T0122398

NEUROMETHODS

Series Editor
Wolfgang Walz
University of Saskatchewan
Saskatoon, SK, Canada

For further volumes:
http://www.springer.com/series/7657

Neuromethods publishes cutting-edge methods and protocols in all areas of neuroscience as well as translational neurological and mental research. Each volume in the series offers tested laboratory protocols, step-by-step methods for reproducible lab experiments and addresses methodological controversies and pitfalls in order to aid neuroscientists in experimentation. *Neuromethods* focuses on traditional and emerging topics with wide-ranging implications to brain function, such as electrophysiology, neuroimaging, behavioral analysis, genomics, neurodegeneration, translational research and clinical trials. *Neuromethods* provides investigators and trainees with highly useful compendiums of key strategies and approaches for successful research in animal and human brain function including translational "bench to bedside" approaches to mental and neurological diseases.

Brain Tumors

Edited by

Giorgio Seano

INSERM U1021 - CNRS UMR3347, Institut Curie Research Center,
Paris-Orsay, France

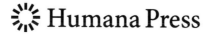

Editor
Giorgio Seano
INSERM U1021 - CNRS UMR3347 Institut Curie Research Center
Orsay, France

ISSN 0893-2336   ISSN 1940-6045   (electronic)
Neuromethods
ISBN 978-1-0716-0858-6   ISBN 978-1-0716-0856-2   (eBook)
https://doi.org/10.1007/978-1-0716-0856-2

© Springer Science+Business Media, LLC, part of Springer Nature 2021
This work is subject to copyright. All rights are reserved by the Publisher, whether the whole or part of the material is concerned, specifically the rights of translation, reprinting, reuse of illustrations, recitation, broadcasting, reproduction on microfilms or in any other physical way, and transmission or information storage and retrieval, electronic adaptation, computer software, or by similar or dissimilar methodology now known or hereafter developed.
The use of general descriptive names, registered names, trademarks, service marks, etc. in this publication does not imply, even in the absence of a specific statement, that such names are exempt from the relevant protective laws and regulations and therefore free for general use.
The publisher, the authors, and the editors are safe to assume that the advice and information in this book are believed to be true and accurate at the date of publication. Neither the publisher nor the authors or the editors give a warranty, expressed or implied, with respect to the material contained herein or for any errors or omissions that may have been made. The publisher remains neutral with regard to jurisdictional claims in published maps and institutional affiliations.

Cover Caption: High-resolution 3D intravital imaging of brain vasculature. Dextran perfused vasculature of the brain cortex imaged thought chronic cranial window (see Chapter 4 for technical details). Cover by Dr. Seano Giorgio.

This Springer imprint is published by the registered company Springer Science+Business Media, LLC part of Springer Nature.
The registered company address is: 1 New York Plaza, New York, NY 10004, U.S.A.

Preface to the Series

Experimental life sciences have two basic foundations: concepts and tools. The *Neuromethods* series focuses on the tools and techniques unique to the investigation of the nervous system and excitable cells. It will not, however, shortchange the concept side of things as care has been taken to integrate these tools within the context of the concepts and questions under investigation. In this way, the series is unique in that it not only collects protocols but also includes theoretical background information and critiques which led to the methods and their development. Thus, it gives the reader a better understanding of the origin of the techniques and their potential future development. The *Neuromethods* publishing program strikes a balance between recent and exciting developments like those concerning new animal models of disease, imaging, in vivo methods, and more established techniques, including, for example, immunocytochemistry and electrophysiological technologies. New trainees in neurosciences still need a sound footing in these older methods in order to apply a critical approach to their results.

Under the guidance of its founders, Alan Boulton and Glen Baker, the *Neuromethods* series has been a success since its first volume published through Humana Press in 1985. The series continues to flourish through many changes over the years. It is now published under the umbrella of Springer Protocols. While methods involving brain research have changed a lot since the series started, the publishing environment and technology have changed even more radically. Neuromethods has the distinct layout and style of the Springer Protocols program, designed specifically for readability and ease of reference in a laboratory setting.

The careful application of methods is potentially the most important step in the process of scientific inquiry. In the past, new methodologies led the way in developing new disciplines in the biological and medical sciences. For example, Physiology emerged out of Anatomy in the nineteenth century by harnessing new methods based on the newly discovered phenomenon of electricity. Nowadays, the relationships between disciplines and methods are more complex. Methods are now widely shared between disciplines and research areas. New developments in electronic publishing make it possible for scientists that encounter new methods to quickly find sources of information electronically. The design of individual volumes and chapters in this series takes this new access technology into account. Springer Protocols makes it possible to download single protocols separately. In addition, Springer makes its print-on-demand technology available globally. A print copy can therefore be acquired quickly and for a competitive price anywhere in the world.

Saskatoon, SK, Canada   *Wolfgang Walz*

Preface

This volume explores the latest models and methods used to study brain tumor biology.

Brain tumors are deadly and unfortunately not many improvements have been achieved to improve the survival of patients with brain tumors. Only accurate models, as well as preclinical and clinical methods will allow us to reach a deep knowledge of the complex molecular, cellular, and anatomic alterations involved in the brain tumors. This level of knowledge may be able to expose vulnerabilities that could be exploited to efficiently treat brain tumors.

The pathology and biology of brain tumors is highly complex and multifaceted. Thus, in order to cover the multiple and different strategies employed to study the multifaceted biology of brain tumors, this book is composed of four distinct sections: (a) In vivo models, (b) Ex vivo models, (c) Treatments in mice, and (d) Clinical imaging.

This collection of 14 interdisciplinary articles lets the readers shed lights on the heterogeneity of the methods used to study the complexity of brain tumors. In the first section, chapters cover topics on in vivo preclinical models of lower grade gliomas, medulloblastoma, and brain metastases as well as intravital imaging of brain tumors. In the second section, we describe ex vivo methods for glioblastoma patient-derived cell lines, organotypic brain cultures for metastasis, human glioblastoma organoids, and mechanobiology of brain tumors. The third section touches topics on in vivo treatments of preclinical models, such as how to assess neurological function, how to dynamically study immunotherapy, and how to explore neurological impacts of brain irradiation. And lastly, we cover some aspects of clinical imaging and modeling, such as biomechanics, vascular perfusion, and magnetic resonance morphometry.

Paris-Orsay, France   *Giorgio Seano*

Acknowledgments

I thank the Tumor Microenvironment Laboratory at Institut Curie for critical reading and insightful suggestions, specifically Drs. Boris Julien, Océane Anézo, Cathy Pichol-Thievend, David Wasilewski, Aafrin Pettiwala, and Charita Furumaya.

Contents

Contributors

FRANCESCO ACERBI • *Department of Neurosurgery, Fondazione IRCCS Instituto Neurologico Carlo Besta, Milan, Italy*

ABRAMO AGOSTI • *MOX, Dipartimento di Matematica, Politecnico di Milano, Milan, Italy*

OCÉANE ANÉZO • *Tumor Microenvironment Laboratory, Institut Curie Research Center, Paris Saclay University, PSL Research University, Inserm U1021, CNRS UMR3347, Orsay, France*

SOFIA ARCHONTIDI • *Inserm U 1127, CNRS UMR 7225, Sorbonne Universités, UPMC Univ Paris 06 UMR S 1127, Institut du Cerveau et de la Moelle épinière, ICM, Paris, France*

CHIARA BARDELLA • *Institute of Cancer and Genomic Sciences, University of Birmingham, Birmingham, UK*

NIHA BEIG • *Department of Biomedical Engineering, Case Western Reserve University, Cleveland, OH, USA*

KAUSTAV BERA • *Department of Biomedical Engineering, Case Western Reserve University, Cleveland, OH, USA*

ANDREAS BIKFALVI • *INSERM U1029, University Bordeaux, Pessac, France*

ALBERTO BIZZI • *Neuroradiology Unit, Fondazione IRCCS Instituto Neurologico Carlo Besta, Milan, Italy*

FAWZI BOUMEZBEUR • *NeuroSpin, CEA, Université Paris-Saclay, Gif-sur-Yvette, France*

FRANÇOIS D. BOUSSIN • *Laboratoire de RadioPathologie, UMRE008/U1274, Inserm, Université de Paris, Université Paris-Saclay, CEA, Fontenay-aux Roses, France*

JIE CHEN • *Department of Radiation Oncology, Edwin L. Steele Laboratories, Massachusetts General Hospital and Harvard Medical School, Boston, MA, USA; Department of Oral and Maxillofacial Surgery, Xiangya Hospital, Central South University, Changsha, Hunan, China*

TIFFANIE CHOULEUR • *INSERM U1029, University Bordeaux, Pessac, France*

PASQUALE CIARLETTA • *MOX, Dipartimento di Matematica, Politecnico di Milano, Milan, Italy*

MICHELE CRESTANI • *IFOM, FIRC Institute of Molecular Oncology, Milan, Italy*

THOMAS DAUBON • *INSERM U1029, University Bordeaux, Pessac, France; CNRS UMR5095, IBGC, University of Bordeaux, Bordeaux, France*

FRANCESCO DIMECO • *Department of Neurosurgery, Fondazione IRCCS Instituto Neurologico Carlo Besta, Milan, Italy; Department of Pathophysiology and Transplantation, Università degli Studi di Milano, Milan, Italy*

BERTRAND DUVILLIÉ • *Institut Curie, Orsay, France; INSERM U1021, Centre Universitaire, Orsay, France; CNRS UMR 3347, Centre Universitaire, Orsay, France; University Paris Sud—Paris Saclay, Orsay, France; PSL Research University, Orsay, France*

KYRRE EEG EMBLEM • *Department of Diagnostic Physics, Oslo University Hospital, Oslo, Norway*

ALAIN EYCHÈNE • *Institut Curie, Orsay, France; INSERM U1021, Centre Universitaire, Orsay, France; CNRS UMR 3347, Centre Universitaire, Orsay, France; University Paris Sud—Paris Saclay, Orsay, France; PSL Research University, Orsay, France*

JACOPO FALCO • *Department of Neurosurgery, Fondazione IRCCS Instituto Neurologico Carlo Besta, Milan, Italy*

PAOLO FERROLI • *Department of Neurosurgery, Fondazione IRCCS Instituto Neurologico Carlo Besta, Milan, Italy*

KATRINA FIFE • *Department of Biomedical Engineering, Cleveland Clinic, Cleveland, OH, USA*

CHLOÉ FORAY • *Institut Curie, Orsay, France; INSERM U1021, Centre Universitaire, Orsay, France; CNRS UMR 3347, Centre Universitaire, Orsay, France; University Paris Sud—Paris Saclay, Orsay, France; PSL Research University, Orsay, France*

XING GAO • *Department of Radiation Oncology, Edwin L. Steele Laboratories, Massachusetts General Hospital and Harvard Medical School, Boston, MA, USA; Department of Oral and Maxillofacial Surgery, Xiangya Hospital, Central South University, Changsha, Hunan, China*

NILS C. GAUTHIER • *IFOM, FIRC Institute of Molecular Oncology, Milan, Italy*

LUIZ HENRIQUE MEDEIROS GERALDO • *INSERM U970, Paris Center for Cardiovascular Research (PARCC), Paris, France*

JORIS GUYON • *INSERM U1029, University Bordeaux, Pessac, France*

CHRISTOPHER G. HUBERT • *Department of Biomedical Engineering, Cleveland Clinic, Cleveland, OH, USA*

EMMANUELLE HUILLARD • *Inserm U 1127, CNRS UMR 7225, Sorbonne Universités, UPMC Univ Paris 06 UMR S 1127, Institut du Cerveau et de la Moelle épinière, ICM, Paris, France*

SANDRA JOPPÉ • *Inserm U 1127, CNRS UMR 7225, Sorbonne Universités, UPMC Univ Paris 06 UMR S 1127, Institut du Cerveau et de la Moelle épinière, ICM, Paris, France*

BORIS JULIEN • *Tumor Microenvironment Laboratory, Institut Curie Research Center, Paris Saclay University, PSL Research University, Inserm U1021, CNRS UMR3347, Orsay, France*

YANIS KHENNICHE • *Inserm U 1127, CNRS UMR 7225, Sorbonne Universités, UPMC Univ Paris 06 UMR S 1127, Institut du Cerveau et de la Moelle épinière, ICM, Paris, France*

MAGALIE LARCHER • *Institut Curie, Orsay, France; INSERM U1021, Centre Universitaire, Orsay, France; CNRS UMR 3347, Centre Universitaire, Orsay, France; University Paris Sud—Paris Saclay, Orsay, France; PSL Research University, Orsay, France*

DENIS LE BIHAN • *NeuroSpin, CEA, Université Paris-Saclay, Gif-sur-Yvette, France*

GRACE Y. LEE • *St. Mark's School, Southborough, MA, USA*

STEFANO MARCHESI • *MOX, Dipartimento di Matematica, Politecnico di Milano, Milan, Italy; IFOM, FIRC Institute of Molecular Oncology, Milan, Italy*

THOMAS MATHIVET • *INSERM U970, Paris Center for Cardiovascular Research (PARCC), Paris, France; Laboratoire de Neurosciences Expérimentales et Cliniques, Université de Poitiers, INSERM U1084, Poitiers, France*

LILIANA MIRABAL-ORTEGA • *Institut Curie, Orsay, France; INSERM U1021, Centre Universitaire, Orsay, France; CNRS UMR 3347, Centre Universitaire, Orsay, France; University Paris Sud—Paris Saclay, Orsay, France; PSL Research University, Orsay, France*

PASCALE MONZO • *IFOM, FIRC Institute of Molecular Oncology, Milan, Italy*

MORGANE MORABITO • *Institut Curie, Orsay, France; INSERM U1021, Centre Universitaire, Orsay, France; CNRS UMR 3347, Centre Universitaire, Orsay, France; University Paris Sud—Paris Saclay, Orsay, France; PSL Research University, Orsay, France*

MARC-ANDRÉ MOUTHON • *Laboratoire de RadioPathologie, UMRE008/U1274, Inserm, Université de Paris, Université Paris-Saclay, CEA, Fontenay-aux Roses, France*

LAURA MOUTON • *Laboratoire de RadioPathologie, UMRE008/U1274, Inserm, Université de Paris, Université Paris-Saclay, CEA, Fontenay-aux Roses, France; NeuroSpin, CEA, Université Paris-Saclay, Gif-sur-Yvette, France*

LINE BRENNHAUG NILSEN • *Department of Diagnostic Physics, Oslo University Hospital, Oslo, Norway*

BEATRICE PHILIP • *Tumor Microenvironment Laboratory, Institut Curie Research Center, Paris Saclay University, PSL Research University, Inserm U1021, CNRS UMR3347, Orsay, France*

CATHY PICHOL-THIEVEND • *Tumor Microenvironment Laboratory, Institut Curie Research Center, Paris Saclay University, PSL Research University, Inserm U1021, CNRS UMR3347, Orsay, France*

MARCO C. PINHO • *Department of Radiology and Advanced Imaging Research Center, UT Southwestern Medical Center, Dallas, TX, USA*

SCOTT R. PLOTKIN • *Department of Neurology, Stephen E. and Catherine Pappas Center for Neuro-Oncology, Massachusetts General Hospital, Boston, MA, USA*

W. DEAN PONTIUS • *Department of Genetics and Genome Sciences, Case Western Reserve University, Cleveland, OH, USA; Department of Molecular Medicine, Cleveland Clinic, Cleveland, OH, USA*

CELIO POUPONNOT • *Institut Curie, Orsay, France; INSERM U1021, Centre Universitaire, Orsay, France; CNRS UMR 3347, Centre Universitaire, Orsay, France; University Paris Sud—Paris Saclay, Orsay, France; PSL Research University, Orsay, France; Institut Curie—Recherche, Centre Universitaire, Orsay Cedex, France*

MONICA RIBEIRO • *Centre Borelli, Université Paris-Saclay, Ecole Normale Supérieure Paris-saclay, Service de Santé des Armées, CNRS, Université de Paris, Paris, France*

DAMIEN RICARD • *Centre Borelli, Université Paris-Saclay, Ecole Normale Supérieure Paris-saclay, Service de Santé des Armées, CNRS, Université de Paris, Paris, France; Service de Neurologie, Service de Santé des Armées, Hôpital d'Instruction des Armées Percy, Clamart, France; Service de Santé des Armées, Ecole du Val-de-Grâce, Paris, France*

GIORGIO SCITA • *IFOM, FIRC Institute of Molecular Oncology, Milan, Italy; Department of Oncology and Hemato-Oncology, Università degli Studi di Milano, Milan, Italy*

GIORGIO SEANO • *Tumor Microenvironment Laboratory, Institut Curie Research Center, Paris Saclay University, PSL Research University, Inserm U1021, CNRS UMR3347, Orsay, France*

YAO SUN • *Department of Radiation Oncology, Edwin L. Steele Laboratories, Massachusetts General Hospital and Harvard Medical School, Boston, MA, USA; Department of*

Radiation Oncology, Tianjin Medical University Cancer Institute and Hospital, National Clinical Research Center for Cancer, Key Laboratory of Cancer Prevention and Therapy, Tianjin Clinical Research for Cancer, Tianjin, China

RAQUEL D. THALHEIMER • *Department of Neurology, Stephen E. and Catherine Pappas Center for Neuro-Oncology, Massachusetts General Hospital, Boston, MA, USA*

PALLAVI TIWARI • *Department of Biomedical Engineering, Case Western Reserve University, Cleveland, OH, USA*

MANUEL VALIENTE • *Brain Metastasis Group, Spanish National Cancer Research Center (CNIO), Madrid, Spain*

PIERRE VERRELLE • *Département d'Oncologie Radiothérapie, Institut Curie, Paris, France; UMR 9187/U1196 (CMIB), Institut Curie, Orsay, France*

IGNAZIO G. VETRANO • *Department of Neurosurgery, Fondazione IRCCS Instituto Neurologico Carlo Besta, Milan, Italy*

LISA C. WALLACE • *Department of Biomedical Engineering, Cleveland Clinic, Cleveland, OH, USA*

CHRISTINA S. WONG • *Merrimack Pharmaceuticals, Inc., Cambridge, MA, USA*

LIMENG WU • *Department of Radiation Oncology, Edwin L. Steele Laboratories, Massachusetts General Hospital and Harvard Medical School, Boston, MA, USA; Department of Oral and Maxillofacial Surgery, Xiangya Hospital, Central South University, Changsha, Hunan, China*

LEI XU • *Department of Radiation Oncology, Edwin L. Steele Laboratories, Massachusetts General Hospital and Harvard Medical School, Boston, MA, USA*

YUNLING XU • *INSERM U970, Paris Center for Cardiovascular Research (PARCC), Paris, France*

LUCÍA ZHU • *Brain Metastasis Group, Spanish National Cancer Research Center (CNIO), Madrid, Spain*

Part I

In Vivo Models

Chapter 1

Mouse Models of Diffuse Lower-Grade Gliomas of the Adult

Sofia Archontidi, Sandra Joppé, Yanis Khenniche, Chiara Bardella, and Emmanuelle Huillard

Abstract

Diffuse gliomas of the adult are common primary brain tumors. Glioblastomas are the most aggressive type of gliomas. Lower-grade gliomas (astrocytomas and oligodendrogliomas) are less aggressive yet can progress to glioblastomas. Whereas our understanding of glioblastoma biology has increased dramatically in the last few years, the cellular and molecular mechanisms underlying the initiation and development of lower-grade gliomas are less understood. This is partly due to the lack of relevant models for this disease. The goal of this chapter is to provide the reader with a review of existing tools to model diffuse gliomas in mice, to discuss the current models and perspectives on modeling lower-grade gliomas, with a particular focus on oligodendrogliomas.

Key words Lower-grade glioma, Astrocytoma, Oligodendroglioma, Mouse model

1 Classification, Genetics, and Origin of Adult Diffuse Gliomas

Diffuse gliomas are the most common malignant primary brain tumors in adults, representing approximately 80% of the malignant cases [1]. As their name suggests, gliomas exhibit cellular characteristics of glial cells, such as astrocytes and oligodendrocytes. As of 2016, diffuse gliomas are classified by the World Health Organization (WHO) based on its histology and specific molecular parameters [2] (Fig. 1.1). Diffuse gliomas can be divided into three general entities: (1) oligodendrogliomas (Grade II and Grade III), (2) astrocytomas (Grade II and Grade III), and (3) glioblastomas (Grade IV, formerly known as astrocytoma Grade IV, also known as glioblastoma multiforme and commonly abbreviated as GBM). The hallmark alterations underlying the distinction of glioma entities are the presence of mutations in Isocitrate Dehydrogenase genes (mostly *IDH1* or less frequently *IDH2*) and the status of the loss of the chromosomal arms 1p and 19q (termed 1p/19q codeletion). Tumors lacking mutations in the *IDH* genes

Giorgio Seano (ed.), *Brain Tumors*, Neuromethods, vol. 158,
https://doi.org/10.1007/978-1-0716-0856-2_1, © Springer Science+Business Media, LLC, part of Springer Nature 2021

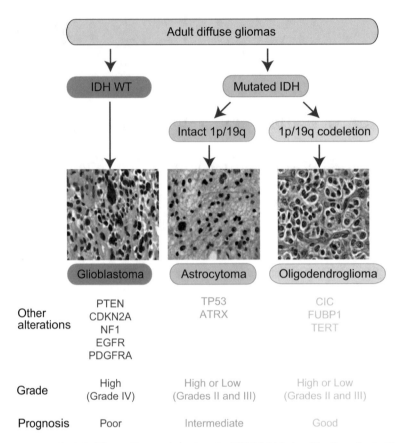

Fig. 1.1 Major classes of adult diffuse gliomas, based on the WHO 2016 classification of primitive brain tumors. Diffuse gliomas are mainly based on the presence or absence of *IDH* mutations, 1p19q codeletion, and *TP53* mutations. Typical histologies (hematoxylin and eosin stainings) of the three main glioma types are displayed

generally fall into the category of glioblastomas, with the exception of secondary glioblastomas (10% of glioblastomas) that harbor *IDH* mutations. Within the *IDH*-mutant category, two subgroups are distinguished based on the status of 1p/19q codeletion: oligodendrogliomas, characterized by the presence of 1p/19q codeletion, and astrocytomas in which 1p/19 codeletion is absent [2] (Fig. 1.1).

1.1 Glioblastomas (GBM)

GBMs represent about half of all diffuse gliomas. They represent the most aggressive types of gliomas, with a median survival of 11–16 months from diagnosis and predominating in patients over 55 [3]. These tumors mostly develop rapidly de novo, without a preexisting lesion. At the molecular level, GBMs are characterized by multiple alterations related with inactivation of tumor suppressor and activation of pro-survival oncogenic pathways. In particular, GBMs can exhibit inactivation of the TP53 and retinoblastoma (RB) tumor suppressor pathways through *TP53* mutations and homozygous deletion of the cyclin-dependent kinases

CDKN2A/CDKN2B. In addition, GBMs show dysregulation of the receptor tyrosine kinase (RTK), Ras, and phosphatidylinositol 3-kinase (PI3K) pathways, primarily via amplification and mutational activation of growth factor receptors (EGFR, PDGFRA) and through Neurofibromin 1 (NF1) and Phosphatase and Tensin homolog (PTEN) deletion [4]. Although indistinguishable on histological criteria, GBMs are heterogeneous at the molecular level. Four GBM molecular subgroups (proneural, classical, neural, mesenchymal) have been proposed on the basis of their genomic alterations and gene expression profiles, but the natural history and clinical relevance of these subgroups are not known. For instance, it is not clear to what extent genetic alterations and the lineage of origin determine subgroups. Initially thought as independent tumor subtypes, multiple GBM subgroups can be detected within the same tumor, suggesting that they may be related to each other [5, 6].

1.2 Diffuse Lower-Grade Gliomas (Astrocytoma and Oligodendroglioma)

Gliomas with *IDH* mutations define an entity associated with a more favorable prognosis compared to *IDH* non-mutated tumors and are referred to as lower-grade gliomas. *IDH* mutations are considered to be initiating events in gliomagenesis [7]. The most frequent alteration is a heterozygote mutation of Isocitrate Dehydrogenase 1 (IDH1) gene on Arg132 (R132H). The mutated IDH enzyme reduces α-Ketoglutarate (α-KG) to D-2-Hydroxyglutarate (D-2HG) [8]. One of the best characterized effects of D-2HG is the competitive inhibition of α-KG-dependent dioxygenases, including histone demethylases and the TET family of 5-methylcytosine hydroxylases, resulting in genome-wide modifications in histone and DNA methylation and inhibition of differentiation [9, 10]. A striking example of the consequences of *IDH1* mutations on DNA methylation is the disruption of chromatin domains, leading to aberrant expression of oncogenes such as *PDGFRA* [11].

Diffuse astrocytomas are characterized by *IDH* mutations and the absence of the 1p19q codeletion. They display *TP53* mutations and *ATRX* loss [2, 12]. Oligodendrogliomas are characterized by *IDH* mutations and 1p19q codeletion, which result from an unbalanced translocation and are associated with a better outcome for patients [13–15]. However, its functional impact on tumor development remains unknown. In addition, mutations within the core promoter of *TERT*, encoding the catalytic subunit of telomerase, are in all 1p19q codeleted gliomas [16]. These mutations result in increased telomerase activity, leading to cell immortalization [17, 18]. In addition, mutations in the *CIC* and *FUBP1* genes are frequent events in oligodendrogliomas, accounting for about 60% and 30% of oligodendrogliomas, respectively [7, 12].

1.3 Glioma Cells of Origin

Although the genetics of diffuse gliomas is relatively well characterized, treatment is currently limited to conventional radio- and chemotherapies. Identifying the lineages that are responsible for tumor growth and understanding how mutations affect their biology are key to understand the different steps of the tumorigenic process and to design rational therapeutic strategies. During brain development, neural stem cells (NSCs) divide to generate neural progenitor cells, giving rise to neurons, oligodendrocytes, and astrocytes in a sequential manner [19]. Oligodendrocyte precursor cells (OPCs) are generated from neural stem/progenitors at late embryonic stages and postnatal stages and start differentiating into oligodendrocytes, the myelin-producing cells, after birth. Neural stem cells persist in the postnatal and adult brain, where they generate subsets of neurons and oligodendrocytes from the subventricular zone (SVZ), the largest neurogenic niche of the brain. OPCs also persist into adulthood and continue to divide, accounting for the majority of proliferating cells in the adult brain [20].

Evidence from clinical studies and experimental models suggest that glioblastomas may originate from the transformation of neural stem cells [21–23] (Fig. 1.2). In addition, evidence from mouse models indicate that cells with stem-like and OPC-like features are responsible for tumor growth in GBMs [22]. Other studies have implicated more differentiated cell types as glioma cells of origin. For instance, mature astrocytes and neurons have been reported to undergo dedifferentiation and generate malignant gliomas [24], but this observation has been challenged in a recent study from the Parada group [25]. In the case of lower-grade gliomas, the nature of the cell population that is initially targeted by mutations and the tumor-propagating cells is less characterized. OPCs have been proposed to be the cells of origin for oligodendrogliomas [26–28]. Human oligodendrogliomas have been found to be associated with white matter tracts and demonstrate an OPC-like expression profile rather than a NSC profile [28]. Cells expressing the OPC marker NG2 have been demonstrated to exhibit high tumorigenic potential, proposedly by losing their ability to divide asymmetrically and generate more differentiated cellular types [26–28]. Nonetheless, OPCs may not be the sole cell type acting as cells of origin for oligodendrogliomas. Single-cell RNA sequencing from *IDH*-mutant human oligodendrogliomas pinpointed the existence of a rare subpopulation of cells displaying an undifferentiated program resembling neural stem cell expression signature [29]. Similar observations were drawn in an independent study, where oligodendroglioma tumor cells displayed morphological, immunohistochemical, and transcriptomic similarities with embryonic neural stem/progenitor cells [30].

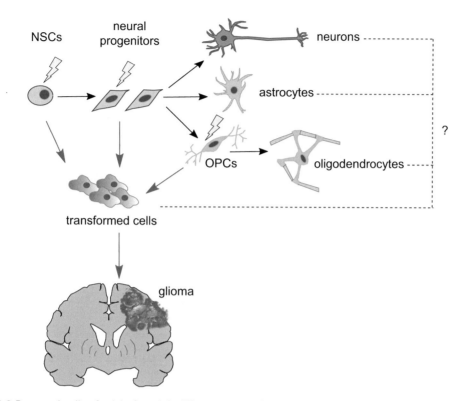

Fig. 1.2 Proposed cells of origin for adult diffuse gliomas. Current evidence from clinical studies and mouse models indicates neural stem/progenitor cells and oligodendrocyte precursor cells (OPCs) as potential cells of origin (indicated by a yellow lightening). It remains unclear whether in some cases mature neural cell types such as neurons, astrocytes, and oligodendrocytes may be transformed by genetic alterations (dotted gray line)

2   Methods to Model Gliomas in Mice

Mouse models represent essential tools to understand tumor biology. Mouse models permit to investigate the consequences of a single or a combination of genetic alterations, in the whole organism or in a given cellular lineage. They provide an opportunity to monitor the whole tumorigenic process, from the first transformed cells to the full-blown tumor. Importantly, they allow to address the contribution of the brain microenvironment to the tumorigenic process. Finally, they represent useful platforms for preclinical testing. In order to be the most relevant to the human tumor, mouse models have to recapitulate as much as possible the histopathological and molecular features found in the patients. We review below the various approaches that can be used to generate mouse models of gliomas, emphasizing on genetically engineered mouse models, and discuss their advantages and limitations.

2.1   Xenograft Models

Xenograft models (PDXs: patient-derived xenografts) are based on the transplantation of patient-derived cells into mice that are immunocompromised, in order to reduce the rejection of the tumor cells by the host. A patient's tumor can be chopped and grafted or, alternatively, amplified in culture before being injected. An important advantage of PDXs is that they retain histological and molecular characteristics of the patient's tumor, therefore providing a clinically relevant model. For this reason, these models are particularly suited for exploring the mechanisms that underlie drug resistance. In PDXs, the number of injected tumor cells is particularly high, which does not represent the natural history of the tumor that arises from a limited number of cells. A major limitation of this approach is that the host mouse has a deficient immunity, making the model not appropriate for investigating the contribution of the immune system to tumor development and for testing immunotherapies. To circumvent this major issue, mice that express human immune cells (humanized mice) have been developed by transplanting human total blood or CD34+ hematopoietic stem cells in immunocompromised mice [31]. A recent study compared the growth of GBM PDXs under different conditions of immunodeficiency, such as immunosuppression (by dexamethasone injection) or immunotolerance (by blocking the recruitment of T-cells) [32]. This study showed that tumors generated in immunotolerant mice bear more features of human tumors including, among others, strong neovascularization and blood-brain barrier leakage [32]. These data highlight the importance to maintain an intact immune system when modeling glioma development.

2.2 Genetically Engineered Mouse Models (GEMMs)

In genetically engineered mouse models (GEMMs), defined gene alterations identified in human tumors are introduced in the germline (knockout, knock-in, transgenic models) or in somatic cells (via viral-mediated gene delivery), in order to allow for de novo tumor formation (Fig. 1.3) [33]. GEMMs represent the best systems for monitoring the initiation and progression steps of the gliomagenic process, in an organism with an intact immune system. GEMMs are critical in assessing biomarker expression, drug delivery, and therapeutic responses and even testing prevention therapies. Several studies using GEMMs have been successfully translated into clinical trials (for a review, see ref. [34]). Prior to developing a GEMM, it is essential to define the scientific or medical question and determine the different parameters that are relevant to address this question, that is, the nature of the genetic alterations and the cell population(s) to be targeted with these mutations. The developmental time frame (whether mutations are introduced in the developing or mature brain) has also to be determined. Finally, the use of a reporter system is useful for tracking the fate of mutated cells and their progeny. We review below the main strategies that are used to generate germline and somatic mutations and what these approaches have taught us on glioma development.

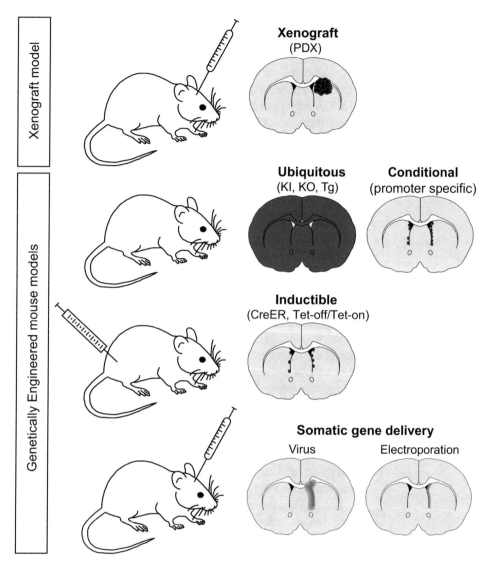

Fig. 1.3 Overview of xenograft models and the major types of genetically engineered mouse models. *PDXs:* patient-derived xenografts, *KI:* knock-in, *KO:* knockout; *Tg:* transgenic

2.2.1 Germline and Conditional Models

In knockout (and knock-in) strategies, the gene of interest is targeted by homologous recombination in embryonic stem cells, which are then implanted into recipient blastocysts. In the transgenic approaches, a DNA sequence consisting of a defined promoter (to target expression to a specific cell population) and a cDNA of interest is injected into the pronucleus of a fertilized egg and results in random genomic integration. Whole-body knockouts or transgenic approaches result in mice that are deficient for, or overexpress, a given gene in all cells of the organism. The first mouse model of glioma was generated from mice knockout for both *Nf1* and *Trp53* tumor suppressors [35]. However, such germline models, because of the ubiquitous inactivation of genes of interest, can lead to embryonic lethality. Furthermore, these models are not representative of tumor initiation in humans, which occurs from a limited number of cells. To provide higher specificity in the cell populations to be targeted, conditional models, such as the *Cre/loxP* system, were developed [36]. In this system, the Cre recombinase enables the excision of gene sequences (typically exons) located between two *loxP* sites. Targeted deletion or activation in a defined cell type is achieved by inducing Cre expression under the control of cell-specific promoters. This approach was initially used to inactivate *Trp53* and *Nf1* in radial glial cells by crossing *hGFAP-cre* mice (expressing Cre under the human GFAP promoter, enabling targeting of radial glia) to *Trp53+/−; Nf1+/fl* mice. The resulting tumors displayed characteristics of GBM, further demonstrating the cooperation between the two tumor suppressors Trp53 and Nf1 in gliomagenesis [37]. This approach was also used to conditionally express an activated version of EGFR (EGFRvIII), frequently mutated in GBMs, upon infection with an adenovirus encoding Cre [38]. An advantage of the *Cre/loxP* system is the possibility, when used in combination with a reporter gene (i.e., fluorescent reporter) preceded by a *lox-STOP-lox* cassette, to track the behavior and progeny of Cre-expressing cells. The generation of mouse lines in which the activity of Cre is dependent on tamoxifen administration (CreER, CreERT2) allows the temporal control of Cre expression, which is crucial for addressing, for instance, whether postnatal and/or adult NSCs can serve as cells of origin for gliomas. This system was used to show that embryonic and adult Nestin+ NSCs can generate tumors when inactivated for the tumor suppressors *Trp53*, *Nf1*, and *Pten* [39].

Other inducible systems include the Tet-on and Tet-off systems, which allow doxycycline-dependent gene expression or repression, respectively [36]. These "Tet-on" and "Tet-off" models are particularly interesting since they allow acute induction of the gene of interest for a desired period of time. However, doxycycline can be toxic at high doses. This type of model has been used by Robinson et al., who investigated the role of Akt signaling, in combination with activated KRas and loss of *Ink4a/Arf*, on glioma growth and recurrence following suppression of KRas expression [40]. This study showed that abrogation of KRas signaling resulted in significant tumor regression and that tumor recurrence occurred more rapidly in the presence of activated Akt signaling [40].

2.2.2 Somatic Gene Transfer

Many models deliver the gene of interest to chosen somatic cells, using viral or nonviral vectors (Table 1.1). Viral constructs encoding the transgene of interest (i.e., oncogene, reporter gene, short hairpin RNA, guide RNA, Cre recombinase, under the control of either viral or specific promoters) are relatively easy, fast, and inexpensive to generate. Several families of virus can be used, depending

Table 1.1
Overview of vectors used for gene delivery in mouse glioma models

	Vector type	Advantages	Limitations	Examples
Viral	Retrovirus	Long-term gene expression	Genome integration may disrupt gene expression	[6, 123]
		Selectively infects dividing cells	Safety concerns when transducing oncogenes	
		Cell-specific infection (RCAS-TVA system)		
		Useful for lineage tracing		
	Lentivirus	Long-term gene expression	Genome integration may disrupt gene expression	[24]
		Infects both dividing and nondividing cells	Safety concerns when transducing oncogenes	
		Useful for lineage tracing		
	Adenovirus	No genome integration (largely episomal)	High immunogenicity	[125]
		Large insert size	Short-term expression	
	Adeno-associated virus (AAV)	Long-term gene expression	Limited transgene capacity	[47, 48]
		Infects both dividing and nondividing cells		
		Low immunogenicity		
		Cell type-specific infection depending on the serotype		
Nonviral	transposon (Sleeping Beauty, PiggyBac)	Long-term gene expression	Genome integration may disrupt gene expression	[55, 124]
		Large insert size (PiggyBac)	Transduction method (electroporation) is mostly done at embryonic and postnatal stages	
		Low immunogenicity		
		Suitable for discovery of tumor drivers		

on the question needs: in the *Retroviridae* family, lentiviruses infect all cells, including quiescent cells, whereas retroviruses infect only dividing cells. Both viruses integrate into the mouse genome of the targeted cells, allowing permanent expression of the transduced gene [41]. The RCAS-TVA system is particularly interesting as it allows the delivery of genes of interest by the RCAS avian retrovirus into defined cell populations engineered to express the RCAS receptor TVA [42]. This model permits the infection of a low number of cells adjacent to the injection site. In addition, it is possible to modulate the expression of several genes by co-injecting different RCAS. The use of this model is therefore easier to implement and less expensive than the production of mice with multiple mutations [43]. This approach was originally used to demonstrate the role of combined Ras and Akt pathway activation in the formation of GBMs from neural stem/progenitors [44] and is still used in numerous glioma models, in combination with the Cre/lox system [6]. In contrast to retrovirus, adenovirus enables to express transiently and efficiently the transgene without genome integration, limiting randomly insertion with the murine genome that may result in modulation of endogenous gene expression [45, 46]. Adeno-associated viruses (AAVs) are widely used in neuroscience research, as they offer selective targeting depending on the serotype and mediate long-term gene expression with no apparent toxicity [41]. The use of these viruses to target tumor cells is gaining interest in preclinical models of gliomas [47]. By combining CRISPR/Cas9 gene editing to AAV transduction, Chow et al. developed an in vivo screen to identify co-occurring tumor drivers [48]. They designed AAV-CRISPR vectors encoding Cre recombinase under a GFAP promoter, to target neural stem cells and astrocytes and guide RNAs targeting genes significantly mutated in human cancer. When injected into Cre-inducible conditional Cas9 mice, this library robustly generated tumors with features of human GBM. This strategy, combined with co-mutation analysis, led to the identification of co-occurring driver combinations (i.e., B2m-Nf1 or Zc3h13-Rb1). These data are particularly interesting since they reveal the relationships of specific mutations during gliomagenesis. Finally, the authors showed that resistance to the first-line chemotherapy used in the clinics (temozolomide) may be associated with co-occurrence of certain genes, for example, *Zc3h13* or *Pten* mutations in *Rb1*-mutant glioma cells [48].

Nonviral vectors are also used for gene delivery in glioma models. Plasmid DNA can be directly injected to the lateral ventricles of the brain and subjected to electroporation. The electroporation technique uses electric pulses to generate transient pores in the cell membranes, which allows transduction of the negatively charged DNA in neural stem/progenitor cells [49, 50]. This technique therefore has a temporal specificity, since DNA delivery can be achieved only during electric pulses, but the electroporated cell types are limited to those close to the ventricles (ependymal cells, neural stem/progenitors). Thus, similar to the CreER system, this approach offers the advantage to target specific cell populations in a time-specific manner. Transposon-based plasmids have gained interest as vectors to deliver a transgene of interest or as tools for cancer gene discovery in the mouse [51]. Transposons are DNA sequences that can move from one site to another site in the genome. Two transposition systems (*Sleeping Beauty* and *PiggyBac*) are predominantly used [52]. *Sleeping Beauty* (SB) is based on the use of a nonautonomous transposon (*T2/Onc*) that requires a transposase element (*SB11*) in order to activate and reintegrate at a different site. The transposon can deregulate expression of a putative oncogene, when integrated upstream in the same transcriptional orientation, and can also terminate transcription of a putative tumor suppressor gene. This results in loss or gain of function of genes that can further be identified by the inserted transposon sequence. The *SB* system can be used with Cre/*loxP*, in order to conditionally activate the transposase element. This mutagenesis strategy is particularly appropriate for high-throughput screen for driver genes in gliomas [53, 54]. A recent study used cotransduction of transposon vectors encoding *PDGFA* and shRNAs against *Nf1* and *Trp53* (shp53/shNf1) to initiate gliomagenesis from neural stem/progenitors in postnatal mice [55].

3 Models for Diffuse Lower-Grade Gliomas

The vast majority of glioma models that have been generated and characterized are models of high-grade gliomas such as GBM. Models for lower-grade gliomas are rare and have been developed only recently, following the identification of *IDH* mutations in these tumors a decade ago.

3.1 Models for Diffuse Astrocytomas

3.1.1 Cell Lines and Xenograft Models

In 1992, Weiss and Reynolds described that neural stem cells can be stably maintained and propagated as neurosphere cultures by using a medium deprived of serum, in the presence of specific growth factors [56]. Accordingly, by growing tumor spheroids in a serum-free medium containing basic fibroblast growth factor (bFGF), epidermal growth factor (EGF), and the neuronal cell culture supplement B-27, it is possible to select and expand glioma stem-like cells in vitro from explanted gliomas [57–59]. Indeed, GBM tumor stem cells cultured with bFGF and EGF more faithfully retain the phenotype and genotype of the primary tumor compared to cell lines cultured in serum [60]. Nevertheless, the successful outcome of the neurosphere cultures depends on the histological grade and genetic background of the tumor analyzed [57]. Cultures from high-grade gliomas such as GBMs are relatively easy to establish using these protocols. On the contrary, it has been difficult to derive cultures from glioma patients with *IDH* mutation, since the *IDH*-mutant cells are easily eliminated during the first passages of the culture [57, 61]. Only a small number of studies have been successful in establishing cell lines from *IDH*-mutant lower-grade tumors [62–67], suggesting that the existing in vitro culture conditions are suboptimal for these tumors and/or additional genetic aberrations are required to bypass the metabolic stress of the standard culture methods (discussed below). *IDH*-mutant cultures derived from human gliomas grow slowly in vitro, and culture protocols using various suspension or adherent conditions [61–63, 66] or coculture with neurons [66] were not useful to improve the propagation of these cells. Thus, to expand these cells, a more responsive microenvironment is required, and to circumvent this problem, some researchers have subsequently implanted *IDH*-mutant neurospheres orthotopically in severe immunocompromised mice [63, 66]. When glioma cells are cultured as spheroids, once implanted as xenografts in mice, they are able to retain the tumor-initiating capacity and aggressive behavior of the original tumor cell population [63, 66, 68]. A better understanding of the biology of lower-grade gliomas is nonetheless required to improve culture protocols specific for the derivation of cellular models from these tumors.

3.1.2 Mouse Models for IDH Mutation

Mutant *IDH1* has been identified as a main driver gene in gliomas and other malignancies. How *IDH* mutations contribute to tumorigenesis is only starting to be deciphered. IDH1 enzyme converts isocitrate to α-ketoglutarate (αKG), but when mutated, it possesses a novel enzymatic function that reduces αKG to D-2-hydroxyglutarate (D-2HG). D-2HG is thought to act as an oncometabolite, by inhibiting αKG-dependent enzymes, involved in DNA and histone demethylation among others [69]. To study the role of mutant *IDH1* in the onset and progression of gliomas, various genetically engineered *Idh1*-mutated mouse models have been developed (Table 1.2). Sasaki et al. were the first ones to generate a brain-specific conditional *Idh1* knock-in model, where mutant *Idh1* was targeted to neural stem/progenitor cells of the embryonic brain by using a *Nestin-Cre* transgenic mouse. Expression of *Idh1*^{R132H} at embryonic day E10.5 resulted in perinatal lethality of the mice from cerebral hemorrhages without evidence of underlying malignancy. *Idh1*-mutant brains accumulated high levels of D-2HG, which was associated with the inhibition of the prolyl-hydroxylation of hypoxia-inducible transcription factor-1α (HIF1α) and activation of HIF1α transcriptional target genes. D-2HG also caused the inhibition of the prolyl-hydroxylation of collagen, resulting in defects in collagen maturation and extracellular matrix aberrations [70]. The same group expressed mutant *Idh1* by using *GFAP-Cre*: starting from E15.5, around 60% of the animals exhibited brain hemorrhages and 92% of mice died perinatally; the surviving mice, however, did not develop

Table 1.2
Genetically engineered mouse models of diffuse lower-grade gliomas

Gene targeted	Mouse model	Genetics	Cell of origin	Phenotype	References
Idh1	Knock-in (Mak lab)	*Nestin-Cre;Idh1*[R132H/+]	Neural stem/progenitor cells (starting at E10.5)	Perinatal lethality (no surviving mice)	[70]
				Brain hemorrhage	
				D-2HG accumulation	
				Decreased ROS levels	
				No change in histone H3 methylation	
				Reduced levels of 5-hmC in NSCs and glial progenitors	
				Increased HIF1a stability and VEGF expression	
				Defective collagen maturation and basement membrane	
	Knock-in (Mak lab)	*GFAP-cre;Idh1*[R132H/+]	Neural progenitors and astrocytes (starting at E14.5)	Perinatal lethality (8% surviving mice)	[70]
				Brain hemorrhage	

(continued)

Table 1.2 (continued)

Gene targeted	Mouse model	Genetics	Cell of origin	Phenotype	References
	Knock-in (Tomlinson lab)	*Nestin-CreERT2;Idh1fl*[R132H]/+	Adult neural stem/ progenitor cells	Death 4–6 weeks post tamoxifen administration	[71]
				Hydrocephalus	
				Proliferative subventricular nodules	
				Expansion of the subventricular zone, increased proliferation of neural stem/progenitor cells, increased Olig2+ cell numbers	
				Parenchyma infiltration	
				Increase in D-2HG and decrease in αKG levels	
				No change in collagen expression, ROS levels, histone methylation, and HIF stability	
				Increase in stem, Wnt, and telomere pathway activity	
				Decrease in 5-hmC and increase in 5mC levels	

Gene targeted	Mouse model	Genetics	Cell of origin	Phenotype	References
	Knock-in (Yan lab)	*hGFAP-Cre; Idh1*[LSL:R132H/WT]	Neural stem/progenitor cells (starting at E13.5)	Median survival around 25 days	[72]
				Hydrocephalus	
				Increase in D-2HG	
				Hemorrhagic foci	
				Decreased proliferation in SVZ	
		Nestin-CreERT2; Idh1[LSL:R132H/WT]	Neural stem/progenitor cells (tamoxifen administration at E18.5)	Asymptomatic 1 year post induction	[72]
		Nestin-CreERT2; Idh1[LSL:R132H/WT]*;Trp53*[fl/fl]	Neural stem/progenitor cells (tamoxifen administration at E18.5)	Increased survival compared to *Nestin::CreERT2;Tp53*[fl/fl]mice	[72]
				Increased NSC proliferation in vitro	
				Increased D-2HG levels	
				Increased 5-mC levels	

(continued)

Table 1.2 (continued)

Gene targeted	Mouse model	Genetics	Cell of origin	Phenotype	References
	Viral delivery (RCAS-TVA system)	*Nestin-TVA* + RCAS-IDH1^{R132H}/shTrp53	Neural progenitors of newborn mice	No tumor formation	[75]
	Viral delivery (RCAS-TVA system)	*Nestin-TVA* + RCAS-PDGFA + RCAS-IDH1^{R132H}/shTrp53	Neural progenitors of newborn mice	Glioma formation Increased median survival conferred by Idh1 mutation	[75]
		Nestin-TVA;Cdkn2a−/− + RCAS-PDGFA + RCAS-IDH1^{R132H}/shTrp53	Neural progenitors of newborn mice	Glioma formation No effect of Idh1 mutation on median survival Increased D-2HG levels Increased DNA methylation of CpG islands and 5-hmC Decreased immune infiltration in tumor	[75]
		Nestin-TVA;Cdkn2a$^{lox/lox}$,Atrx$^{lox/lox}$,Pten$^{lox/lox}$ + RCAS PDGFA + RCAS cre + RCAS IDH1^{R132H}	Neural progenitors of newborn mice	High-grade glioma (proneural subtype) Decreased survival compared to IDH1 wild-type tumors Increased proliferation in vitro Increase in D-2HG Decreased 5-hmC levels and methylation changes on CpG regions	[73]

Gene targeted	Mouse model	Genetics	Cell of origin	Phenotype	References
	Transposon (Sleeping Beauty)	shTrp53 + shAtrx + NRas G12V + Idh1^{R132H}	Neural stem/progenitors of newborn mice	Increased median survival conferred by Idh1 mutation	[74]
				Increased D-2HG levels	
				Inhibition of cell differentiation	
				Increase in histone H3 methylation	
				Increase in DNA damage response via upregulation of ATM signaling	
Cic	Knockout (Cictm2a(KOMP) Wtsi)	$Cic^{-/-}$	Germline	Embryonically lethal	[94]
				Hyperproliferation in subventricular zone and Olig2+ cells in septum	
				Defective oligodendrocyte differentiation in vitro	
	Conditional knockout; orthotopic transplantations in NSG mice	$Cic^{fl/fl}$ + Ad:cre + PDGFB retrovirus	Neural progenitors from postnatal brain	Low-grade infiltrating tumors	[94]
				Shorter survival compared to $Cic^{+/+}$ cells	
				Decreased mitotic activity and vascular proliferation	

(continued)

Table 1.2 (continued)

Gene targeted	Mouse model	Genetics	Cell of origin	Phenotype	References
	Conditional knockout	$FoxG1$-$Cre;Cic^{fl/fl}$	Neural progenitors of the forebrain (starting at E10.5)	Lethal after weaning Smaller brain Decreased neuron and increased glial numbers in cortex Increased NSC proliferation and self-renewal in vitro Increased specification toward oligodendrocytes	[104]
Fubp1	Knockout	$Actin$-$Cre;Fubp1^{fl/fl}$	Germline	Embryonically lethal Decreased body weight Hypercellularity in brain parenchyma Defective hematopoiesis	[106]
	Orthotopic transplantations in NSG mice	shFubp1 + IDH1^{R132H} + PIK3CA^{H1047R} viruses	cdkn2a$^{-/-}$ newborn neural progenitors	Tumor growth (glioma?)	[107]

Gene targeted	Mouse model	Genetics	Cell of origin	Phenotype	References
Others	Transgenic	*S100b-v-erbB*	Glial progenitors	Low-grade oligodendrogliomas localized mainly to hindbrain	[84]
		S100b-v-erbB;Trp53⁻ᐟ⁻	Glial progenitors	High-grade oligodendrogliomas, often supratentorial	[28]
	Viral delivery	PDGFB retrovirus	Oligodendrocyte precursor cells of newborn mice	Oligodendrogliomas Grade II Tumor incidence: 33%	[79]
	Transposon (PiggyBac)	RasV12	Embryonic oligodendrocyte lineage cells (starting at E16.5)	Oligodendrogliomas	[124]

ROS: reactive oxygen species, *5-hmC*: 5-hydroxymethylcytosine, *NSCs*: neural stem cells, *NSG*: NOD/scid-Il2Rgc knockout, *Ad:cre*: Cre-expressing adenovirus

any tumors [70]. To avoid embryonic lethality, Bardella et al. [71] targeted $Idh1^{R132H}$ to neural stem and progenitor cells in the two neurogenic niches of the adult mouse brain, the SVZ and dentate gyrus of the hippocampus, using a tamoxifen-inducible *Nestin-CreERT2* mouse line. *Nestin-CreERT2;Idh1^{R132H}* mice developed a number of abnormalities relevant to IDH1-driven gliomas, such as increased proliferation and self-renewal of stem/progenitors cells, cellular infiltration into surrounding brain regions, formation of subventricular nodules, and a gene expression profile overlapping that of human gliomas. This mouse model was the first one demonstrating that expression of Idh1^{R132H} in mouse SVZ induces a phenotype clearly related to gliomagenesis. In 2017, Pirozzi and coauthors [72] investigated the consequences of mutant Idh1 expression on developing mouse brains. Expression of *Idh1*-mutant allele was induced from E13.5 through the CNS by using *hGFAP-Cre* transgenic mice. Mutant brains showed hemorrhagic lesions, reduced proliferation of neural stem cells, and a disorganized SVZ. High levels of D-2HG were measured in the brain, possibly being partially responsible for these defects [70, 71]. Pirozzi and coauthors also targeted *Idh1* mutant allele to neural stem/progenitors cells of embryonic mouse brains, by using a tamoxifen-inducible *Nestin-CreERT2* line. Upon administration of tamoxifen to the pregnant mothers, recombination was observed from E18.5. However, the resulting animals were asymptomatic even after 1 year of observation [72]. These data were to some extent in contrast with the previous studies. The experimental discrepancies and the phenotypic differences observed in *Idh1*-mutant mice described above suggest that the timing in which this mutant gene is induced is important, as well as the cell types targeted. Also the levels of D-2HG produced need to be considered, which additionally seem to give rise to different consequences on developing compared to adult tissues.

While mutant *IDH1* is a main driver gene and an early event in gliomagenesis, these studies demonstrated that alone it is not sufficient to initiate gliomagenesis in vivo. Indeed, it has recently been demonstrated that mutant IDH1 requires multiple additional genetic alterations to determine astrocytoma development in vivo [73]. However, depending on the genetic background and the presence of co-occurring alterations (i.e., *Atrx, Trp53, Cdkn2a, Pten*), IDH1 mutation promotes or antagonizes tumor development [73–75] (see Table 1.2). Although recent studies seem to indicate that IDH-mutated gliomas probably originate from the transformation of a stem/progenitor cell or glial progenitor endowed with proliferative capacity [29, 71], the candidate cell of origin of these tumors is still elusive at the present, and more research is required to established whether mutant IDH targets neural stem/progenitor or more differentiated lineages.

3.2 Oligodend roglioma Models

3.2.1 Cell Lines and Xenograft Models

Oligodendrogliomas are slow-growing tumors compared to GBMs. Whereas the success rate for establishing GBM-derived cultures is about 50% in many laboratories, this value barely reaches 10% for *IDH*-mutated lower-grade gliomas and even less for oligodendrogliomas. As a result, just a few laboratories have successfully derived oligodendroglioma cultures, grown in serum-free conditions [62, 64, 65, 76]. These cell lines are slow-growing (needing passage every 14–28 days), are sensitive to chemotherapy treatment, and retain *IDH1^{R132H}* mutation and the 1p19q codeletion [62]. The fact that lower-grade gliomas contain less cells with stem-like properties compared to higher-grade gliomas may account for this behavior. Moreover, the nature of the tumor-driving population may be different in lower-grade gliomas (neural progenitor, OPC-like cells) compared to GBM (neural stem cell-like cells).

Very few oligodendroglioma intracranial xenograft models have been developed. Tumors were generated following grafting intracranially 1p19q codeleted cell lines in immunocompromised mice. However, in one study, the cell line did not maintain *IDH* mutation [62], while the other cell line that was implanted carried *IDH1* mutation but did not engraft. A few *IDH*-mutated xenografts were generated [63, 65], but in the case of the Luchman study (2012), the authors did not show that the cells harbored the 1p19q codeletion. More recently, Klink and colleagues were able to propagate oligodendrogliomas in mice that displayed hallmark genetic changes of oligodendrogliomas, such as *IDH1*, *CIC*, and *FUBP1* mutations, 1p19q codeletion, and activation of PDGFRA [66].

3.2.2 Genetically Engineered Mouse Models of Oligodendrogliomas

The lack of knowledge on the cellular and molecular mechanisms implicated in oligodendroglioma development is partly due to the lack of genetically relevant mouse models. Only a few genetically engineered mouse models of oligodendrogliomas have been developed. Most models use activation of the PDGF signaling pathway to generate oligodendrogliomas. Retroviral-mediated transduction of PDGF-B, a ligand for PDGFRα, into neural progenitor cells or OPCs in the developing brain resulted in oligodendrogliomas [77–79]. However, PDGF-B infusion in the lateral ventricles of the adult mouse brain was not sufficient to initiate tumorigenesis [80], suggesting that adult neural progenitor cells may require additional alterations to be transformed. Similarly, transduction of adult OPCs from the subcortical white matter with a PDGF-B retrovirus did not lead to tumor formation; tumors were instead generated only in a *Pten*-null;*Trp53*-null background but these tumors resembled glioblastomas [81]. These data suggest different sensitivities of postnatal and adult OPCs to PDGF signaling, which may be linked to the fact that postnatal and adult OPCs have different transcriptomic and cell cycle kinetics [82, 83].

Another model of oligodendroglioma takes advantage of an activated allele of EGFR (*v-erbB*). Transgenic mice expressing *v-erbB* in glial cells develop lower-grade gliomas that bear histopathological features of oligodendrogliomas, such as infrequent mitoses, rounded nuclei and clear haloes, and branching capillaries [84]. In this model, inactivation of *Cdkn2a* or *Trp53* increased tumor penetrance and grade. In the *v-erbB;Trp53*-null model, tumor cells express high levels of OPC markers, and OPC-like cells, rather than NSCs, drive tumor formation [26, 28, 84]. Accordingly, OPC-like cells from human oligodendrogliomas are highly tumorigenic and preferentially associate with white matter tracts, demonstrating a key contribution of this population to tumor development [28].

A limitation with the above models is that for the most part, they do not recapitulate the genetics of human oligodendrogliomas. First, since these models were created before the identification of *IDH* mutations in gliomas in 2009, it is not known whether this mutation is present in these models. In addition, it is not clear whether they retain Atrx expression and have low Trp53 levels. Second, the *v-erbB* (transforming allele of EGFR) model is based on EGFR overexpression, which is highly expressed in oligodendrogliomas but not specific for this tumor group [85]. The status of the loss of mouse chromosomal regions syntenic to 1p and 19q was assessed in these models: deletions of genomic regions homologous to 1p, 19q, and 10q were not detected in PDGF-induced gliomas [77]. Of note, in the *v-erbB* mouse model, a region on chromosome 4, orthologous with human chromosome 1p, was lost [84].

3.3 Emerging Models of Oligodendrogliomas

There thus is an urgent need to develop models that faithfully recapitulate the genetics of oligodendrogliomas. Among the most frequently mutated genes in oligodendrogliomas, two genes, *CIC* and *FUBP1*, are located on 19q and 1p, respectively. *CIC* is mutated in 58–62% and *FUBP1* in 29–31% of *IDH*-mutated and 1p19q-codeleted gliomas [7, 12]. We review here the models that have been generated for these genes.

3.3.1 CIC Models

First discovered in *Drosophila*, CIC (or Capicua) encodes a transcriptional repressor of the High Mobility Group (HMG) family. CIC is ubiquitously expressed and its function has been demonstrated to be necessary for normal lung alveolarization, bile acid homeostasis, and brain development [86]. In the brain, *Cic* conditional inactivation in the developing forebrain led to reduced cortical and dentate gyrus thickness, which resulted into hyperactivity and impaired learning and memory [87].

CIC protein acts as a transcriptional repressor, by interacting with cofactors of the Ataxin (ATXN1, ATXNL1) and histone deacetylase (SIN3) families [88, 89]. CIC represses expression of its target genes by binding to regulatory sequences in promoter or enhancers containing the octameric binding site of the HMG-box T(G/C)AATG(A/G)A [90]. The most characterized functions of CIC implicate this factor as a key sensor of EGFR/Ras/ERK signaling [86]. In the absence of ERK signaling, CIC represses the expression of RTK-responsive genes such as *ETV1*, *ETV4*, *ETV5*, and *SHC3* [91, 90], which are involved in proliferation and migration. Upon ERK activation, CIC repression is relieved, allowing expression of these target genes [92]. CIC activity appears to be mostly controlled at the posttranslational level. In fact, RTK/Ras/ERK pathway acts at several levels to reduce CIC activity: it mediates CIC ubiquitylation and degradation [93] or promotes its phosphorylation which results in attenuated DNA binding [91]. CIC was shown to modulate expression of genes in an EGFR-independent manner [94], and conversely, other pathways have been described to modulate CIC activity independently of RTK signaling, such as DYRK1A/Mnb [95]. Also, it was shown in *Drosophila* that CIC converges with Hippo signaling to repress and restrict the expression of a gene subset to prevent hyperproliferation [96]. Together these data suggest that CIC is an important integrator of signaling pathways in the cell, directing cell fate depending on the nature and intensity of the stimulus.

CIC gene is mutated in several cancers such as oligodendrogliomas; breast, lung, and gastric cancers [97, 98]; and T-cell lymphoblastic leukemia at a low frequency [99] where it was shown to be a major regulator of tumor progression and metastasis [97, 98]. Interestingly, in GBMs that do not display *CIC* gene alterations, CIC protein is continuously degraded by the proteasome as a consequence of hyperactive RTK/Ras/ERK signaling in these tumors [93]. This study highlights the major role of CIC in modulation of oncogenic pathways and the necessity for the tumor to inactivate it. In brain tumors, *CIC* mutations have been identified in about 60% of *IDH*-mutated, 1p19q codeleted tumors and are exclusive to this subgroup [100–103]. *CIC* mutations are associated with a poor outcome and a shorter time to anaplastic transformation in 1p19q codeleted gliomas [100]. Interestingly, *CIC* maps to chromosome region 19q13.2, suggesting that *CIC* mutations, which affect the unique remaining copy of the gene, may be partly responsible for the tumorigenic effect of the 1p19q codeletion. The majority of *CIC* mutations is located in exon 5, encoding the DNA binding domain of the protein, or in exon 20, which encodes a domain important for DNA binding and CIC repressor activity [101, 102, 104]. About half of the mutations result in a truncated form of the protein. In accordance with this observation, most of the missense mutations were associated with an undetectable expression of the CIC protein in glioma samples. Transcriptomic analysis revealed that *CIC* mutations were associated with an upregulation of genes normally repressed by CIC (*ETV1, ETV4, ETV5, CCND1, ASCL1*), as well as proliferative pathways and OPC genes [100]. Collectively, these data suggest that *CIC* mutations in oligodendrogliomas result in inactivation of the protein.

A few recent studies have given insights into CIC function in glioma development. *Cic* knockout mice are embryonically lethal, and the rare pups that survive show aberrant expansion of neural stem/progenitors in the subventricular zone, reminiscent of a preneoplastic phenotype [94]. In another model, *Cic* inactivation from *hGFAP-Cre*-expressing neural stem/progenitor cells did not induce any detectable lesion in the brain [99], indicating that loss of Cic alone is not sufficient to initiate gliomagenesis. *Cic*−/− NSC cultures fail to differentiate into mature oligodendrocytes and are arrested in an OPC-like state [94, 105] (Table 1.2). Our group has made similar observations from an in vivo conditional model of *Cic* inactivation (Lerond, Khenniche et al., unpublished data). Together, these data suggest that *CIC* mutations in oligodendrogliomas are not initiating events but may promote tumor development through the maintenance of NSC-like and OPC-like glioma progenitors.

3.3.2 FUBP1 Models

Far Upstream Element-Binding Protein 1 (FUBP1) gene is mutated in about 30% of lower-grade gliomas with *IDH* mutation and 1p19 codeletion, and these mutations are exclusive to this tumor subgroup [7, 12]. Since *FUBP1* gene is located on 1p, *FUBP1* mutation results in loss of heterozygosity in oligodendrogliomas. *FUBP1* encodes a multifunctional DNA- and RNA-binding protein that notably regulates *Myc* expression [106]. A limited number of studies have investigated the role of FUBP1 in brain development and tumorigenesis. *Fubp1* knockout mice are embryonically lethal and display hypercellularity in the brain parenchyma at perinatal stages, suggesting specific tumor suppressor function in the brain [107]. A recent study reported that Fubp1 is necessary for cell cycle exit and neuronal differentiation of neural progenitors, through regulation of alternative splicing of Lysine-specific Histone Demethylase 1A (LDS1/KDM1A) mRNA [108]. Interestingly, downregulation of Fubp1 by short hairpin RNA (shRNA) promoted tumor development when combined with expression of *IDH1*[R132H] and *PIK3CA*-activating mutation and loss of *Cdkn2a* [108]. However, the study did not report whether the resulting tumors display features of oligodendrogliomas.

4   Next Challenges in Oligodendroglioma Modeling

Models need to be optimized for both patient-derived cultures and genetic mouse models. We discuss below points to consider to improve current models to better recapitulate human oligodendrogliomas.

4.1   Choose the Right Combination of Mutations

There is currently no model that recapitulates the genetics of oligodendrogliomas. The minimal number of alterations required to initiate tumorigenesis has yet to be determined. Probably one of the biggest challenges is to successfully model the 1p19q codeletion, whether in human neural stem cells or in the mouse. Because the mouse genome is highly syntenic to the human genome, it is possible to engineer chromosomal rearrangements in the mouse genome, as this was done for several human syndromes [109]. However, in the case of oligodendrogliomas, human and mouse genomes have to be compared to ensure human 1p and 19q have syntenic regions. Mutations in the *TERT* promoter are found in nearly all oligodendrogliomas, resulting in increased telomerase expression [12]. Telomerase lengthens chromosome ends, thus providing unlimited proliferative potential. Telomerase is expressed at high levels in the developing brain but at low levels in the adult brain, being restricted to neural stem/progenitor cells niches [110]. Expression in oligodendrocyte lineage cells has not been reported. Future models will need to include activation of telomerase activity in combination with other genetic alterations. *CIC* and *FUBP1* mutations are exclusively found in *IDH*-mutated gliomas, raising the possibility that tumor formation requires synergy between these alterations. One study suggested cooperative effects of *IDH1* and *CIC* mutations in the production of the D2-HG oncometabolite from serum-grown cell lines [111], although these findings need to be replicated in more relevant cell types. Finally, other recurrent mutations have been identified and correspond to mutations in components of the PI3K-mTOR pathway (PIK3CA, PIK3R1, PTEN), inactivating mutations of the NOTCH receptors (NOTCH1, NOTCH2, NOTCH4), components of the SWI/SNF chromatin-remodeling complex (ARID1A, ARID1B, SMARCA4 (6–13%)), and neurodevelopmental transcription factors (TCF12, ZBTB20) [7, 12, 112]. Therefore, there are numerous combinations of mutations that can be envisioned in order to find the right code of alterations that will create oligodendroglioma in mice.

4.2   Choose the Timing of Mutation and the Cell Types to Be Targeted

Lower-grade gliomas are diagnosed at an average age of 30–40 years old [113]. The natural history of lower-grade gliomas is believed to occur in three phases: the occult phase, during which the first transformation events occur but are undetectable by MRI; the silent phase, during which a tumor can be detected but remains asymptomatic; and then the symptomatic phase [114]. The mean duration of the silent phase was estimated at 15 years when considering a linear growth of an average of 4 mm/year [114], suggesting that mutations may first appear in the late childhood/adolescence period. Based on these considerations, it seems reasonable to initiate tumorigenesis in mice at ages corresponding to postnatal/young adult stages rather than adult ages corresponding to the diagnosis age in the patients.

Another important consideration is the nature of the cell that should be initially targeted with mutations. As discussed above, OPCs appear as the main cell population responsible for tumor growth in oligodendrogliomas. However, recent work demonstrated the existence of cells with a neural stem/progenitor features that are cycling in oligodendrogliomas, suggesting that these cells may represent a subpopulation of tumor-driving cells in these tumors [29, 115]. Whether these populations are able to reconstitute oligodendrogliomas upon transplantation has not been addressed. Future models will have to target these cells with oligodendroglioma-relevant alterations to determine whether these lineages can drive tumor development.

4.3 Model Interactions Between Glioma Cells and the Microenvironment

Oligodendroglioma-derived cultures are difficult to establish and maintain. Oligodendroglioma cell lines that were successfully generated were so in the presence of neural stem cell medium (without serum and in the presence of EGF and bFGF). Cells did not grow in the absence of mitogens [62]. Klink et al. reported that tumor cells derived from oligodendroglioma xenografts did not expand in culture, in the presence of EGF, FGF, and/or PDGFA, or on different extracellular matrix substrates (laminin, fibronectin). These cells were highly adherent to newborn rat hippocampal neurons, yet this did not result in enhanced growth [66]. Therefore, culture conditions need to be optimized to increase the take rate for these tumors. An obvious and crucial factor to take into account is the brain microenvironment. Non-neoplastic cells that are located around gliomas represent numerous cell types and molecules, among which are immune cells, extracellular matrix components, neural cells (astrocytes, oligodendrocytes, neurons), and blood vessels. All these cell populations interact with tumor cells to modulate tumor biology (for a review, see [116]). It is therefore crucial to consider these interactions when culturing glioma cells and to develop systems that allow the investigations of these interactions. Several GEMMs have been developed to characterize the interactions between glioblastoma cells and microglia, endogenous neural cell populations, or the vasculature (for review, see [116]). These GEMMs and xenografts are, however, expensive and not ideal for live tracking of tumor cell behavior, such as invasion. Glioma cells migrate along white matter tracts and blood vessels, below meninges, and cluster around neurons (perineuronal satellitosis) [117]. Sophisticated approaches such as live imaging using multiphoton intravital microscopy enabled to demonstrate the in vivo migration behavior of GBM cells, which migrated as single cells along blood vessels [76]. However, this approach is limited by the low imaging depth. Another approach for assessing invasion or tumor-stroma interactions in vivo is to take advantage of new methods that allow optical clearing of the brain. With the iDISCO technique, a whole brain can be immunostained with several markers and imaged using light sheet fluorescence microscopy, yielding 3D information on a particular process [118]. This method was used recently to monitor the migration behavior of GBM cells in different brain regions [119]. Also, organotypic brain slice systems have been developed, in which GBM cells are grafted onto a slice of adult mouse brain and maintained for several weeks [120, 121]. These systems enable monitoring live cell behavior and are particularly suitable for studying tumor cell invasion and interactions between tumor cells and the stroma (e.g., with endothelial cells) [76].

Brain organoids have been recently developed and represent exciting new tools to model neurodevelopmental and neurological disorders. These self-organizing structures reproduce the layered organization of the brain and can be maintained over long periods of time. Brain organoids can be directed to differentiate into dorsal or ventral forebrain regions. A recent study reported the generation of brain organoids containing oligodendrocytes that temporally and spatially overlap with astrocytes and neurons [122]. In this system, stages of OPC development were detected, including mature oligodendrocytes that myelinate neuronal axons. Such a system may be valuable to assess functionally mutations found in oligodendrogliomas. These structures can be modified (by virus-medicated gene transfer or CRISPR/Cas9 genome edition) and be transplanted into immunodeficient mice. Interestingly, Ogawa and colleagues recently demonstrated that brain organoids represent a model for monitoring glioma initiation and development [123]. In this system, organoid cells can be directly modified (by CRISPR/Cas9 genome editing) to express activated Ras and disrupt TP53, resulting in the generation of tumor cells that can be tracked for tumor progression and invasion in the organoid. Patient-derived glioblastoma cells can also be transplanted on the organoid and their proliferative and invasive behavior monitored [123]. Organoid approaches represent a new approach that has been successfully used for glioblastomas; they now need to be optimized to lower-grade gliomas.

5 Concluding Remarks

In conclusion, novel models that faithfully recapitulate the genetics of human lower-grade gliomas, and particularly oligodendrogliomas, are crucially needed. These models may serve as preclinical platforms and accelerate the transition of promising drugs to the clinic. Importantly, integration of data generated from these models and from transcriptomic and genomic studies on patient samples is highly needed to improve our knowledge on oligodendroglioma development and to design better treatments.

Acknowledgment

We thank Giorgios Solomou, Marc Sanson, and Jean-Philippe Hugnot for constructive feedbacks on the manuscript. We thank Karima Mokhtari for providing histological examples in Fig. 1.1. Research in the Huillard team is funded by the Ligue Nationale contre le Cancer, Fondation ARC, OligoNation and Operation OligoCure associations. We acknowledge the contribution of SiRIC CURAMUS (INCA-DGOS-Inserm_12560) which is financially supported by the French National Cancer Institute, the French Ministry of Solidarity and Health, and Inserm. Bardella team is funded by Cancer Research UK and OligoNation and Operation OligoCure associations.

References

1. Ostrom QT, Gittleman H, Liao P, Vecchione-Koval T, Wolinsky Y, Kruchko C, Barnholtz-Sloan JS (2017) CBTRUS statistical report: primary brain and other central nervous system tumors diagnosed in the United States in 2010–2014. Neuro-Oncology 19(suppl_5):v1–v88. https://doi.org/10.1093/neuonc/nox158

2. Louis DN, Perry A, Reifenberger G, von Deimling A, Figarella-Branger D, Cavenee WK, Ohgaki H, Wiestler OD, Kleihues P, Ellison DW (2016) The 2016 World Health Organization classification of tumors of the central nervous system: a summary. Acta Neuropathol 131(6):803–820. https://doi.org/10.1007/s00401-016-1545-1

3. Ohgaki H, Kleihues P (2013) The definition of primary and secondary glioblastoma. Clin Cancer Res 19(4):764–772. https://doi.org/10.1158/1078-0432.CCR-12-3002

4. The Cancer Genome Atlas Research Network (2008) Comprehensive genomic characterization defines human glioblastoma genes and core pathways. Nature 455(7216):1061–1068. https://doi.org/10.1038/nature07385

5. Sottoriva A, Spiteri I, Piccirillo SG, Touloumis A, Collins VP, Marioni JC, Curtis C, Watts C, Tavare S (2013) Intratumor heterogeneity in human glioblastoma reflects cancer evolutionary dynamics. Proc Natl Acad Sci U S A 110(10):4009–4014. https://doi.org/10.1073/pnas.1219747110

6. Ozawa T, Riester M, Cheng YK, Huse JT, Squatrito M, Helmy K, Charles N, Michor F, Holland EC (2014) Most human non-GCIMP glioblastoma subtypes evolve from a common proneural-like precursor glioma. Cancer Cell 26(2):288–300. https://doi.org/10.1016/j.ccr.2014.06.005

7. Suzuki H, Aoki K, Chiba K, Sato Y, Shiozawa Y, Shiraishi Y, Shimamura T, Niida A, Motomura K, Ohka F, Yamamoto T, Tanahashi K, Ranjit M, Wakabayashi T, Yoshizato T, Kataoka K, Yoshida K, Nagata Y, Sato-Otsubo A, Tanaka H, Sanada M,

Kondo Y, Nakamura H, Mizoguchi M, Abe T, Muragaki Y, Watanabe R, Ito I, Miyano S, Natsume A, Ogawa S (2015) Mutational landscape and clonal architecture in grade II and III gliomas. Nat Genet 47(5):458–468. https://doi.org/10.1038/ng.3273

8. Dang L, White DW, Gross S, Bennett BD, Bittinger MA, Driggers EM, Fantin VR, Jang HG, Jin S, Keenan MC, Marks KM, Prins RM, Ward PS, Yen KE, Liau LM, Rabinowitz JD, Cantley LC, Thompson CB, Vander Heiden MG, Su SM (2009) Cancer-associated IDH1 mutations produce 2-hydroxyglutarate. Nature 462(7274):739–744. https://doi.org/10.1038/nature08617

9. Xu W, Yang H, Liu Y, Yang Y, Wang P, Kim SH, Ito S, Yang C, Wang P, Xiao MT, Liu LX, Jiang WQ, Liu J, Zhang JY, Wang B, Frye S, Zhang Y, Xu YH, Lei QY, Guan KL, Zhao SM, Xiong Y (2011) Oncometabolite 2-hydroxyglutarate is a competitive inhibitor of alpha-ketoglutarate-dependent dioxygenases. Cancer Cell 19(1):17–30. https://doi.org/10.1016/j.ccr.2010.12.014

10. Lu C, Ward PS, Kapoor GS, Rohle D, Turcan S, Abdel-Wahab O, Edwards CR, Khanin R, Figueroa ME, Melnick A, Wellen KE, O'Rourke DM, Berger SL, Chan TA, Levine RL, Mellinghoff IK, Thompson CB (2012) IDH mutation impairs histone demethylation and results in a block to cell differentiation. Nature 483(7390):474–478. https://doi.org/10.1038/nature10860

11. Flavahan WA, Drier Y, Liau BB, Gillespie SM, Venteicher AS, Stemmer-Rachamimov AO, Suva ML, Bernstein BE (2016) Insulator dysfunction and oncogene activation in IDH mutant gliomas. Nature 529(7584):110–114. https://doi.org/10.1038/nature16490

12. Cancer Genome Atlas Research N, Brat DJ, Verhaak RG, Aldape KD, Yung WK, Salama SR, Cooper LA, Rheinbay E, Miller CR, Vitucci M, Morozova O, Robertson AG, Noushmehr H, Laird PW, Cherniack AD, Akbani R, Huse JT, Ciriello G, Poisson LM, Barnholtz-Sloan JS, Berger MS, Brennan C, Colen RR, Colman H, Flanders AE, Giannini C, Grifford M, Iavarone A, Jain R, Joseph I, Kim J, Kasaian K, Mikkelsen T, Murray BA, O'Neill BP, Pachter L, Parsons DW, Sougnez C, Sulman EP, Vandenberg SR, Van Meir EG, von Deimling A, Zhang H, Crain D, Lau K, Mallery D, Morris S, Paulauskis J, Penny R, Shelton T, Sherman M, Yena P, Black A, Bowen J, Dicostanzo K, Gastier-Foster J, Leraas KM, Lichtenberg TM, Pierson CR, Ramirez NC, Taylor C, Weaver S, Wise L, Zmuda E, Davidsen T, Demchok JA, Eley G, Ferguson ML, Hutter CM, Mills Shaw KR, Ozenberger BA, Sheth M, Sofia HJ, Tarnuzzer R, Wang Z, Yang L, Zenklusen JC, Ayala B, Baboud J, Chudamani S, Jensen MA, Liu J, Pihl T, Raman R, Wan Y, Wu Y, Ally A, Auman JT, Balasundaram M, Balu S, Baylin SB, Beroukhim R, Bootwalla MS, Bowlby R, Bristow CA, Brooks D, Butterfield Y, Carlsen R, Carter S, Chin L, Chu A, Chuah E, Cibulskis K, Clarke A, Coetzee SG, Dhalla N, Fennell T, Fisher S, Gabriel S, Getz G, Gibbs R, Guin R, Hadjipanayis A, Hayes DN, Hinoue T, Hoadley K, Holt RA, Hoyle AP, Jefferys SR, Jones S, Jones CD, Kucherlapati R, Lai PH, Lander E, Lee S, Lichtenstein L, Ma Y, Maglinte DT, Mahadeshwar HS, Marra MA, Mayo M, Meng S, Meyerson ML, Mieczkowski PA, Moore RA, Mose LE, Mungall AJ, Pantazi A, Parfenov M, Park PJ, Parker JS, Perou CM, Protopopov A, Ren X, Roach J, Sabedot TS, Schein J, Schumacher SE, Seidman JG, Seth S, Shen H, Simons JV, Sipahimalani P, Soloway MG, Song X, Sun H, Tabak B, Tam A, Tan D, Tang J, Thiessen N, Triche T Jr, Van Den Berg DJ, Veluvolu U, Waring S, Weisenberger DJ, Wilkerson MD, Wong T, Wu W, Xi L, Xu AW, Yang L, Zack TI, Zhang J, Aksoy BA, Arachchi H, Benz C, Bernard B, Carlin D, Cho J, DiCara D, Frazer S, Fuller GN, Gao J, Gehlenborg N, Haussler D, Heiman DI, Iype L, Jacobsen A, Ju Z, Katzman S, Kim H, Knijnenburg T, Kreisberg RB, Lawrence MS, Lee W, Leinonen K, Lin P, Ling S, Liu W, Liu Y, Liu Y, Lu Y, Mills G, Ng S, Noble MS, Paull E, Rao A, Reynolds S, Saksena G, Sanborn Z, Sander C, Schultz N, Senbabaoglu Y, Shen R, Shmulevich I, Sinha R, Stuart J, Sumer SO, Sun Y, Tasman N, Taylor BS, Voet D, Weinhold N, Weinstein JN, Yang D, Yoshihara K, Zheng S, Zhang W, Zou L, Abel T, Sadeghi S, Cohen ML, Eschbacher J, Hattab EM, Raghunathan A, Schniederjan MJ, Aziz D, Barnett G, Barrett W, Bigner DD, Boice L, Brewer C, Calatozzolo C, Campos B, Carlotti CG Jr, Chan TA, Cuppini L, Curley E, Cuzzubbo S, Devine K, DiMeco F, Duell R, Elder JB, Fehrenbach A, Finocchiaro G, Friedman W, Fulop J, Gardner J, Hermes B, Herold-Mende C, Jungk C, Kendler A, Lehman NL, Lipp E, Liu O, Mandt R, McGraw M, McLendon R, McPherson C, Neder L, Nguyen P, Noss A, Nunziata R, Ostrom QT, Palmer C, Perin A, Pollo B, Potapov A, Potapova O, Rathmell WK, Rotin D, Scarpace L, Schilero C, Senecal K, Shimmel K, Shurkhay V, Sifri S, Singh R, Sloan AE, Smolenski K, Staugaitis SM, Steele R, Thorne L, Tirapelli DP, Unterberg A, Vallurupalli M, Wang Y, Warnick R, Williams F, Wolinsky Y, Bell S, Rosenberg M, Stewart C, Huang F, Grimsby JL, Radenbaugh AJ, Zhang J (2015)

Comprehensive, integrative genomic analysis of diffuse lower-grade Gliomas. N Engl J Med 372(26):2481–2498. https://doi.org/10.1056/NEJMoa1402121

13. Bromberg JE, van den Bent MJ (2009) Oligodendrogliomas: molecular biology and treatment. Oncologist 14(2):155–163. https://doi.org/10.1634/theoncologist.2008-0248

14. Griffin CA, Burger P, Morsberger L, Yonescu R, Swierczynski S, Weingart JD, Murphy KM (2006) Identification of der(1,19)(q10;p10) in five oligodendrogliomas suggests mechanism of concurrent 1p and 19q loss. J Neuropathol Exp Neurol 65(10):988–994. https://doi.org/10.1097/01.jnen.0000235122.98052.8f

15. Jenkins RB, Blair H, Ballman KV, Giannini C, Arusell RM, Law M, Flynn H, Passe S, Felten S, Brown PD, Shaw EG, Buckner JC (2006) A t(1,19)(q10;p10) mediates the combined deletions of 1p and 19q and predicts a better prognosis of patients with oligodendroglioma. Cancer Res 66(20):9852–9861. https://doi.org/10.1158/0008-5472.CAN-06-1796

16. Killela PJ, Reitman ZJ, Jiao Y, Bettegowda C, Agrawal N, Diaz LA Jr, Friedman AH, Friedman H, Gallia GL, Giovanella BC, Grollman AP, He TC, He Y, Hruban RH, Jallo GI, Mandahl N, Meeker AK, Mertens F, Netto GJ, Rasheed BA, Riggins GJ, Rosenquist TA, Schiffman M, Shih Ie M, Theodorescu D, Torbenson MS, Velculescu VE, Wang TL, Wentzensen N, Wood LD, Zhang M, McLendon RE, Bigner DD, Kinzler KW, Vogelstein B, Papadopoulos N, Yan H (2013) TERT promoter mutations occur frequently in gliomas and a subset of tumors derived from cells with low rates of self-renewal. Proc Natl Acad Sci U S A 110(15):6021–6026. https://doi.org/10.1073/pnas.1303607110

17. Huang FW, Hodis E, Xu MJ, Kryukov GV, Chin L, Garraway LA (2013) Highly recurrent TERT promoter mutations in human melanoma. Science 339(6122):957–959. https://doi.org/10.1126/science.1229259

18. Hanahan D, Weinberg RA (2011) Hallmarks of cancer: the next generation. Cell 144(5):646–674. https://doi.org/10.1016/j.cell.2011.02.013

19. Kriegstein A, Alvarez-Buylla A (2009) The glial nature of embryonic and adult neural stem cells. Annu Rev Neurosci 32:149–184. https://doi.org/10.1146/annurev.neuro.051508.135600

20. Richardson WD, Young KM, Tripathi RB, McKenzie I (2011) NG2-glia as multipotent neural stem cells: fact or fantasy? Neuron 70(4):661–673. https://doi.org/10.1016/j.neuron.2011.05.013

21. Lee JH, Lee JE, Kahng JY, Kim SH, Park JS, Yoon SJ, Um JY, Kim WK, Lee JK, Park J, Kim EH, Lee JH, Lee JH, Chung WS, Ju YS, Park SH, Chang JH, Kang SG, Lee JH (2018) Human glioblastoma arises from subventricular zone cells with low-level driver mutations. Nature 560(7717):243–247. https://doi.org/10.1038/s41586-018-0389-3

22. Zong H, Parada LF, Baker SJ (2015) Cell of origin for malignant Gliomas and its implication in therapeutic development. Cold Spring Harb Perspect Biol 7(5). https://doi.org/10.1101/cshperspect.a020610

23. Bardella C, Al-Shammari AR, Soares L, Tomlinson I, O'Neill E, Szele FG (2018) The role of inflammation in subventricular zone cancer. Prog Neurobiol 170:37–52. https://doi.org/10.1016/j.pneurobio.2018.04.007

24. Friedmann-Morvinski D, Bushong EA, Ke E, Soda Y, Marumoto T, Singer O, Ellisman MH, Verma IM (2012) Dedifferentiation of neurons and astrocytes by oncogenes can induce gliomas in mice. Science 338(6110):1080–1084. https://doi.org/10.1126/science.1226929

25. Llaguno SA, Sun D, Pedraza AM, Vera E, Wang Z, Burns DK, Parada LF (2019) Cell-of-origin susceptibility to glioblastoma formation declines with neural lineage restriction. Nat Neurosci. https://doi.org/10.1038/s41593-018-0333-8

26. Sugiarto S, Persson AI, Munoz EG, Waldhuber M, Lamagna C, Andor N, Hanecker P, Ayers-Ringler J, Phillips J, Siu J, Lim DA, Vandenberg S, Stallcup W, Berger MS, Bergers G, Weiss WA, Petritsch C (2011) Asymmetry-defective oligodendrocyte progenitors are glioma precursors. Cancer Cell 20(3):328–340. https://doi.org/10.1016/j.ccr.2011.08.011

27. Liu C, Sage JC, Miller MR, Verhaak RG, Hippenmeyer S, Vogel H, Foreman O, Bronson RT, Nishiyama A, Luo L, Zong H (2011) Mosaic analysis with double markers reveals tumor cell of origin in glioma. Cell 146(2):209–221. https://doi.org/10.1016/j.cell.2011.06.014

28. Persson AI, Petritsch C, Swartling FJ, Itsara M, Sim FJ, Auvergne R, Goldenberg DD, Vandenberg SR, Nguyen KN, Yakovenko S, Ayers-Ringler J, Nishiyama A, Stallcup WB, Berger MS, Bergers G, McKnight TR, Goldman SA, Weiss WA (2010) Non-stem cell origin for oligodendroglioma. Cancer Cell 18(6):669–682. https://doi.org/10.1016/j.ccr.2010.10.033

29. Tirosh I, Venteicher AS, Hebert C, Escalante LE, Patel AP, Yizhak K, Fisher JM, Rodman C, Mount C, Filbin MG, Neftel C, Desai N, Nyman J, Izar B, Luo CC, Francis JM, Patel AA, Onozato ML, Riggi N, Livak KJ, Gennert D, Satija R, Nahed BV, Curry WT, Martuza RL, Mylvaganam R, Iafrate AJ, Frosch MP, Golub TR, Rivera MN, Getz G, Rozenblatt-Rosen O, Cahill DP, Monje M, Bernstein BE, Louis DN, Regev A, Suva ML (2016) Single-cell RNA-seq supports a developmental hierarchy in human oligodendroglioma. Nature 539(7628):309–313. https://doi.org/10.1038/nature20123

30. Bielle F, Ducray F, Mokhtari K, Dehais C, Adle-Biassette H, Carpentier C, Chanut A, Polivka M, Poggioli S, Rosenberg S, Giry M, Marie Y, Duyckaerts C, Sanson M, Figarella-Branger D, Idbaih A, Pola N (2016) Tumor cells with neuronal intermediate progenitor features define a subgroup of 1p/19q co-deleted anaplastic gliomas. Brain Pathol. https://doi.org/10.1111/bpa.12434

31. Choi Y, Lee S, Kim K, Kim SH, Chung YJ, Lee C (2018) Studying cancer immunotherapy using patient-derived xenografts (PDXs) in humanized mice. Exp Mol Med 50(8):99. https://doi.org/10.1038/s12276-018-0115-0

32. Semenkow S, Li S, Kahlert UD, Raabe EH, Xu J, Arnold A, Janowski M, Oh BC, Brandacher G, Bulte JWM, Eberhart CG, Walczak P (2017) An immunocompetent mouse model of human glioblastoma. Oncotarget 8(37):61072–61082. https://doi.org/10.18632/oncotarget.17851

33. Cheon DJ, Orsulic S (2011) Mouse models of cancer. Annu Rev Pathol 6:95–119. https://doi.org/10.1146/annurev.pathol.3.121806.154244

34. Day CP, Merlino G, Van Dyke T (2015) Preclinical mouse cancer models: a maze of opportunities and challenges. Cell 163(1):39–53. https://doi.org/10.1016/j.cell.2015.08.068

35. Reilly KM, Loisel DA, Bronson RT, McLaughlin ME, Jacks T (2000) Nf1;Trp53 mutant mice develop glioblastoma with evidence of strain-specific effects. Nat Genet 26(1):109–113. https://doi.org/10.1038/79075

36. Dhaliwal J, Lagace DC (2011) Visualization and genetic manipulation of adult neurogenesis using transgenic mice. Eur J Neurosci 33(6):1025–1036. https://doi.org/10.1111/j.1460-9568.2011.07600.x

37. Zhu Y, Guignard F, Zhao D, Liu L, Burns DK, Mason RP, Messing A, Parada LF (2005) Early inactivation of p53 tumor suppressor gene cooperating with NF1 loss induces malignant astrocytoma. Cancer Cell 8(2):119–130. https://doi.org/10.1016/j.ccr.2005.07.004

38. Zhu H, Acquaviva J, Ramachandran P, Boskovitz A, Woolfenden S, Pfannl R, Bronson RT, Chen JW, Weissleder R, Housman DE, Charest A (2009) Oncogenic EGFR signaling cooperates with loss of tumor suppressor gene functions in gliomagenesis. Proc Natl Acad Sci U S A 106(8):2712–2716. https://doi.org/10.1073/pnas.0813314106

39. Alcantara Llaguno S, Chen J, Kwon CH, Jackson EL, Li Y, Burns DK, Alvarez-Buylla A, Parada LF (2009) Malignant astrocytomas originate from neural stem/progenitor cells in a somatic tumor suppressor mouse model. Cancer Cell 15(1):45–56. https://doi.org/10.1016/j.ccr.2008.12.006

40. Robinson JP, Vanbrocklin MW, McKinney AJ, Gach HM, Holmen SL (2011) Akt signaling is required for glioblastoma maintenance in vivo. Am J Cancer Res 1(2):155–167

41. Howarth JL, Lee YB, Uney JB (2010) Using viral vectors as gene transfer tools (cell biology and toxicology special issue: ETCS-UK 1 day meeting on genetic manipulation of cells). Cell Biol Toxicol 26(1):1–20. https://doi.org/10.1007/s10565-009-9139-5

42. Orsulic S (2002) An RCAS-TVA-based approach to designer mouse models. Mamm Genome 13(10):543–547. https://doi.org/10.1007/s00335-002-4003-4

43. Moore LM, Holmes KM, Fuller GN, Zhang W (2011) Oncogene interactions are required for glioma development and progression as revealed by a tissue specific transgenic mouse model. Chin J Cancer 30(3):163–172

44. Holland EC, Celestino J, Dai C, Schaefer L, Sawaya RE, Fuller GN (2000) Combined activation of Ras and Akt in neural progenitors induces glioblastoma formation in mice. Nat Genet 25(1):55–57. https://doi.org/10.1038/75596

45. Lentz TB, Gray SJ, Samulski RJ (2012) Viral vectors for gene delivery to the central nervous system. Neurobiol Dis 48(2):179–188. https://doi.org/10.1016/j.nbd.2011.09.014

46. Manfredsson FP (2016) Introduction to viral vectors and other delivery methods for gene therapy of the nervous system. Methods Mol Biol 1382:3–18. https://doi.org/10.1007/978-1-4939-3271-9_1

47. Zolotukhin I, Luo D, Gorbatyuk O, Hoffman B, Warrington K Jr, Herzog R, Harrison J, Cao O (2013) Improved Adeno-associated viral gene transfer to murine Glioma. J Genet Syndr Gene Ther 4(133). https://doi.org/10.4172/2157-7412.1000133

48. Chow RD, Guzman CD, Wang G, Schmidt F, Youngblood MW, Ye L, Errami Y, Dong MB, Martinez MA, Zhang S, Renauer P, Bilguvar K, Gunel M, Sharp PA, Zhang F, Platt RJ, Chen S (2017) AAV-mediated direct in vivo CRISPR screen identifies functional suppressors in glioblastoma. Nat Neurosci 20(10):1329–1341. https://doi.org/10.1038/nn.4620

49. Bloquel C, Fabre E, Bureau MF, Scherman D (2004) Plasmid DNA electrotransfer for intracellular and secreted proteins expression: new methodological developments and applications. J Gene Med 6(Suppl 1):S11–S23. https://doi.org/10.1002/jgm.508

50. Bigey P, Bureau MF, Scherman D (2002) In vivo plasmid DNA electrotransfer. Curr Opin Biotechnol 13(5):443–447

51. Copeland NG, Jenkins NA (2010) Harnessing transposons for cancer gene discovery. Nat Rev Cancer 10(10):696–706. https://doi.org/10.1038/nrc2916

52. Lampreht Tratar U, Horvat S, Cemazar M (2018) Transgenic mouse models in Cancer Research. Front Oncol 8:268. https://doi.org/10.3389/fonc.2018.00268

53. Koso H, Takeda H, Yew CC, Ward JM, Nariai N, Ueno K, Nagasaki M, Watanabe S, Rust AG, Adams DJ, Copeland NG, Jenkins NA (2012) Transposon mutagenesis identifies genes that transform neural stem cells into glioma-initiating cells. Proc Natl Acad Sci U S A 109(44):E2998–E3007. https://doi.org/10.1073/pnas.1215899109

54. Bender AM, Collier LS, Rodriguez FJ, Tieu C, Larson JD, Halder C, Mahlum E, Kollmeyer TM, Akagi K, Sarkar G, Largaespada DA, Jenkins RB (2010) Sleeping beauty-mediated somatic mutagenesis implicates CSF1 in the formation of high-grade astrocytomas. Cancer Res 70(9):3557–3565. https://doi.org/10.1158/0008-5472.CAN-09-4674

55. Sumiyoshi K, Koso H, Watanabe S (2018) Spontaneous development of intratumoral heterogeneity in a transposon-induced mouse model of glioma. Cancer Sci 109(5):1513–1523. https://doi.org/10.1111/cas.13579

56. Reynolds BA, Weiss S (1992) Generation of neurons and astrocytes from isolated cells of the adult mammalian central nervous system. Science 255(5052):1707–1710

57. Balvers RK, Kleijn A, Kloezeman JJ, French PJ, Kremer A, van den Bent MJ, Dirven CM, Leenstra S, Lamfers ML (2013) Serum-free culture success of glial tumors is related to specific molecular profiles and expression of extracellular matrix-associated gene modules. Neuro-Oncology 15(12):1684–1695. https://doi.org/10.1093/neuonc/not116

58. Conti L, Pollard SM, Gorba T, Reitano E, Toselli M, Biella G, Sun Y, Sanzone S, Ying QL, Cattaneo E, Smith A (2005) Niche-independent symmetrical self-renewal of a mammalian tissue stem cell. PLoS Biol 3(9):e283. https://doi.org/10.1371/journal.pbio.0030283

59. Brewer GJ, Torricelli JR, Evege EK, Price PJ (1993) Optimized survival of hippocampal neurons in B27-supplemented Neurobasal, a new serum-free medium combination. J Neurosci Res 35(5):567–576. https://doi.org/10.1002/jnr.490350513

60. Lee J, Kotliarova S, Kotliarov Y, Li A, Su Q, Donin NM, Pastorino S, Purow BW, Christopher N, Zhang W, Park JK, Fine HA (2006) Tumor stem cells derived from glioblastomas cultured in bFGF and EGF more closely mirror the phenotype and genotype of primary tumors than do serum-cultured cell lines. Cancer Cell 9(5):391–403. https://doi.org/10.1016/j.ccr.2006.03.030

61. Piaskowski S, Bienkowski M, Stoczynska-Fidelus E, Stawski R, Sieruta M, Szybka M, Papierz W, Wolanczyk M, Jaskolski DJ, Liberski PP, Rieske P (2011) Glioma cells showing IDH1 mutation cannot be propagated in standard cell culture conditions. Br J Cancer 104(6):968–970. https://doi.org/10.1038/bjc.2011.27

62. Kelly JJ, Blough MD, Stechishin OD, Chan JA, Beauchamp D, Perizzolo M, Demetrick DJ, Steele L, Auer RN, Hader WJ, Westgate M, Parney IF, Jenkins R, Cairncross JG, Weiss S (2010) Oligodendroglioma cell lines containing t(1,19)(q10;p10). Neuro-Oncology 12(7):745–755. https://doi.org/10.1093/neuonc/noq031

63. Luchman HA, Stechishin OD, Dang NH, Blough MD, Chesnelong C, Kelly JJ, Nguyen SA, Chan JA, Weljie AM, Cairncross JG, Weiss S (2012) An in vivo patient-derived model of endogenous IDH1-mutant glioma. Neuro-Oncology 14(2):184–191. https://doi.org/10.1093/neuonc/nor207

64. Koivunen P, Lee S, Duncan CG, Lopez G, Lu G, Ramkissoon S, Losman JA, Joensuu P, Bergmann U, Gross S, Travins J, Weiss S, Looper R, Ligon KL, Verhaak RG, Yan H, Kaelin WG Jr (2012) Transformation by the (R)-enantiomer of 2-hydroxyglutarate linked to EGLN activation. Nature 483(7390):484–488. https://doi.org/10.1038/nature10898

65. Rohle D, Popovici-Muller J, Palaskas N, Turcan S, Grommes C, Campos C, Tsoi J, Clark O, Oldrini B, Komisopoulou E, Kunii K, Pedraza A, Schalm S, Silverman L, Miller A, Wang F, Yang H, Chen Y, Kernytsky A, Rosenblum MK, Liu W, Biller SA, Su SM, Brennan CW, Chan TA, Graeber TG, Yen

KE, Mellinghoff IK (2013) An inhibitor of mutant IDH1 delays growth and promotes differentiation of glioma cells. Science 340(6132):626–630. https://doi.org/10.1126/science.1236062

66. Klink B, Miletic H, Stieber D, Huszthy PC, Valenzuela JA, Balss J, Wang J, Schubert M, Sakariassen PO, Sundstrom T, Torsvik A, Aarhus M, Mahesparan R, von Deimling A, Kaderali L, Niclou SP, Schrock E, Bjerkvig R, Nigro JM (2013) A novel, diffusely infiltrative xenograft model of human anaplastic oligodendroglioma with mutations in FUBP1, CIC, and IDH1. PLoS One 8(3):e59773. https://doi.org/10.1371/journal.pone.0059773

67. Turcan S, Makarov V, Taranda J, Wang Y, Fabius AWM, Wu W, Zheng Y, El-Amine N, Haddock S, Nanjangud G, LeKaye HC, Brennan C, Cross J, Huse JT, Kelleher NL, Osten P, Thompson CB, Chan TA (2018) Mutant-IDH1-dependent chromatin state reprogramming, reversibility, and persistence. Nat Genet 50(1):62–72. https://doi.org/10.1038/s41588-017-0001-z

68. Claes A, Schuuring J, Boots-Sprenger S, Hendriks-Cornelissen S, Dekkers M, van der Kogel AJ, Leenders WP, Wesseling P, Jeuken JW (2008) Phenotypic and genotypic characterization of orthotopic human glioma models and its relevance for the study of anti-glioma therapy. Brain Pathol 18(3):423–433. https://doi.org/10.1111/j.1750-3639.2008.00141.x

69. Dang L, Yen K, Attar EC (2016) IDH mutations in cancer and progress toward development of targeted therapeutics. Ann Oncol 27(4):599–608. https://doi.org/10.1093/annonc/mdw013

70. Sasaki M, Knobbe CB, Itsumi M, Elia AJ, Harris IS, Chio II, Cairns RA, McCracken S, Wakeham A, Haight J, Ten AY, Snow B, Ueda T, Inoue S, Yamamoto K, Ko M, Rao A, Yen KE, Su SM, Mak TW (2012) D-2-hydroxyglutarate produced by mutant IDH1 perturbs collagen maturation and basement membrane function. Genes Dev 26(18):2038–2049. https://doi.org/10.1101/gad.198200.112

71. Bardella C, Al-Dalahmah O, Krell D, Brazauskas P, Al-Qahtani K, Tomkova M, Adam J, Serres S, Lockstone H, Freeman-Mills L, Pfeffer I, Sibson N, Goldin R, Schuster-Boeckler B, Pollard PJ, Soga T, McCullagh JS, Schofield CJ, Mulholland P, Ansorge O, Kriaucionis S, Ratcliffe PJ, Szele FG, Tomlinson I (2016) Expression of Idh1R132H in the murine subventricular zone stem cell niche recapitulates features of early gliomagenesis. Cancer Cell 30(4):578–594. https://doi.org/10.1016/j.ccell.2016.08.017

72. Pirozzi CJ, Carpenter AB, Waitkus MS, Wang CY, Zhu H, Hansen LJ, Chen LH, Greer PK, Feng J, Wang Y, Bock CB, Fan P, Spasojevic I, McLendon RE, Bigner DD, He Y, Yan H (2017) Mutant IDH1 disrupts the mouse subventricular zone and alters brain tumor progression. Mol Cancer Res 15(5):507–520. https://doi.org/10.1158/1541-7786.MCR-16-0485

73. Philip B, Yu DX, Silvis MR, Shin CH, Robinson JP, Robinson GL, Welker AE, Angel SN, Tripp SR, Sonnen JA, VanBrocklin MW, Gibbons RJ, Looper RE, Colman H, Holmen SL (2018) Mutant IDH1 promotes glioma formation in vivo. Cell Rep 23(5):1553–1564. https://doi.org/10.1016/j.celrep.2018.03.133

74. Nunez FJ, Mendez FM, Kadiyala P, Alghamri MS, Savelieff MG, Garcia-Fabiani MB, Haase S, Koschmann C, Calinescu AA, Kamran N, Saxena M, Patel R, Carney S, Guo MZ, Edwards M, Ljungman M, Qin T, Sartor MA, Tagett R, Venneti S, Brosnan-Cashman J, Meeker A, Gorbunova V, Zhao L, Kremer DM, Zhang L, Lyssiotis CA, Jones L, Herting CJ, Ross JL, Hambardzumyan D, Hervey-Jumper S, Figueroa ME, Lowenstein PR, Castro MG (2019) IDH1-R132H acts as a tumor suppressor in glioma via epigenetic up-regulation of the DNA damage response. Sci Transl Med 11(479). https://doi.org/10.1126/scitranslmed.aaq1427

75. Amankulor NM, Kim Y, Arora S, Kargl J, Szulzewsky F, Hanke M, Margineantu DH, Rao A, Bolouri H, Delrow J, Hockenbery D, Houghton AM, Holland EC (2017) Mutant IDH1 regulates the tumor-associated immune system in gliomas. Genes Dev 31(8):774–786. https://doi.org/10.1101/gad.294991.116

76. Griveau A, Seano G, Shelton SJ, Kupp R, Jahangiri A, Obernier K, Krishnan S, Lindberg OR, Yuen TJ, Tien AC, Sabo JK, Wang N, Chen I, Kloepper J, Larrouquere L, Ghosh M, Tirosh I, Huillard E, Alvarez-Buylla A, Oldham MC, Persson AI, Weiss WA, Batchelor TT, Stemmer-Rachamimov A, Suva ML, Phillips JJ, Aghi MK, Mehta S, Jain RK, Rowitch DH (2018) A glial signature and Wnt7 signaling regulate glioma-vascular interactions and tumor microenvironment. Cancer Cell 33(5):874–889. e877. https://doi.org/10.1016/j.ccell.2018.03.020

77. Dai C, Celestino JC, Okada Y, Louis DN, Fuller GN, Holland EC (2001) PDGF autocrine stimulation dedifferentiates cultured astrocytes and induces oligodendrogliomas and oligoastrocytomas from neural

progenitors and astrocytes in vivo. Genes Dev 15(15):1913–1925. https://doi.org/10.1101/gad.903001

78. Appolloni I, Calzolari F, Tutucci E, Caviglia S, Terrile M, Corte G, Malatesta P (2009) PDGF-B induces a homogeneous class of oligodendrogliomas from embryonic neural progenitors. Int J Cancer 124(10):2251–2259. https://doi.org/10.1002/ijc.24206

79. Lindberg N, Kastemar M, Olofsson T, Smits A, Uhrbom L (2009) Oligodendrocyte progenitor cells can act as cell of origin for experimental glioma. Oncogene 28(23):2266–2275. https://doi.org/10.1038/onc.2009.76

80. Jackson EL, Garcia-Verdugo JM, Gil-Perotin S, Roy M, Quinones-Hinojosa A, VandenBerg S, Alvarez-Buylla A (2006) PDGFR alpha-positive B cells are neural stem cells in the adult SVZ that form glioma-like growths in response to increased PDGF signaling. Neuron 51(2):187–199. https://doi.org/10.1016/j.neuron.2006.06.012

81. Lei L, Sonabend AM, Guarnieri P, Soderquist C, Ludwig T, Rosenfeld S, Bruce JN, Canoll P (2011) Glioblastoma models reveal the connection between adult glial progenitors and the proneural phenotype. PLoS One 6(5):e20041. https://doi.org/10.1371/journal.pone.0020041

82. Psachoulia K, Jamen F, Young KM, Richardson WD (2009) Cell cycle dynamics of NG2 cells in the postnatal and ageing brain. Neuron Glia Biol 5(3–4):57–67. https://doi.org/10.1017/S1740925X09990354

83. Moyon S, Dubessy AL, Aigrot MS, Trotter M, Huang JK, Dauphinot L, Potier MC, Kerninon C, Melik Parsadaniantz S, Franklin RJ, Lubetzki C (2015) Demyelination causes adult CNS progenitors to revert to an immature state and express immune cues that support their migration. J Neurosci Off J Soc Neurosci 35(1):4–20. https://doi.org/10.1523/JNEUROSCI.0849-14.2015

84. Weiss WA, Burns MJ, Hackett C, Aldape K, Hill JR, Kuriyama H, Kuriyama N, Milshteyn N, Roberts T, Wendland MF, DePinho R, Israel MA (2003) Genetic determinants of malignancy in a mouse model for oligodendroglioma. Cancer Res 63(7):1589–1595

85. Horbinski C, Hobbs J, Cieply K, Dacic S, Hamilton RL (2011) EGFR expression stratifies oligodendroglioma behavior. Am J Pathol 179(4):1638–1644. https://doi.org/10.1016/j.ajpath.2011.06.020

86. Simon-Carrasco L, Jimenez G, Barbacid M, Drosten M (2018) The Capicua tumor suppressor: a gatekeeper of Ras signaling in development and cancer. Cell Cycle 17(6):702–711. https://doi.org/10.1080/15384101.2018.1450029

87. Lu HC, Tan Q, Rousseaux MW, Wang W, Kim JY, Richman R, Wan YW, Yeh SY, Patel JM, Liu X, Lin T, Lee Y, Fryer JD, Han J, Chahrour M, Finnell RH, Lei Y, Zurita-Jimenez ME, Ahimaz P, Anyane-Yeboa K, Van Maldergem L, Lehalle D, Jean-Marcais N, Mosca-Boidron AL, Thevenon J, Cousin MA, Bro DE, Lanpher BC, Klee EW, Alexander N, Bainbridge MN, Orr HT, Sillitoe RV, Ljungberg MC, Liu Z, Schaaf CP, Zoghbi HY (2017) Disruption of the ATXN1-CIC complex causes a spectrum of neurobehavioral phenotypes in mice and humans. Nat Genet 49(4):527–536. https://doi.org/10.1038/ng.3808

88. Lam YC, Bowman AB, Jafar-Nejad P, Lim J, Richman R, Fryer JD, Hyun ED, Duvick LA, Orr HT, Botas J, Zoghbi HY (2006) ATAXIN-1 interacts with the repressor Capicua in its native complex to cause SCA1 neuropathology. Cell 127(7):1335–1347. https://doi.org/10.1016/j.cell.2006.11.038

89. Weissmann S, Cloos PA, Sidoli S, Jensen ON, Pollard S, Helin K (2018) The tumor suppressor CIC directly regulates MAPK pathway genes via histone deacetylation. Cancer Res 78(15):4114–4125. https://doi.org/10.1158/0008-5472.CAN-18-0342

90. Kawamura-Saito M, Yamazaki Y, Kaneko K, Kawaguchi N, Kanda H, Mukai H, Gotoh T, Motoi T, Fukayama M, Aburatani H, Takizawa T, Nakamura T (2006) Fusion between CIC and DUX4 up-regulates PEA3 family genes in Ewing-like sarcomas with t(4;19)(q35;q13) translocation. Hum Mol Genet 15(13):2125–2137. https://doi.org/10.1093/hmg/ddl136

91. Dissanayake K, Toth R, Blakey J, Olsson O, Campbell DG, Prescott AR, MacKintosh C (2011) ERK/p90(RSK)/14-3-3 signalling has an impact on expression of PEA3 Ets transcription factors via the transcriptional repressor capicua. Biochem J 433(3):515–525. https://doi.org/10.1042/BJ20101562

92. Jimenez G, Shvartsman SY, Paroush Z (2012) The Capicua repressor—a general sensor of RTK signaling in development and disease. J Cell Sci 125(Pt 6):1383–1391. https://doi.org/10.1242/jcs.092965

93. Bunda S, Heir P, Metcalf J, Li ASC, Agnihotri S, Pusch S, Yasin M, Li M, Burrell K, Mansouri S, Singh O, Wilson M, Alamsahebpour A, Nejad R, Choi B, Kim D, von Deimling A, Zadeh G, Aldape K (2019) CIC protein instability contributes to tumorigenesis in glioblastoma. Nat Commun 10(1):661. https://doi.org/10.1038/s41467-018-08087-9

94. Yang R, Chen LH, Hansen LJ, Carpenter AB, Moure CJ, Liu H, Pirozzi CJ, Diplas BH,

Waitkus MS, Greer PK, Zhu H, McLendon RE, Bigner DD, He Y, Yan H (2017) Cic loss promotes gliomagenesis via aberrant neural stem cell proliferation and differentiation. Cancer Res 77(22):6097–6108. https://doi.org/10.1158/0008-5472.CAN-17-1018

95. Yang L, Paul S, Trieu KG, Dent LG, Froldi F, Fores M, Webster K, Siegfried KR, Kondo S, Harvey K, Cheng L, Jimenez G, Shvartsman SY, Veraksa A (2016) Minibrain and wings apart control organ growth and tissue patterning through down-regulation of Capicua. Proc Natl Acad Sci U S A 113(38):10583–10588. https://doi.org/10.1073/pnas.1609417113

96. Pascual J, Jacobs J, Sansores-Garcia L, Natarajan M, Zeitlinger J, Aerts S, Halder G, Hamaratoglu F (2017) Hippo reprograms the transcriptional response to Ras signaling. Dev Cell 42(6):667–680. e664. https://doi.org/10.1016/j.devcel.2017.08.013

97. Sjoblom T, Jones S, Wood LD, Parsons DW, Lin J, Barber TD, Mandelker D, Leary RJ, Ptak J, Silliman N, Szabo S, Buckhaults P, Farrell C, Meeh P, Markowitz SD, Willis J, Dawson D, Willson JK, Gazdar AF, Hartigan J, Wu L, Liu C, Parmigiani G, Park BH, Bachman KE, Papadopoulos N, Vogelstein B, Kinzler KW, Velculescu VE (2006) The consensus coding sequences of human breast and colorectal cancers. Science 314(5797):268–274. https://doi.org/10.1126/science.1133427

98. Okimoto RA, Breitenbuecher F, Olivas VR, Wu W, Gini B, Hofree M, Asthana S, Hrustanovic G, Flanagan J, Tulpule A, Blakely CM, Haringsma HJ, Simmons AD, Gowen K, Suh J, Miller VA, Ali S, Schuler M, Bivona TG (2017) Inactivation of Capicua drives cancer metastasis. Nat Genet 49(1):87–96. https://doi.org/10.1038/ng.3728

99. Simon-Carrasco L, Grana O, Salmon M, Jacob HKC, Gutierrez A, Jimenez G, Drosten M, Barbacid M (2017) Inactivation of Capicua in adult mice causes T-cell lymphoblastic lymphoma. Genes Dev 31(14):1456–1468. https://doi.org/10.1101/gad.300244.117

100. Gleize V, Alentorn A, Connen de Kerillis L, Labussiere M, Nadaradjane A, Mundwiller E, Ottolenghi C, Mangesius S, Rahimian A, Ducray F, Behalf Of The Pola Network O, Mokhtari K, Villa C, Sanson M (2015) CIC inactivating mutations identify aggressive subset of 1p19q codeleted gliomas. Ann Neurol. https://doi.org/10.1002/ana.24443

101. Bettegowda C, Agrawal N, Jiao Y, Sausen M, Wood LD, Hruban RH, Rodriguez FJ, Cahill DP, McLendon R, Riggins G, Velculescu VE, Oba-Shinjo SM, Marie SK, Vogelstein B, Bigner D, Yan H, Papadopoulos N, Kinzler KW (2011) Mutations in CIC and FUBP1

contribute to human oligodendroglioma. Science 333(6048):1453–1455. https://doi.org/10.1126/science.1210557

102. Yip S, Butterfield YS, Morozova O, Chittaranjan S, Blough MD, An J, Birol I, Chesnelong C, Chiu R, Chuah E, Corbett R, Docking R, Firme M, Hirst M, Jackman S, Karsan A, Li H, Louis DN, Maslova A, Moore R, Moradian A, Mungall KL, Perizzolo M, Qian J, Roldan G, Smith EE, Tamura-Wells J, Thiessen N, Varhol R, Weiss S, Wu W, Young S, Zhao Y, Mungall AJ, Jones SJ, Morin GB, Chan JA, Cairncross JG, Marra MA (2012) Concurrent CIC mutations, IDH mutations, and 1p/19q loss distinguish oligodendrogliomas from other cancers. J Pathol 226(1):7–16. https://doi.org/10.1002/path.2995

103. Wesseling P, van den Bent M, Perry A (2015) Oligodendroglioma: pathology, molecular mechanisms and markers. Acta Neuropathol 129(6):809–827. https://doi.org/10.1007/s00401-015-1424-1

104. Fores M, Simon-Carrasco L, Ajuria L, Samper N, Gonzalez-Crespo S, Drosten M, Barbacid M, Jimenez G (2017) A new mode of DNA binding distinguishes Capicua from other HMG-box factors and explains its mutation patterns in cancer. PLoS Genet 13(3):e1006622. https://doi.org/10.1371/journal.pgen.1006622

105. Ahmad ST, Rogers AD, Chen MJ, Dixit R, Adnani L, Frankiw LS, Lawn SO, Blough MD, Alshehri M, Wu W, Marra MA, Robbins SM, Cairncross JG, Schuurmans C, Chan JA (2019) Capicua regulates neural stem cell proliferation and lineage specification through control of Ets factors. Nat Commun 10(1):2000. https://doi.org/10.1038/s41467-019-09949-6

106. Debaize L, Troadec MB (2019) The master regulator FUBP1: its emerging role in normal cell function and malignant development. Cell Mol Life Sci 76(2):259–281. https://doi.org/10.1007/s00018-018-2933-6

107. Zhou W, Chung YJ, Parrilla Castellar ER, Zheng Y, Chung HJ, Bandle R, Liu J, Tessarollo L, Batchelor E, Aplan PD, Levens D (2016) Far upstream element binding protein plays a crucial role in embryonic development, Hematopoiesis, and stabilizing Myc expression levels. Am J Pathol 186(3):701–715. https://doi.org/10.1016/j.ajpath.2015.10.028

108. Hwang I, Cao D, Na Y, Kim DY, Zhang T, Yao J, Oh H, Hu J, Zheng H, Yao Y, Paik J (2018) Far upstream element-binding protein 1 regulates LSD1 alternative splicing to promote terminal differentiation of neural progenitors. Stem Cell Reports 10(4):1208–

1221. https://doi.org/10.1016/j.stemcr.2018.02.013

109. van der Weyden L, Bradley A (2006) Mouse chromosome engineering for modeling human disease. Annu Rev Genomics Hum Genet 7:247–276. https://doi.org/10.1146/annurev.genom.7.080505.115741

110. Liu MY, Nemes A, Zhou QG (2018) The emerging roles for telomerase in the central nervous system. Front Mol Neurosci 11:160. https://doi.org/10.3389/fnmol.2018.00160

111. Chittaranjan S, Chan S, Yang C, Yang KC, Chen V, Moradian A, Firme M, Song J, Go NE, Blough MD, Chan JA, Cairncross JG, Gorski SM, Morin GB, Yip S, Marra MA (2014) Mutations in CIC and IDH1 cooperatively regulate 2-hydroxyglutarate levels and cell clonogenicity. Oncotarget 5(17):7960–7979. https://doi.org/10.18632/oncotarget.2401

112. Labreche K, Simeonova I, Kamoun A, Gleize V, Chubb D, Letouze E, Riazalhosseini Y, Dobbins SE, Elarouci N, Ducray F, de Reynies A, Zelenika D, Wardell CP, Frampton M, Saulnier O, Pastinen T, Hallout S, Figarella-Branger D, Dehais C, Idbaih A, Mokhtari K, Delattre JY, Huillard E, Mark Lathrop G, Sanson M, Houlston RS (2015) TCF12 is mutated in anaplastic oligodendroglioma. Nat Commun 6:7207. https://doi.org/10.1038/ncomms8207

113. Delgado-Lopez PD, Corrales-Garcia EM, Martino J, Lastra-Aras E, Duenas-Polo MT (2017) Diffuse low-grade glioma: a review on the new molecular classification, natural history and current management strategies. Clin Transl Oncol 19(8):931–944. https://doi.org/10.1007/s12094-017-1631-4

114. Pallud J, Capelle L, Taillandier L, Badoual M, Duffau H, Mandonnet E (2013) The silent phase of diffuse low-grade gliomas. Is it when we missed the action? Acta Neurochir 155(12):2237–2242. https://doi.org/10.1007/s00701-013-1886-7

115. Bielle F, Ducray F, Mokhtari K, Dehais C, Adle-Biassette H, Carpentier C, Chanut A, Polivka M, Poggioli S, Rosenberg S, Giry M, Marie Y, Duyckaerts C, Sanson M, Figarella-Branger D, Idbaih A, Pola N (2017) Tumor cells with neuronal intermediate progenitor features define a subgroup of 1p/19q co-deleted anaplastic gliomas. Brain Pathol 27(5):567–579. https://doi.org/10.1111/bpa.12434

116. Simeonova I, Huillard E (2014) In vivo models of brain tumors: roles of genetically engineered mouse models in understanding tumor biology and use in preclinical studies. Cell

Mol Life Sci 71(20):4007–4026. https://doi.org/10.1007/s00018-014-1675-3

117. Cuddapah VA, Robel S, Watkins S, Sontheimer H (2014) A neurocentric perspective on glioma invasion. Nat Rev Neurosci 15(7):455–465. https://doi.org/10.1038/nrn3765

118. Renier N, Wu Z, Simon DJ, Yang J, Ariel P, Tessier-Lavigne M (2014) iDISCO: a simple, rapid method to immunolabel large tissue samples for volume imaging. Cell 159(4):896–910. https://doi.org/10.1016/j.cell.2014.10.010

119. Lagerweij T, Dusoswa SA, Negrean A, Hendrikx EML, de Vries HE, Kole J, Garcia-Vallejo JJ, Mansvelder HD, Vandertop WP, Noske DP, Tannous BA, Musters RJP, van Kooyk Y, Wesseling P, Zhao XW, Wurdinger T (2017) Optical clearing and fluorescence deep-tissue imaging for 3D quantitative analysis of the brain tumor microenvironment. Angiogenesis 20(4):533–546. https://doi.org/10.1007/s10456-017-9565-6

120. Pencheva N, de Gooijer MC, Vis DJ, Wessels LFA, Wurdinger T, van Tellingen O, Bernards R (2017) Identification of a druggable pathway controlling glioblastoma invasiveness. Cell Rep 20(1):48–60. https://doi.org/10.1016/j.celrep.2017.06.036

121. Marques-Torrejon MA, Gangoso E, Pollard SM (2018) Modelling glioblastoma tumour-host cell interactions using adult brain organotypic slice co-culture. Dis Model Mech 11(2). https://doi.org/10.1242/dmm.031435

122. Marton RM, Miura Y, Sloan SA, Li Q, Revah O, Levy RJ, Huguenard JR, Pasca SP (2019) Differentiation and maturation of oligodendrocytes in human three-dimensional neural cultures. Nat Neurosci 22(3):484–491. https://doi.org/10.1038/s41593-018-0316-9

123. Ogawa J, Pao GM, Shokhirev MN, Verma IM (2018) Glioblastoma model using human cerebral organoids. Cell Rep 23(4):1220–1229. https://doi.org/10.1016/j.celrep.2018.03.105

124. Assanah M, Lochhead R, Ogden A, Bruce J, Goldman J, Canoll P (2006) Glial progenitors in adult white matter are driven to form malignant gliomas by platelet-derived growth factor-expressing retroviruses. J Neurosci Off J Soc Neurosci 26(25):6781–6790. https://doi.org/10.1523/JNEUROSCI.0514-06.2006

125. Glasgow SM, Zhu W, Stolt CC, Huang TW, Chen F, LoTurco JJ, Neul JL, Wegner M, Mohila C, Deneen B (2014) Mutual antagonism between Sox10 and NFIA regulates diversification of glial lineages and glioma subtypes. Nat Neurosci 17(10):1322–1329. https://doi.org/10.1038/nn.3790

Chapter 2

In Vivo Medulloblastoma Modeling

Liliana Mirabal-Ortega, Magalie Larcher, Morgane Morabito, Chloé Foray, Bertrand Duvillié, Alain Eychène, and Celio Pouponnot

Abstract

Medulloblastoma (MB), the most frequent malignant brain tumor in children, is localized in the cerebellum. The standard care includes surgery, radiotherapy, and chemotherapy leading to an overall survival (OS) of 70–80%, but survivors suffer from severe side effects. Based on gene expression, MB is divided in four different molecular subgroups—WNT, SHH, Group 3 (G3), and Group 4 (G4)—which differ in terms of clinics, prognosis, genetic alterations, and cell of origin. The WNT group is characterized by the activation of the WNT/β-catenin signaling pathway and displays the best prognosis. The SHH group is driven by deregulation of the SHH signaling pathway and has an intermediate prognosis. G3 and G4 are less characterized. Contrary to the SHH and WNT groups, no specific alteration of a given signaling pathway has been described. G3 is the group with the worse prognosis. Few recurrent genetic alterations have been characterized including MYC amplification in less than 20% of G3 tumors. Nevertheless, all G3-MBs overexpress MYC through mechanisms not completely understood. G3-MBs also express an abnormal photoreceptor differentiation program found in the retina but not in the cerebellum during normal development. It has been shown that NRL and CRX, two master transcription factors (TF) of the photoreceptor lineage, are required for the establishment of this program as well as for G3 tumor maintenance. G4 has an intermediate prognosis, and the most frequent alteration is the overexpression of PRDM6. It has been recently proposed that this group could be driven by activation of an ERBB4-SRC signaling.

Established cell lines and patient-derived xenografts (PDXs) are available to study MB. The different groups of MB have also been modeled in vivo using either genetically engineered mouse models (GEMMs) or orthotopic transplantation of mouse cerebellar progenitors modified to overexpress oncogenes and/or to inactivate tumor suppressors. Here, we provide the readers with tools and information that allow MB modeling in vivo. We describe how to purify granular cell cerebellar progenitors or PDXs and to culture them in vitro in order to modulate gene expression by lentiviral infection. We provide protocols for the retrovirus production and infection. We also describe the experimental procedures for orthotopic grafting in the cerebellum, which is used to assess how genetic modifications alter in vivo tumor formation of reinjected modified PDX cells or GCPs.

Key words Medulloblastoma, Animal models, Patient-derived xenograft (PDX), Cerebellar progenitors, Cerebellum, Orthotopic grafting

Giorgio Seano (ed.), *Brain Tumors*, Neuromethods, vol. 158,
https://doi.org/10.1007/978-1-0716-0856-2_2, © Springer Science+Business Media, LLC, part of Springer Nature 2021

1   Introduction

In developing countries, pediatric tumors are the second most frequent cause of death in children older than 1. Among childhood cancers, leukemia is the most prevalent, representing about 30% of the cases, while tumors of the central nervous system (CNS) are the most common solid tumors (20%) [1].

Medulloblastoma (MB) is a pediatric CNS tumor originating in the cerebellum. Although rare, it represents the most malignant brain tumor in childhood. It arises at a median age of 7, with more than 70% of patients that are below the age of 10, but it can be also diagnosed in adolescents and adults. Patients with MB present symptoms such as intracranial pressure, hydrocephalus, nausea, vomiting, and balance or motor coordination problems [2]. Different histological variants have been described including classic, desmoplastic, and large cells/anaplastic (LC/A) MB [3]. Metastases are found in ~30% of patients at diagnosis and are associated with a bad prognosis. They are mainly found in leptomeninges, although systemic metastases can be observed in very rare patients with advanced and highly treated disease [3].

The current standard care for MB patients is composed of surgery, followed by radiotherapy of the entire cranio-spinal axis and association of different chemotherapies. This treatment scheme is applied to all patients except infants below the age of 3, who are spared of radiotherapy due to the high toxicity on the developing brain. This heavy multimodal treatment has allowed reaching an overall survival (OS) of 80% but at the cost of severe side effects including neurological and cognitive deficits, endocrine disorders, hearing loss, and, possibly, secondary cancers [3]. When relapse is observed, very few therapeutic options remain and the outcome is almost always fatal [3].

1.1   The MB Groups

To better understand the disease and stratify patients, molecular analyses have established a novel classification of the disease. As expected and observed in many different cancers, inter-tumor heterogeneity has been uncovered. MB does not represent a single entity but gathers at least four distinct groups differing in their molecular and clinical characteristics, prognosis, and cell of origin [3, 4]. These four groups—WNT, SHH, Group 3 (G3), and Group 4 (G4)—have first been identified through transcriptomic analyses [5–8] and further validated by DNA methylation profiling [9] and proteomic analyses [10–12] (Table 2.1). Recently, three different studies have shown that each group could be further subdivided into subtypes providing a better prognostic value. However, the number and nature of these subtypes are variable between these studies, and no consensus has yet been reached [13–15].

Table 2.1

Main clinical and molecular features of medulloblastoma subgroups

	WNT	SHH	G3	G4
Age at diagnosis	Children and adults	Infants, children, and adults	Infants and children	Children and adults
Incidence	10–15%	25–30%	20–25%	35–40%
Metastasis at diagnosis	5–10%	20%	40–50%	35–40%
Overall survival (5 years)	Very good/95%	Intermediate/75%	Poor/50%	Intermediate/75%
Histology	Classic	Classic, desmoplastic, LC/A	Classic, LC/A	Classic, LC/A
Proposed cell of origin	Lower rhombic lip progenitors	Granule cell progenitors	Neural stem cells or Granule cell progenitors	Not yet determined
Main molecular drivers	*CTNNB1*	*PTCH1*	*MYC, GFI1 and GFIB*	PRDM6, SRC
Expression signature	WNT signaling	SHH signaling	Photoreceptor/ GABAergic	Neuronal/ glutamergic

1.1.1 The WNT Group

The WNT group, as highlighted by its name, is driven by the activation of the WNT/β-catenin signaling pathway. It represents 10–15% of all MBs and is usually found in older patients, adolescents and adults. WNT group displays the most favorable outcome with an OS at 5 years over 95%. It is in most cases of classic histology and is very rarely metastatic at diagnosis. Clinical trials are ongoing to evaluate whether treatment de-escalation could be an option in order to decrease side effects. It should be noticed that adult WNT-MBs have a less favorable outcome [3, 16]. Different pieces of evidence, including a genetically engineered mouse model (GEMM) based on the concurrent conditional expression of a mutant Ctnnb1 (stabilized form of β-catenin) and inactivation of TP53, led to propose that the cell of origin of the WNT group is a lower rhombic lip progenitor residing in the brainstem outside the cerebellum [17].

1.1.2 The SHH Group

The SHH-MBs represent 25–30% of MB cases and are driven by deregulation of the cognate SHH signaling pathway. Metastases are found in 20% of SHH-MB at diagnosis. It is associated with the three main types of histology: desmoplastic, classic, and LC/A [3].

It should be emphasized that desmoplastic histology is mainly found in SHH-MBs and virtually absent in other groups and presents a better prognosis. This group is the most prevalent in infants below the age of 3 and in adults. It is less frequent in children and adolescents. Overall, it has an intermediate prognosis around 75% of survival. Within this group, OS can be refined according to the different subtypes recently described [13, 15]. Some SHH subtypes display an extremely bad prognosis, while others have a predictive very good outcome. The granule cell progenitors (GCPs), which are highly proliferative during postnatal cerebellar development, have been clearly demonstrated to be the cell of origin of SHH-MB [18, 19], and several mouse models have been generated by targeting genetic alterations in this cell type (see below).

While SHH and WNT groups are driven by the deregulation of specific signaling pathways, alterations that specifically drive G3 and G4 are much less characterized. Moreover, much less mutations and recurrent alterations are found in these two latter groups.

1.1.3 Group 3

G3 represents 20–25% of all MB cases and has the worse prognosis with a 5-year OS of 60%. These tumors usually occur in infants or young children. They display mainly a classic or LC/A histology and are highly metastatic at diagnosis (40–50%). Tumor recurrence is usually not found at the primary tumor bed but rather at metastatic sites [3]. While MYC amplification is found in less than 20% of G3, all G3-MBs express high levels of MYC through unknown mechanisms. Accordingly, G3 tumors are characterized by a MYC target gene signature with high expression of ribosomal genes, genes involved in mRNA processing, transcription, and translation [5, 6, 10, 11]. MYCN is also amplified in 5% of cases. Beside MYC amplification, GFI1 transcription factors (GFI1 and GFI1B) are overexpressed in 15% of patients, mostly through enhancer hijacking mechanisms [20]. Their role as a codriver together with MYC has been validated using animal models [20, 21]. Moreover, G3 tumors display very surprising characteristics. They express a set of genes usually turned on specifically in the photoreceptor cells of the retina but never in the cerebellum [5, 6, 22]. Thus, G3 shows an aberrant identity unrelated to its tissue of origin, the cerebellum. We recently showed that NRL and CRX, two master transcription factors (TF) of the photoreceptor lineage, establish this aberrant identity. Importantly, they are also required for MB growth [22]. This challenges the widely accepted concept of lineage addiction, which postulates that cancer cells depend on the identity of a given cell lineage from which cancer grows. This work shows that cancer cells can be driven by an abnormal identity unrelated to its tissue of origin. As the SHH-MB, G3 have been further divided into subtypes with different clinical outcomes. Indeed, it has been shown that one subtype characterized by high MYC expression, including MYC amplified tumors, shows the worse prognosis [13]. The cell of origin of G3 tumors remains a matter of debate. It has been suggested that G3 cells could arise either from CD133+ cerebellar neural stem cells [23] or from the GCPs (see below).

1.1.4 Group 4

G4 is the most prevalent MB group found in 35–40% of patients. Its histology is most frequently classic, but LC/A histology is also encountered at lower frequency. It is found in older patients, mainly older children, adolescents, and adults. It is of intermediate prognosis with an OS at 5 years around 75%. It is also frequently metastatic at diagnosis (35–40%), and, as in G3-MB, relapses are more frequent at metastatic sites [3]. The most frequent alteration, found in around 20% of cases, is an enhancer hijacking mechanism that leads to strong overexpression of PRDM6, a putative lysine-methyl transferase. PRDM6 alteration is considered as a putative driver event in G4, but its exact role has not been clearly demonstrated yet [14]. Recently, proteomic and phospho-proteomic studies have proposed that G4 could be driven by activation of an ERBB4-SRC signaling. The relevance of SRC activation has been further validated with the development of an animal model based on in utero electroporation and that displays some characteristics of G4 tumors [11].

1.2 MB Models

As in other cancers, established cell lines and patient-derived xenografts (PDX) are frequently used in the MB field. The different groups of MB have been modeled in vivo either by genetically engineered mouse models (GEMMs) or by orthotopic transplantation of modified mouse cerebellar progenitors manipulated to overexpress oncogenes and/or to inactivate tumor suppressors. In contrast, no patient-derived organoid has been established for MB so far, and, in contrast to glioblastoma, primary cultures from patient samples can be maintained for only very short term in MB and are not commonly used.

Interestingly, the stable expression of a Luciferase gene in all these models allowed a longitudinal follow-up of tumor growth upon orthotopic transplantation into the cerebellum of immuno-deficient mice by bioluminescence quantification. This allows assessing the impact of gene overexpression, gene extinction, or drug treatment on tumor growth. We have used this system on cell lines or MB-PDX to demonstrate the role of the transcription factor NRL, a master regulator of photoreceptor development, on Group 3-MB growth [22].

In the following section, we will give a brief overview of some in vivo models used in the MB field. We do not attempt to provide an exhaustive description of them.

1.2.1 Patient-Derived Xenografts

Patient-derived xenografts (PDX) have become a widely used model in cancer research since they are presumed to faithfully recapitulate the original tumors from which they derived. PDXs are established from fresh surgical MB material that is grafted subcutaneously or orthotopically directly into the cerebellum of immuno-compromised mice (Nod Scid Gamma mice for the establishment). The PDXs are maintained in vivo by serial passages in immunode-ficient mice either subcutaneously or in the cerebellum. It has been demonstrated that PDXs can be established from the different MB groups and remain stable across serial sub-transplantations [24] although some subclonal selection can occur [25]. Very recently, a biobank has been established allowing the availability of 15 MB PDXs including 1 WNT-MB, 4 SHH-MBs, 7 G3-MBs, and 3 G4-MBs [25]. Five out of the seven G3 PDXs harbor MYC ampli-fication. The establishment rate for MB was around 35%. It should be noticed that PDXs have been mostly established from high-risk MBs indicating that aggressive tumors are more favorable to grow as PDXs. When compared to subcutaneous grafting, initial grafting in the cerebellum may be more efficient and might allow a better grafting efficiency for less aggressive tumors. Important informa-tion is provided on these PDXs including transcriptomic and whole-exome sequencing data: https://research.fhcrc.org/olson/en/btrl.html and https://hgserver1.amc.nl/cgi-bin/r2/main.cgi?&dscope=PDX_OLSON&option=about_dscope. PDX mod-els have been used, for example, to validate important players in MB biology [22] or to investigate different potential therapies such as smoothened inhibitors in the SHH group [26], the CDK4/CDK6 Palbociclib inhibitor for SHH and G3 groups [27], or anti-BCL therapy in the G3 [22]. In Sect. 2.1, we provide the experi-mental procedure to short-term culture these PDXs in order to overexpress or downregulate gene expression by retroviral infec-tion and to investigate how this genetic manipulation can affect tumor growth orthotopically.

1.2.2 Genetically Engineered Mouse Models (GEMMs)

Although highly time-consuming, laborious, and expensive, genet-ically engineered mouse models (GEMMs) have proven to be one of the most valuable tools in cancer research. Such models are based on the editing of the mouse genome, including gene dele-tion (knockout mice), gene mutation (knock-in mice), or gene overexpression (transgenic mice). In contrast to other models, GEMMs develop de novo tumors in their natural and immune proficient environment allowing to carefully study the different steps in tumorigenesis, from initiation to advanced cancer states [28]. GEMMs have allowed to validate different genetic drivers in MB as well as to identify the cell of origin of different groups, espe-cially for the SHH and WNT groups.

These models have been widely used to study the SHH group. The first MB GEMM was described in 1997. It consists of $PTCH1^{-/+}$ mice that develop medulloblastoma at low frequency, in about 20% of cases. Noteworthy, these MBs have lost the second allele of PTCH1 [29]. This and other ubiquitous or conditional knockout mouse models have been widely used to validate different cooperating oncogenic events in this group. Moreover, different conditional GEMMs have allowed to firmly establish that SHH-MB originates from GCPs and that a GCP identity is crucial for tumor formation [18, 19].

A unique GEMM has been described for the WNT tumors. Several observations support the fact that WNT-MB might arise outside the cerebellum, from cells in the dorsal brainstem, which originate from the lower rhombic lip [17]. Accordingly, a knock-in mice, in which the expression of an activated mutated form of β-catenin can be conditionally induced in different cell progenitors, have shown that activated β-catenin has no effect on cerebellar progenitors but induces abnormal accumulation of cells in the dorsal brainstem.

Using transgenic mice, it has been shown that NMYC overexpression in the cerebellum induces different types of MBs including G3, G4, and SHH [30, 31].

1.2.3 Orthotopic Transplantation of Modified/Transformed Cerebellar Progenitors

To get insights into the cell of origin of the G3 group, a model originally developed for SHH-MB by Dr. Roussel's laboratory has been used. This model relies on the modification of murine cerebellar progenitors that are then orthotopically grafted in animals. For the original SHH model, GCPs were purified from $p53^{-/-}::Ink4C^{-/-}$ mice at early postnatal stages between p2 and p8, when these progenitors are still proliferating. Oncogenic hits such as overexpression of NMYC were introduced by retroviral infection. The resulting modified GCPs, when grafted into the cerebellum of immunocompromised mice, led to the formation of SHH-MB [32]. A modified experimental procedure allowed G3 tumor formation when other oncogenic combinations were used. For instance, the loss of p53 together with MYC (c-MYC) overexpression in GCPs forms G3-MB when transplanted into the cerebellum of nude mice. These results led to propose that GCPs may be the cell of origin of G3 [33].

Another model described for G3 is based on the retroviral transduction of another cerebellar cell population that expresses the neural stem cell (NSC) marker CD133 (prominin-1) [23]. It has been subsequently demonstrated that G3 tumors can be modeled when these cerebellar stem cells are engineered to overexpress a stabilized form of MYC together with a dominant negative form of p53 (DN-p53) and are subsequently transplanted in the cerebellum of NSG mice [23]. This model supports the idea that G3 may arise from CD133+ cerebellar NSCs.

Noticeably, while these two types of modified progenitors could be in principle transplanted back into syngeneic mice, only immunodeficient animals were used as recipients in published reports. Indeed, our own data tend to indicate that transplantation in syngeneic animals is much less efficient for unknown reasons.

Moreover, since G3 can be modeled from different cerebellar progenitors or NSC, its cell of origin remains elusive. Even more provokingly, using in utero electroporation leading to overexpression of MYC and a dominant negative form of TP53, it has been shown that G3-MB can arise from different cerebellar progenitors. Thus, it has been proposed that G3-MB is mainly driven by specific oncogenic hits, in particular MYC overexpression, rather than its deregulation in a specific cell of origin [34].

It is important to mention that while all the above models indeed display G3 characteristics, they all combine MYC overexpression with p53 inactivation. However, neither p53 loss nor its mutation is found in G3 at diagnosis [3], questioning the relevance of these models. Nevertheless, these models were used to validate the driving role of GFI-1 TF in G3 tumors (see above). It has been shown that retroviral mediated overexpression of MYC together with GFI-1 both in CD133+ cerebellar NSC [20] and in GCP [21] can induce G3-MB when transplanted. These models based on the in vitro transformation of GCPs by MYC and GFI1 are described in the following section.

2   Experimental Tools to Model and Study the Biology of MB In Vivo

In the following sections, we describe the short-term in vitro culture of PDXs that allows gene expression manipulation by retroviral infection. We next provide information on MB modeling using the modified GCP culture. We provide protocols for the purification of these cells and their culture conditions. We also describe retroviral production and infection allowing gene expression manipulation in PDXs and GCPs. Finally, we provide a description of the experimental procedure to perform orthotopic grafting in the cerebellum, which is used to reinject both PDXs cells and GCPs.

2.1 Patient-Derived Xenografts (PDXs)

Patient-derived xenograft (PDX) model constitutes an essential tool to study MB. It better recapitulates heterogeneity and molecular features of patient tumors compared to in vitro models. For PDX establishment, tumor tissues freshly isolated from untreated patients are inoculated into immunodeficient mice, usually NSG mice, where they can be maintained and serially passaged either subcutaneously or orthotopically in the cerebellum. Cells from these PDXs, when cultured in vitro, can be engineered to knock down (KD) or overexpress a given gene, whose role in MB can be subsequently studied. Once these modified cells are grafted either subcutaneously or orthotopically into the cerebellum of nude mice, it is possible to evaluate the effect of the expression of this specific gene in MB tumor biology. As an example of such an approach, we have shown that NRL, a master transcription factor of the photoreceptor lineage in the retina, is critical for G3-MB maintenance [22]. To this end, cultures of different G3 PDXs were established, and cells from PDXs were manipulated using retroviruses to induce either NRL KD or overexpression. The effect of these modifications on MB tumor growth was then evaluated in vivo, by performing orthotopic grafting of PDX cells in nude mice, allowing us to establish NRL requirement for MB growth.

2.1.1 Short-Term In Vitro Cultures from PDXs

PDXs extracted from animals are processed to be short-term cultured to allow genetic modifications of the cells. PDXs are first grown in the fat pad neck of mice (NMRI nude mice, Janvier Labs) until they reach a volume of around 1.5 cm^3. PDXs are extracted from freshly euthanized animals using sterile forceps, scalpels, and scissors. Once extracted, the PDX is placed in a Petri dish and any adjacent non-tumor or necrotic tissue should be removed. Using a scalpel, the tumor is cut in small pieces (around 3–4 mm^3) that are then disaggregated by enzymatic digestion. To that end, pieces of tumor are covered by the dissociation buffer (2.5 mL of buffer containing Neurobasal medium (Gibco) supplemented with 1 mg/mL DNase (Worthington), 2.5 mg/mL Collagenase P (Roche), 2.5 mg/mL Collagenase/Dispase (Roche) and B27 supplement without vitamin A (Gibco)) and incubated at 37 °C for 30 min[1]. The suspension is filtered using a 40 μm cell strainer (Sigma) to remove debris. The cell strainer is then rinsed with 10 mL of Neurobasal medium and the filtrate is centrifuged at 520 g for 5 min. After centrifugation, the cell pellet is resuspended in a buffer containing 1 mL of Neurobasal medium supplemented with 0.5 mg/mL DNase, 0.35% D-Glucose (Sigma), and 2 mL of CMF-PBS pH 7.4 buffer (NaCl 0.14 M, KCl 4 mM, glucose 11.1 mM, NaH2PO4 H2O 3.2 mM, KH2PO4 3.2 mM, NaHCO3 0.004%). The different cell populations are separated using discontinuous Percoll gradient (Sigma), constituted of two phases, Percoll 35% and Percoll 60%. Percoll dilutions are prepared in CMF-PBS-EDTA (2.5 mM final) buffer. Trypan blue solution 0.4% (Gibco) is added to the Percoll 60% solution to better visualize the cells. The 3 mL cell suspension is deposited on the top of the Percoll gradient and centrifuged at 1800 g for 13 min with minimal acceleration and break (Eppendorf Centrifuge 5810R). At the end of centrifugation, cells at the interface of Percoll 60–35% are collected and transferred into a 15 mL tube filled with CMF-PBS to obtain a final volume of 15 mL. Cells are then centrifuged at 520 g for 5 min to remove the remaining Percoll solution. Finally, cells are seeded at a concentration of 2×10^6 cells/mL in low attachment Corning® flasks or plates to prevent adhesion since PDX cells are grown in neurosphere conditions. PDX MB primary cultures are maintained in Neurobasal medium (Gibco) supplemented with B27 supplement without vitamin A (Gibco), 0.012% BSA (Sigma), 1% penicillin/streptomycin, 1% L-glutamine

(Invitrogen), 12.5 ng/mL of human bFGF (Peprotech), and 12.5 ng/mL of human EGF (Peprotech)[2]. This culture can be subjected to retroviral infection after 2 h of incubation at 37 °C (see below for the description of the protocol).

2.2 Culture of Granular Cell Progenitors (GCPs)

MYC is overexpressed in all G3 tumors, but, although necessary, it is not sufficient to give rise to G3-MB and therefore requires additional oncogenic hits. This was shown using the G3-MB mouse models based on orthotopic grafting of modified GCPs or cerebellar stem cells. Indeed, overexpression of MYC itself in these cells does not induce MB formation, while its combination with p53 inactivation or GFI-1 overexpression does [20, 21, 23, 33]. As described for the GFI-1 TFs, these models are particularly suited to validate novel G3 driver genes by testing their ability to cooperate with MYC to induce MB. For these models, GCPs need to be isolated, shortly cultured in vitro, and genetically manipulated to be then grafted into the cerebellum of mice. Here, we described how to perform these different steps.

2.2.1 Cerebella Dissection and GCP Isolation

For cerebella dissection, pups (aged between P5 and P8) are decapitated, and the skin from the dorsal part of the head is removed using sterile scissors and forceps. The brain along the skull is dissociated from the rest of the head using forceps. The skull is then carefully removed by pulling it out from the front to the cerebellum, and the brain is transferred into a plate containing cold CMF-PBS. Meninges are scratched using very thin forceps under binocular loupes, paying attention not to damage the cerebellum, which is then separated from the rest of the brain. It is important to fully remove the meninges to avoid contamination of the GCP culture. All the manipulations should be performed at 4 °C.

After dissection, cerebella are transferred into polypropylene conical tubes (15 mL) (two or three cerebella per tube) containing 1 mL of a Trypsin/DNase solution (10 mg/mL Trypsin (Gibco), 1 mg/mL DNase, 0.006 M NaOH, 1.5 mg/mL $MgSO_4$ $7H_2O$) and incubated at 37 °C for 5 min. Then, the Trypsin/DNase solution is carefully removed, and the cerebella are resuspended in 1 mL of DNase solution (Neurobasal medium supplemented with 0.5 mg/mL DNase and 0.35% D-Glucose), and the tissue is mechanically dissociated by successively pipetting up and down with a 1000 μl micropipette and then with a syringe with 20G and, then, 23G needles. Each step is done approximately 15 times. The suspension is then centrifuged at 720 g for 5 min at 4 °C, the cell pellet is resuspended in 1 mL of the DNase solution described above, and 2 mL of CMF-PBS pH 7.4 buffer is then added. To separate the different cell types, the 3 mL cell suspension is centrifuged through a 35–60% Percoll gradient as described above (Sect. 2.1.1). The cell pellet is resuspended in 50 μL of DNase solution, and then GCP culture medium (Neurobasal medium supplemented with B27 supplement, 1% penicillin/streptomycin, 0.2% Fungizone, 1% L-glutamine (Invitrogen), SPITE medium supplement (Sigma), 0.1 mM N-acetyl cysteine (Sigma), 0.45% D-glucose, oleic acid albumin/linoleic acid (Sigma)) is added. Cells are then plated in a 35-mm-diameter culture dish and incubated at 37 °C during a period of time between 45 min and 2 h. This step allows eliminating adherent cells, such as astrocytes, microglia, etc. Then, plates are gently flushed with a 1000 μL micropipette in order to recover cells that are in suspension and slightly adherent. Cells are plated at 2×10^6 GCPs/mL in culture medium containing 0.2 μg/mL Sonic hedgehog (R&D Systems). Of note, GCPs in culture are quite fragile, making it difficult to keep viable cultures for several days[3].

2.3 Retroviral Production

PDX cells or GCP cultures can be subjected to lentiviral infection in order to overexpress or KD (shRNA-mediated gene silencing) genes as previously described [22]. The use of lentiviral systems for modifying cells involves the production of retroviral particles that encode the gene of interest. We use retroviral defective systems that allow gene overexpression (pMIGR (also named MIGR1) (addgene, https://www.addgene.org/27490/) or pMSCV vectors (addgene, https://www.addgene.org/86537/) or shRNA-mediated gene knockdown (pLKO vector (a shRNA library is available at SIGMA https://www.sigmaaldrich.com/life-science/functional-genomics-and-rnai/shrna/library-information/vector-map.html), a packaging vector (encoding the gag/pol gene) and a plasmid encoding the envelope gene. The pMIGR or pMSCV vectors contain an IRES that allows the translation of the gene of interest and the GFP gene from the same cistron. Infected cells can be FACS-sorted using this marker. Other derivatives of these vectors can be found with different markers including the luciferase gene allowing a noninvasive follow-up of tumor growth by bioluminescence imaging and quantification. This tracer is particularly useful when cells are orthotopically grafted in the brain or the cerebellum (see below). On the other hand, to induce stable KD of genes of interest, lentiviral particles are produced using the pLKO.1-TRC vector, commonly used for the expression of shRNAs. pLKO-based vectors contain a selectable marker, the puromycin resistance gene (other resistant genes being also available), making possible the selection of infected cells by adding the antibiotic to the culture media. The choice of the packaging vector and of the envelope encoding plasmid depends on the backbone vector and the species of the cells to be infected, respectively. For example, when using the pLKO vector to infect human PDXs, we used the psPAX and pMD2/VSVG plasmids. The latter allows producing amphotropic retroviral particles that can infect both human and rodent cells. For murine GCP culture, the pMIGR (or MIGR1) together with the pCMV-Gag/Pol [35] and the pSV-E-MLV plasmids [36] is used. The latest plasmid encodes an ecotropic envelope allowing infection of mouse and rat cells. To produce retroviruses, HEK293T cells seeded at a density of 1×10^5 cells/cm^2 are co-transfected with the three plasmids at a concentration of 0.066 µg/cm^2 (retroviral backbone vector), 0.09 µg/cm^2 (packaging vector), and 0.044 µg/cm^2 (env plasmid) using Invitrogen™ Lipofectamine™ 2000 Transfection Reagent, following manufacturer indications. Twenty-four hour post transfection, the media is replaced by a fresh harvest media that is normally used to cultivate the cells that will be infected. Viral particles are harvested at 48, 60, and 72 h after transfection by collecting the medium followed by filtration.

2.4 Retroviral Infection of PDX Cells and GCPs

To infect PDXs, 9 mL of a cell suspension is prepared at 4×10^6 cells/mL and plated in T75 low-attaching flasks. Then, 1 mL of viral particles is added to cell culture.

For GCP infection, a cell suspension at 2×10^6 cells/mL is plated at 500 µL/well in a 12-well plate and incubated 30 min at 37 °C. Then, 250 µL of viral suspension/well is added and plates are incubated 1 h and 30 min at 37 °C, followed by the addition of another 250 µL of viral suspension/well and an incubation for 1 h and 30 min at 37 ° C. Then, 500 µL of media is removed and 250 µL of viral suspension/well is again added. SHH is added to cells at a final concentration of 0.2 µg/mL, and plates are incubated overnight at 37 °C. GCP media are then removed and replaced by fresh media. Cells can be further processed for transplantation in mice.

2.5 Preparing Culture of PDX Cells and GCPs for Orthotopic Grafting

Cells to be orthotopically grafted are flushed and harvested. They are washed twice by centrifugation and resuspended in CMF-PBS at an adequate concentration to be injected. For GCP, 1×10^6 cells/5 µL/mice are used. For each PDX, adequate concentration shall be determined for reproducible tumor growth. We usually use 3×10^5 cells/5 µL/mice for the G3 PDXs available in our lab.

2.6 Orthotopic Grafting

2.6.1 Animal Preparation

Orthotopic transplants are performed in 7–8-weeks-old NMRI nude female mice (Janvier Labs). Mice are acclimated to the animal facility at least 1 week before surgery. The grafting of tumor cells is performed under anesthesia. First, animals are anesthetized in an induction chamber (Anesthesia workstation AST-00, Anestéo) supplied with isoflurane (5%), compressed air (1 L/min), and O_2 (0.8 L/min). For surgery, mice are placed in a stereotaxic frame supplied with a microinjection system (PHYMEP), while they are kept in inhalation masks (isoflurane 3–5%) (Fig. 2.1). Local anesthetic and analgesic are supplied before surgery such as bupivacaine (5 mg/mL, Aguettant) and Buprecare (0.3 mg/mL,

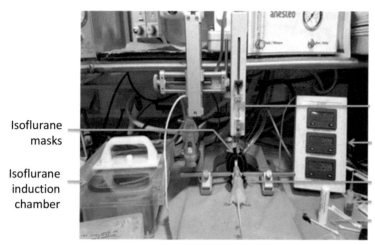

Isoflurane masks

Isoflurane induction chamber

Syringe containing tumoral cells

Stereotaxic coordenates

Anesthetized mouse

Tissue glu

Horsley' bone wax

Fig. 2.1 Stereotaxic surgery area and animal preparation. For anesthesia, mice are placed in an induction chamber supplied with isoflurane (5%), compressed air (1 L/min), and O_2 (0.8 L/min). For the surgery, anesthetized animals are placed in a stereotaxic frame while kept in inhalation masks (isoflurane 3–5%), and the exact coordinates of injection are determined

Axience), respectively. During anesthesia, an adequate body temperature of animals is maintained using heated pads (Anestéo), while heated lamps (Anestéo) are used during the awakening phase.

2.6.2 Stereotaxic Surgery

As part of the preoperative care, mice receive an intraperitoneal injection of 50 µL of Buprecare. The animal is positioned by hooking its incisors in the frame hold using ears bars. Once the animal is well fixed, additional local anesthetic is supplied: two injections of 30 µL of bupivacaine is performed subcutaneously at the level of the skull around the site of incision. The skin of this area is disinfected with a Betadine solution (10%, MEDA Pharma), and then, using surgical sterile scissors, an incision is performed along the midline to expose the skull. The pericranial transparent tissues are scraped with the help of a sterile cotton tip. Then, the exact area of injection is determined using stereotaxic referent coordinates: a small hole in the skull is made using a 25G needle at 2 mm lateral and 2 mm posterior to the Lambda (Fig. 2.2). Then, a Hamilton syringe 1700 (10 µL, 26 G) coupled to an automate injector is inserted in this emplacement, first to a depth of 3 mm and then raised to 2.5 mm to proceed with the injection of 5 µL of cell suspension at a rate of 2.5 µL/min. The syringe should remain in place for two additional minutes following injection and, then, gently removed[4]. The scission is closed using Horsley bone wax (SMI) and the skin is closed with tissue glue (Surgibond)[5].

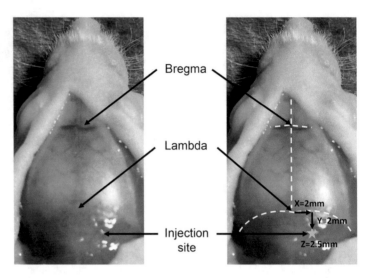

Fig. 2.2 Site of injection for orthotopic grafting in the cerebellum. The site of injection is determined using the Lambda point as a reference. From here, stereotaxic coordinates are determined and cells are injected at 2 mm lateral (*X*) and 2 mm posterior (*Y*) to the Lambda and at a depth of 2.5 mm (*Z*)

2.7 In Vivo Bioluminescence Imaging Using IVIS

Cells orthotopically grafted in mice are engineered to express the luciferase gene, following infection with retrovirus encoding this gene (see above). This allows to follow tumor growth in a noninvasive manner using bioluminescence. Animals bearing tumor cells expressing the luciferase gene are injected with luciferin, the substrate of the luciferase enzyme. During oxidation of luciferin catalyzed by Luciferase, light is emitted. The photons are captured using an IVIS spectrum in vivo imaging system (Perkin Elmer), and images are subsequently analyzed using the Living Image software. The intensity of the signal is a direct measure of tumor size since Luciferase concentration (number of cells expressing the reporter gene) is linearly correlated to photon emission. Bioluminescence imaging and quantification are performed at different time points to follow tumor growth in a noninvasive manner. IVIS imaging of animals is performed under isoflurane anesthesia (2%), 15 min after an IP administration of luciferin (30 mg/mL 50 μL, Perkin Elmer).

As mentioned above, genetic modifications of GCPs have been used in the field to investigate the different oncogenic hits cooperating with MYC to promote G3-MB formation. As an example, we present here the results obtained when GCPs overexpressing a stabilized form of MYC and GFI1 using two pMIGR constructs are orthotopically grafted in the cerebellum of immunodeficient mice. One of the pMIGR construct coexpresses the Luciferase gene, allowing noninvasive follow-up of tumor growth by bioluminescence imaging (Fig. 2.3). Thus, grafted mice develop tumors that are detectable by bioluminescence from day 20 after surgery with

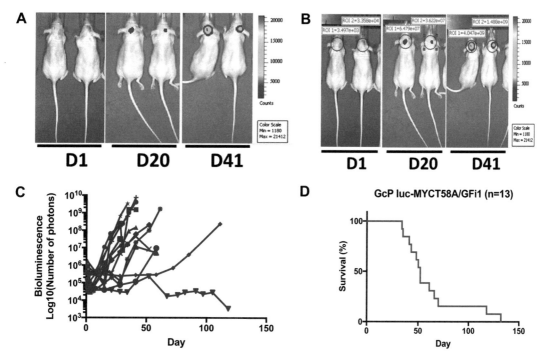

Fig. 2.3 Orthotopic grafting of GCPs overexpressing MYC and GFI1 leads to medulloblastoma formation in immunodeficient mice. (**a, b**) Tumor growth was followed by IVIS bioluminescence imaging. A. Images of luciferase signal in representative animals at the indicated days. (**b**) Representative images shown in A with the selected area of photon measurement (region of interest, ROI). (**c**) Number of photons measured as indicated in B for all the animals in the experiment ($n = 12$) during time. (**d**) Kaplan-Meier survival curves of these animals

a signal increasing in time (Fig. 2.3a). Quantification of bioluminescent signal is performed using the Living Image software. All images obtained from each animal at the different time points are analyzed by measuring the amount of photons emitted. To that end, an area of quantification (red circle—region of interest (ROI)—Fig. 2.3b) is similarly applied to all mice in all images. This allows to measure the amount of photons emitted (Fig. 2.3b).

Photon emission is then plotted at given time points, to obtain the variation of the bioluminescence signal, indicating the size of the developing tumors (Fig. 2.3c). These types of experiments allow the follow-up of tumor growth without animal euthanasia. Animals are spared until they show clinical endpoints requiring euthanasia (Fig. 2.3d).

3 Troubleshooting

1. If tissue disaggregation is not satisfying after this step, the time of incubation could be increased and, in any case, should be optimized for each PDX.

2. PDX MB primary cultures can be maintained in culture for a limited period of time; any experimental procedure with these cultures should be planned taking into account this feature.

3. In order to obtain an adequate number of GCPs, the exact number of processed cerebella should be determined and optimized considering that cell yield depends on the mouse strain and pup age. In our conditions, the best results are obtained at P7. The dissection part could also be optimized to improve the final result.

4. Animal respiration should be monitored all along the surgery. If respiration seems laborious, with an inefficient rhythm or if it stops completely, remove the animal rapidly from the stereotaxic frame and install it in a warm surface. A tail massage, from the end of the tail to the body, can be performed in order to help blood circulation. The thorax can be also compressed repeatedly.

5. After surgery, recovery from anesthesia should be carefully monitored to detect any sign of suffering/distress. Possible symptoms after brain stereotaxic surgery could involve bowed head, reduced locomotion, or distress behaviors such as mice running in circle. In this case, animals should be isolated with minimal disturbance. If signs of pain are still detected the day after surgery, additional analgesic could be provided. If the mouse does not recover from these symptoms, euthanasia is required.

4   Conclusion

In conclusion, the experimental procedures presented here allow to study the role of different players in MB biology in vivo. For example, potential oncogenic hits that are susceptible to cooperate with MYC to induce MB can be validated and, more generally, specific groups of MB can be modeled.

Acknowledgment

We thank members of our laboratory for helpful advice and comments. This work was supported by grants from Ligue Nationale Contre le Cancer (Val d'Oise 2019 - Oise-Yvelines #M18759, #M16649 and Legs Chovet), Institut National du Cancer (INCa, Pair Pediatrie, Mr. ROBOT), the IRS "NanoTheRad" of University U-PSUD (Paris-Saclay), and Gefluc Ile de France, Association AIDA. LMO and MM were supported by a fellowship from the Ministère Français de l'Enseignement Supérieur, de la Recherche et de l'Innovation and Fondation ARC (fourth-year PhD fellowship).

References

1. Ward E, DeSantis C, Robbins A, Kohler B, Jemal A (2014) Childhood and adolescent cancer statistics, 2014. CA Cancer J Clin 64:83–103. https://doi.org/10.3322/caac.21219

2. Dörner L, Fritsch MJ, Stark AM, Mehdorn HM (2007) Posterior fossa tumors in children: how long does it take to establish the diagnosis? Childs Nerv Syst 23:887–890. https://doi.org/10.1007/s00381-007-0323-8

3. Northcott PA, Robinson GW, Kratz CP, Mabbott DJ, Pomeroy SL, Clifford SC, Rutkowski S, Ellison DW, Malkin D, Taylor MD, Gajjar A, Pfister SM (2019) Medulloblastoma. Nat Rev Dis Primers 5:11. https://doi.org/10.1038/s41572-019-0063-6

4. Taylor MD, Northcott PA, Korshunov A, Remke M, Cho YJ, Clifford SC, Eberhart CG, Parsons DW, Rutkowski S, Gajjar A, Ellison DW, Lichter P, Gilbertson RJ, Pomeroy SL, Kool M, Pfister SM (2011) Molecular subgroups of medulloblastoma: the current consensus. Acta Neuropathol 123:465–472. https://doi.org/10.1007/s00401-011-0922-z

5. Cho YJ, Tsherniak A, Tamayo P, Santagata S, Ligon A, Greulich H, Berhoukim R, Amani V, Goumnerova L, Eberhart CG, Lau CC, Olson JM, Gilbertson RJ, Gajjar A, Delattre O, Kool M, Ligon K, Meyerson M, Mesirov JP, Pomeroy SL (2010) Integrative genomic analysis of medulloblastoma identifies a molecular subgroup that drives poor clinical outcome. J Clin Oncol 29:1424–1430. https://doi.org/10.1200/JCO.2010.28.5148

6. Kool M, Koster J, Bunt J, Hasselt NE, Lakeman A, van Sluis P, Troost D, Meeteren NS, Caron HN, Cloos J, Mrsic A, Ylstra B, Grajkowska W, Hartmann W, Pietsch T, Ellison D, Clifford SC, Versteeg R (2008) Integrated genomics identifies five medulloblastoma subtypes with distinct genetic profiles, pathway signatures and clinicopathological features. PLoS One 3:e3088. https://doi.org/10.1371/journal.pone.0003088

7. Northcott PA, Korshunov A, Witt H, Hielscher T, Eberhart CG, Mack S, Bouffet E, Clifford SC, Hawkins CE, French P, Rutka JT, Pfister S, Taylor MD (2010) Medulloblastoma comprises four distinct molecular variants. J Clin Oncol 29:1408–1414. https://doi.org/10.1200/JCO.2009.27.4324

8. Thompson MC, Fuller C, Hogg TL, Dalton J, Finkelstein D, Lau CC, Chintagumpala M, Adesina A, Ashley DM, Kellie SJ, Taylor MD, Curran T, Gajjar A, Gilbertson RJ (2006) Genomics identifies medulloblastoma subgroups that are enriched for specific genetic alterations. J Clin Oncol 24:1924–1931. https://doi.org/10.1200/JCO.2005.04.4974

9. Schwalbe EC, Williamson D, Lindsey JC, Hamilton D, Ryan SL, Megahed H, Garami M, Hauser P, Dembowska-Baginska B, Perek D, Northcott PA, Taylor MD, Taylor RE, Ellison DW, Bailey S, Clifford SC (2013) DNA methylation profiling of medulloblastoma allows robust subclassification and improved outcome prediction using formalin-fixed biopsies. Acta Neuropathol (Berl) 125:359–371. https://doi.org/10.1007/s00401-012-1077-2

10. Archer TC, Ehrenberger T, Mundt F, Gold MP, Krug K, Mah CK, Mahoney EL, Daniel CJ, LeNail A, Ramamoorthy D, Mertins P, Mani DR, Zhang H, Gillette MA, Clauser K, Noble M, Tang LC, Pierre-François J, Silterra J, Jensen J, Tamayo P, Korshunov A, Pfister SM, Kool M, Northcott PA, Sears RC, Lipton JO, Carr SA, Mesirov JP, Pomeroy SL, Fraenkel E (2018) Proteomics, post-translational modifications, and integrative analyses reveal molecular heterogeneity within medulloblastoma subgroups. Cancer Cell 34:396–410.e8. https://doi.org/10.1016/j.ccell.2018.08.004

11. Forget A, Martignetti L, Puget S, Calzone L, Brabetz S, Picard D, Montagud A, Liva S, Sta A, Dingli F, Arras G, Rivera J, Loew D, Besnard A, Lacombe J, Pagès M, Varlet P, Dufour C, Yu H, Mercier AL, Indersie E, Chivet A, Leboucher S, Sieber L, Beccaria K, Gombert M, Meyer FD, Qin N, Bartl J, Chavez L, Okonechnikov K, Sharma T, Thatikonda V, Bourdeaut F, Pouponnot C, Ramaswamy V, Korshunov A, Borkhardt A, Reifenberger G, Poullet P, Taylor MD, Kool M, Pfister SM, Kawauchi D, Barillot E, Remke M, Ayrault O (2018) Aberrant ERBB4-SRC signaling as a hallmark of group 4 medulloblastoma revealed by integrative phosphoproteomic profiling. Cancer Cell 34:379–395.e7. https://doi.org/10.1016/j.ccell.2018.08.002

12. Rivero-Hinojosa S, Lau LS, Stampar M, Staal J, Zhang H, Gordish-Dressman H, Northcott PA, Pfister SM, Taylor MD, Brown KJ, Rood BR (2018) Proteomic analysis of medulloblastoma reveals functional biology with translational potential. Acta Neuropathol Commun 6:48. https://doi.org/10.1186/s40478-018-0548-7

13. Cavalli FMG, Remke M, Rampasek L, Peacock J, Shih DJH, Luu B, Garzia L, Torchia J, Nor C, Morrissy AS, Agnihotri S, Thompson YY, Kuzan-Fischer CM, Farooq H, Isaev K, Daniels

C, Cho B-K, Kim S-K, Wang K-C, Lee JY, Grajkowska WA, Perek-Polnik M, Vasiljevic A, Faure-Conter C, Jouvet A, Giannini C, Nageswara Rao AA, Li KKW, Ng H-K, Eberhart CG, Pollack IF, Hamilton RL, Gillespie GY, Olson JM, Leary S, Weiss WA, Lach B, Chambless LB, Thompson RC, Cooper MK, Vibhakar R, Hauser P, van Veelen M-LC, Kros JM, French PJ, Ra YS, Kumabe T, López-Aguilar E, Zitterbart K, Sterba J, Finocchiaro G, Massimino M, Van Meir EG, Osuka S, Shofuda T, Klekner A, Zollo M, Leonard JR, Rubin JB, Jabado N, Albrecht S, Mora J, Van Meter TE, Jung S, Moore AS, Hallahan AR, Chan JA, Tirapelli DPC, Carlotti CG, Fouladi M, Pimentel J, Faria CC, Saad AG, Massimi L, Liau LM, Wheeler H, Nakamura H, Elbabaa SK, Perezpeña-Diazconti M, Chico Ponce de León F, Robinson S, Zapotocky M, Lassaletta A, Huang A, Hawkins CE, Tabori U, Bouffet E, Bartels U, Dirks PB, Rutka JT, Bader GD, Reimand J, Goldenberg A, Ramaswamy V, Taylor MD (2017) Intertumoral heterogeneity within medulloblastoma subgroups. Cancer Cell 31:737–754.e6. https://doi.org/10.1016/j.ccell.2017.05.005

14. Northcott PA, Buchhalter I, Morrissy AS, Hovestadt V, Weischenfeldt J, Ehrenberger T, Gröbner S, Segura-Wang M, Zichner T, Rudneva VA, Warnatz H-J, Sidiropoulos N, Phillips AH, Schumacher S, Kleinheinz K, Waszak SM, Erkek S, Jones DTW, Worst BC, Kool M, Zapatka M, Jäger N, Chavez L, Hutter B, Bieg M, Paramasivam N, Heinold M, Gu Z, Ishaque N, Jäger-Schmidt C, Imbusch CD, Jugold A, Hübschmann D, Risch T, Amstislavskiy V, Gonzalez FGR, Weber UD, Wolf S, Robinson GW, Zhou X, Wu G, Finkelstein D, Liu Y, Cavalli FMG, Luu B, Ramaswamy V, Wu X, Koster J, Ryzhova M, Cho Y-J, Pomeroy SL, Herold-Mende C, Schuhmann M, Ebinger M, Liau LM, Mora J, McLendon RE, Jabado N, Kumabe T, Chuah E, Ma Y, Moore RA, Mungall AJ, Mungall KL, Thiessen N, Tse K, Wong T, Jones SJM, Witt O, Milde T, Von Deimling A, Capper D, Korshunov A, Yaspo M-L, Kriwacki R, Gajjar A, Zhang J, Beroukhim R, Fraenkel E, Korbel JO, Brors B, Schlesner M, Eils R, Marra MA, Pfister SM, Taylor MD, Lichter P (2017) The whole-genome landscape of medulloblastoma subtypes. Nature 547:311–317. https://doi.org/10.1038/nature22973

15. Schwalbe EC, Lindsey JC, Nakjang S, Crosier S, Smith AJ, Hicks D, Rafiee G, Hill RM, Iliasova A, Stone T, Pizer B, Michalski A, Joshi A, Wharton SB, Jacques TS, Bailey S, Williamson D, Clifford SC (2017) Novel molecular subgroups for clinical classification

and outcome prediction in childhood medulloblastoma: a cohort study. Lancet Oncol 18:958–971. https://doi.org/10.1016/S1470-2045(17)30243-7

16. Wang J, Garancher A, Ramaswamy V, Wechsler-Reya RJ (2018) Medulloblastoma: from molecular subgroups to molecular targeted therapies. Annu Rev Neurosci 41:207–232. https://doi.org/10.1146/annurev-neuro-070815-013838

17. Gibson P, Tong Y, Robinson G, Thompson MC, Currle DS, Eden C, Kranenburg TA, Hogg T, Poppleton H, Martin J, Finkelstein D, Pounds S, Weiss A, Patay Z, Scoggins M, Ogg R, Pei Y, Yang ZJ, Brun S, Lee Y, Zindy F, Lindsey JC, Taketo MM, Boop FA, Sanford RA, Gajjar A, Clifford SC, Roussel MF, McKinnon PJ, Gutmann DH, Ellison DW, Wechsler-Reya R, Gilbertson RJ (2010) Subtypes of medulloblastoma have distinct developmental origins. Nature 468:1095–1099. https://doi.org/10.1038/nature09587

18. Schuller U, Heine VM, Mao J, Kho AT, Dillon AK, Han YG, Huillard E, Sun T, Ligon AH, Qian Y, Ma Q, Alvarez-Buylla A, McMahon AP, Rowitch DH, Ligon KL (2008) Acquisition of granule neuron precursor identity is a critical determinant of progenitor cell competence to form Shh-induced medulloblastoma. Cancer Cell 14:123–134. https://doi.org/10.1016/j.ccr.2008.07.005

19. Yang ZJ, Ellis T, Markant SL, Read TA, Kessler JD, Bourboulas M, Schuller U, Machold R, Fishell G, Rowitch DH, Wainwright BJ, Wechsler-Reya RJ (2008) Medulloblastoma can be initiated by deletion of patched in lineage-restricted progenitors or stem cells. Cancer Cell 14:135–145. https://doi.org/10.1016/j.ccr.2008.07.003

20. Northcott PA, Lee C, Zichner T, Stütz AM, Erkek S, Kawauchi D, Shih DJH, Hovestadt V, Zapatka M, Sturm D, Jones DTW, Kool M, Remke M, Cavalli FMG, Zuyderduyn S, Bader GD, VandenBerg S, Esparza LA, Ryzhova M, Wang W, Wittmann A, Stark S, Sieber L, Seker-Cin H, Linke L, Kratochwil F, Jäger N, Buchhalter I, Imbusch CD, Zipprich G, Raeder B, Schmidt S, Diessl N, Wolf S, Wiemann S, Brors B, Lawerenz C, Eils J, Warnatz H-J, Risch T, Yaspo M-L, Weber UD, Bartholomae CC, von Kalle C, Turányi E, Hauser P, Sanden E, Darabi A, Siesjö P, Sterba J, Zitterbart K, Sumerauer D, van Sluis P, Versteeg R, Volckmann R, Koster J, Schuhmann MU, Ebinger M, Grimes HL, Robinson GW, Gajjar A, Mynarek M, von Hoff K, Rutkowski S, Pietsch T, Scheurlen W, Felsberg J, Reifenberger G, Kulozik AE, von Deimling A, Witt O, Eils R, Gilbertson RJ, Korshunov A, Taylor MD,

Lichter P, Korbel JO, Wechsler-Reya RJ, Pfister SM (2014) Enhancer hijacking activates GFI1 family oncogenes in medulloblastoma. Nature 511:428–434. https://doi.org/10.1038/nature13379

21. Vo BT, Li C, Morgan MA, Theurillat I, Finkelstein D, Wright S, Hyle J, Smith SMC, Fan Y, Wang Y-D, Wu G, Orr BA, Northcott PA, Shilatifard A, Sherr CJ, Roussel MF (2017) Inactivation of Ezh2 upregulates Gfi1 and drives aggressive Myc-driven group 3 medulloblastoma. Cell Rep 18:2907–2917. https://doi.org/10.1016/j.celrep.2017.02.073

22. Garancher A, Lin CY, Morabito M, Richer W, Rocques N, Larcher M, Bihannic L, Smith K, Miquel C, Leboucher S, Herath NI, Dupuy F, Varlet P, Haberler C, Walczak C, El Tayara N, Volk A, Puget S, Doz F, Delattre O, Druillennec S, Ayrault O, Wechsler-Reya RJ, Eychène A, Bourdeaut F, Northcott PA, Pouponnot C (2018) NRL and CRX define photoreceptor identity and reveal subgroup-specific dependencies in Medulloblastoma. Cancer Cell 33:435–449.e6. https://doi.org/10.1016/j.ccell.2018.02.006

23. Pei Y, Moore CE, Wang J, Tewari AK, Eroshkin A, Cho YJ, Witt H, Korshunov A, Read TA, Sun JL, Schmitt EM, Miller CR, Buckley AF, McLendon RE, Westbrook TF, Northcott PA, Taylor MD, Pfister SM, Febbo PG, Wechsler-Reya RJ (2012) An animal model of MYC-driven medulloblastoma. Cancer Cell 21:155–167. https://doi.org/10.1016/j.ccr.2011.12.021

24. Zhao X, Liu Z, Yu L, Zhang Y, Baxter P, Voicu H, Gurusiddappa S, Luan J, Su JM, Leung HE, Li X-N (2012) Global gene expression profiling confirms the molecular fidelity of primary tumor-based orthotopic xenograft mouse models of medulloblastoma. Neuro Oncol 14:574–583. https://doi.org/10.1093/neuonc/nos061

25. Brabetz S, Leary SES, Gröbner SN, Nakamoto MW, Şeker-Cin H, Girard EJ, Cole B, Strand AD, Bloom KL, Hovestadt V, Mack NL, Pakiam F, Schwalm B, Korshunov A, Balasubramanian GP, Northcott PA, Pedro KD, Dey J, Hansen S, Ditzler S, Lichter P, Chavez L, Jones DTW, Koster J, Pfister SM, Kool M, Olson JM (2018) A biobank of patient-derived pediatric brain tumor models. Nat Med 24:1752–1761. https://doi.org/10.1038/s41591-018-0207-3

26. Kool M, Jones DTW, Jäger N, Northcott PA, Pugh TJ, Hovestadt V, Piro RM, Esparza LA, Markant SL, Remke M, Milde T, Bourdeaut F, Ryzhova M, Sturm D, Pfaff E, Stark S, Hutter S, Seker-Cin H, Johann P, Bender S, Schmidt C, Rausch T, Shih D, Reimand J, Sieber L, Wittmann A, Linke L, Witt H, Weber UD, Zapatka M, König R, Beroukhim R, Bergthold G, van Sluis P, Volckmann R, Koster J, Versteeg R, Schmidt S, Wolf S, Lawerenz C, Bartholomae CC, von Kalle C, Unterberg A, Herold-Mende C, Hofer S, Kulozik AE, von Deimling A, Scheurlen W, Felsberg J, Reifenberger G, Hasselblatt M, Crawford JR, Grant GA, Jabado N, Perry A, Cowdrey C, Croul S, Zadeh G, Korbel JO, Doz F, Delattre O, Bader GD, McCabe MG, Collins VP, Kieran MW, Cho Y-J, Pomeroy SL, Witt O, Brors B, Taylor MD, Schüller U, Korshunov A, Eils R, Wechsler-Reya RJ, Lichter P, Pfister SM, PedBrain Tumor Project ICGC (2014) Genome sequencing of SHH medulloblastoma predicts genotype-related response to smoothened inhibition. Cancer Cell 25:393–405. https://doi.org/10.1016/j.ccr.2014.02.004

27. Cook Sangar ML, Genovesi LA, Nakamoto MW, Davis MJ, Knobluagh SE, Ji P, Millar A, Wainwright BJ, Olson JM (2017) Inhibition of CDK4/6 by Palbociclib significantly extends survival in medulloblastoma patient-derived Xenograft mouse models. Clin Cancer Res 23:5802–5813. https://doi.org/10.1158/1078-0432.CCR-16-2943

28. Kersten K, de Visser KE, van Miltenburg MH, Jonkers J (2017) Genetically engineered mouse models in oncology research and cancer medicine. EMBO Mol Med 9:137–153. https://doi.org/10.15252/emmm.201606857

29. Goodrich LV, Milenković L, Higgins KM, Scott MP (1997) Altered neural cell fates and medulloblastoma in mouse patched mutants. Science 277:1109–1113. https://doi.org/10.1126/science.277.5329.1109

30. Swartling FJ, Grimmer MR, Hackett CS, Northcott PA, Fan QW, Goldenberg DD, Lau J, Masic S, Nguyen K, Yakovenko S, Zhe XN, Gilmer HC, Collins R, Nagaoka M, Phillips JJ, Jenkins RB, Tihan T, Vandenberg SR, James CD, Tanaka K, Taylor MD, Weiss WA, Chesler L (2010) Pleiotropic role for MYCN in medulloblastoma. Genes Dev 24:1059–1072. https://doi.org/10.1101/gad.1907510

31. Swartling FJ, Savov V, Persson AI, Chen J, Hackett CS, Northcott PA, Grimmer MR, Lau J, Chesler L, Perry A, Phillips JJ, Taylor MD, Weiss WA (2012) Distinct neural stem cell populations give rise to disparate brain tumors in response to N-MYC. Cancer Cell 21:601–613. https://doi.org/10.1016/j.ccr.2012.04.012

32. Zindy F, Uziel T, Ayrault O, Calabrese C, Valentine M, Rehg JE, Gilbertson RJ, Sherr CJ, Roussel MF (2007) Genetic alterations in mouse medulloblastomas and generation of tumors de novo from primary cerebellar gran-

ule neuron precursors. Cancer Res 67:2676–2684. https://doi.org/10.1158/0008-5472.CAN-06-3418

33. Kawauchi D, Robinson G, Uziel T, Gibson P, Rehg J, Gao C, Finkelstein D, Qu C, Pounds S, Ellison DW, Gilbertson RJ, Roussel MF (2012) A mouse model of the most aggressive subgroup of human medulloblastoma. Cancer Cell 21:168–180. https://doi.org/10.1016/j.ccr.2011.12.023

34. Kawauchi D, Ogg RJ, Liu L, Shih DJH, Finkelstein D, Murphy BL, Rehg JE, Korshunov A, Calabrese C, Zindy F, Phoenix T, Kawaguchi Y, Gronych J, Gilbertson RJ, Lichter P, Gajjar A, Kool M, Northcott PA, Pfister SM, Roussel MF (2017) Novel MYC-driven medulloblastoma models from multiple embryonic cerebellar cells. Oncogene 36:5231–5242. https://doi.org/10.1038/onc.2017.110

35. Reya T, Duncan AW, Ailles L, Domen J, Scherer DC, Willert K, Hintz L, Nusse R, Weissman IL (2003) A role for Wnt signalling in self-renewal of haematopoietic stem cells. Nature 423:409–414. https://doi.org/10.1038/nature01593

36. Landau NR, Littman DR (1992) Packaging system for rapid production of murine leukemia virus vectors with variable tropism. J Virol 66:5110–5113

Chapter 3

In Vivo Models of Brain Metastases

Christina S. Wong

Abstract

Modeling brain metastasis in mice has been an ongoing challenge for the field of cancer research due to the complex anatomy of this morphologically distinct organ in the CNS. Furthermore, as the command center of various physiological functions of the body as a whole, artificial experimental manipulation of the brain comes with caveats and assumptions that cannot be lightly ignored. There is no true one-size-fits-all model to study brain metastasis, and depending on the hypothesis, each model needs to be adjusted and optimized to answer the scientific question at hand as unbiased and accurately as possible. This chapter will cover the different methods currently used to establish *orthotopic* models of brain metastasis, which include (1) intracranial, (2) intracarotid, (3) intracardiac, and (4) spontaneous (resected) routes of cancer cell inoculation, with specific human or murine-derived cells or cell lines tailored for different purposes.

Key words Brain metastasis, In vivo model, Orthotopic, Intracranial, Intracarotid, Intracardiac, Spontaneous model, Fluorescence, Bioluminescence, In vivo imaging

1 Introduction: Brain Metastases

Brain colonization of circulating tumor cells that have spread from the primary tumor site including lung, breast, skin, and renal cancers occurs frequently in patients with metastatic disease. Unfortunately, the diagnosis of this fatal progression of the disease is usually too late for successful treatment intervention, with a median survival of 2–3 months from diagnosis [1]. Furthermore, as treatment strategies continue to improve for primary cancers and their extracranial metastases, extended patient survival may partially contribute to the increased incidence of brain metastases, also fueled in part by gradually improving protocols to screen for brain metastases in cancer subtypes that are known to have higher-risk patients. However, the brain microenvironment proves to be a challenging avenue for drug delivery in the clinic to target brain metastases [2]. Current treatment options include whole-brain radiotherapy, stereotactic radiosurgery, and fractionated radiation. A better understanding of the unique characteristics of the brain

Giorgio Seano (ed.), *Brain Tumors*, Neuromethods, vol. 158,
https://doi.org/10.1007/978-1-0716-0856-2_3, © Springer Science+Business Media, LLC, part of Springer Nature 2021

microenvironment and how it affects migrating cancer cells that eventually colonize it is vital for the successful treatment of this disease. Concurrently, the intrinsic changes in biology behind the micropopulation of malignant cells that are attracted to and flourish in this unique microenvironment also need to be delineated. However, although many models for brain metastasis exist and are currently used in the field, it has been a challenge both biologically and technically to generate brain metastasis models that consistently reproduce the clinical presentation of the disease (particularly spontaneous migration of metastatic clones from the primary tumor site to the brain). This chapter aims to give an overview of the different models available to study brain metastasis, comprehensive but non-encompassing protocols on how to generate each of the different brain metastasis models, and discussion regarding the advantages and disadvantages of each for studying different steps in the metastasis pathway specifically to the brain, depending on the hypothesis put forth.

1.1 Methods to Study Brain Metastases In Vivo

For a particular brain metastasis model to be relevant and practical, at least one of two criteria needs to be fulfilled: either comprehensive imaging facilities (ultrasound, Optical Coherence Tomography, MRI, etc.) are available to monitor tumor take and progression, or a labeled cell line needs to be generated or acquired (i.e., cells with fluorescent or bioluminescent properties), and even then, a bioluminescence/fluorescence small animal imaging machine and/or a plate reader capable for reading blood-soluble bioluminescent enzyme activity are needed. Additionally, depending on which step of the metastasis pathway the hypothesis has been generated to investigate, further engineering of candidate tumor cell lines may be needed, in particular the creation or enrichment of brain-tropic metastatic cell lines (usually for syngeneic models, although applicable to xenograft models as well) and/or the acquisition of tumor material from patient brain metastasis samples for xenograft models. This chapter will discuss the methods to generate four different in vivo models of brain metastasis in mice, categorized largely based on the route of administration of cancer cells to generate the metastatic lesions in the brain. These include the (1) intracranial, (2) intracarotid, (3) intracardiac, and (4) spontaneous (resected) models of brain metastasis. The methods discussed are intended to cover "artificial" models of brain metastasis, which will not include truly spontaneous models of brain metastasis from primary tumors that arise "naturally" (without physical human intervention, namely, genetically modified mouse models of cancer) in their orthotopic sites of origin.

2   Materials

2.1   Tools and Preparation of Surgical Stage

Prior to the day of surgery, all non-disposable surgical tools are cleaned, dried, and autoclaved to ensure sterility for the aseptic surgical process. To enable a smooth and efficient surgical session, it is highly recommended to package sets of instruments based on specialized parts of the surgical procedure, i.e., all surgical tools for a particular procedure—a pair of scissors, a pair of forceps, and a pair of Kelly forceps—are all placed in one autoclave bag/box and autoclaved as a set to maintain sterility when opened in the relevant biosafety cabinet. Please refer to the relevant tables in the sections below for a comprehensive list of materials and tools needed for each part of the procedure. Apparatus and surfaces that cannot be autoclaved but will come into contact with the animals are cleaned with 70% ethanol, Clidox, or an equivalent cleaning agent as deemed suitable for the relevant material. Furthermore, a glass bead sterilizer (VWR Micro Sterilizer, VWR International) can be used to eliminate bacteria, spores, and microorganisms on small surgical instruments in between uses, including forceps, scissors, and needles. Biosafety cabinets are put through at least one 30-min UV cycle to prepare an aseptic environment for the surgical process.

2.2   Animal Preparations

Ensure all animal experiments are approved by the relevant Institutional Animal Care and Use Committee (IACUC) with regular veterinary check-ins to monitor the health and condition of the animals.

For these experiments, mice from a range of different strains were used, e.g., C57Bl/6, Balb/c, FVB, Nu/Nu, and NSG mice, at the optimal weight for imaging between 20 and 25 g. The first step of the model establishment process is the preparation of the animals. Mice need to acclimatize for approximately 3 days (or more, ideally 7 days) before surgery, to the area where they will be housed while recovering from surgery and during the brain metastasis establishment or growth phase. Recent reports have highlighted the differences in brain tumor (including secondary brain tumors such as brain metastases) response to certain therapies in patients of different genders; hence, the gender of mice cohorts should be taken into consideration [3, 4]. Of note, breast cancer brain metastasis is almost always modeled in female mice, although there are cases of male breast cancer as well.

Prior to any surgery, an appropriate dose of carprofen (5 mg/kg) is administered subcutaneously to the animal at least half an hour before the planned start time of surgery to minimize postoperative inflammation, followed by an intraperitoneal injection of a freshly prepared mixture of ketamine and xylazine (100 mg/kg and 20 mg/kg, respectively, using a 27 gauge needle). Depending on the procedure of brain metastasis inoculation, it may be necessary to shave the fur at the site of surgery, particularly for intracranial and intracarotid models of brain metastasis. To do this accurately without causing accidental shaving wounds, the skin on either side of the planned site of incision is gently pulled taut with one hand while carefully shaving as close as possible to the skin surface using a small animal shaver (Wahl Pocket Pro Universal Trimmer, Wahl). The surface area shaved should be slightly larger than the perimeter and/or area of any incisions made, as fur-free skin ensures that the sutures made during wound closure do not trap any fur within the wound, which may otherwise risk infection.

Additionally, eye ointment (Puralube Vet Ointment, Dechra Pharmaceuticals) should be applied to the eyes of the animals prior to the start of surgery and reapplied as needed during the procedure, to protect the eyes from dehydration. For more details on the materials and tools needed, please refer to Table 3.1.

2.3 Surgical Procedures to Expose the Brain

For direct inoculation of brain metastatic cell lines into the brain, an opening in the skull of the animal is necessary, and this is created using a small electric drill (Bone Micro Drill System, Harvard Apparatus) with the relevant drill bits (0.9 and 1.4 mm burrs for micro drill, Fine Science Tools). In between short bouts of drilling, it is advisable to moisten the surface of the skull being drilled with a drop or two of sterile saline for injection to dissipate the heat caused by the drilling motion and also to clean the drill bit to remove bone residue.

After the skull bone is penetrated and the surface of the brain is exposed, the brain surface is kept moist by placing a small piece of sponge (absorbable gelatin sponge, Surgifoam; cut into 2 mm × 2 mm pieces when dry, 4 mm × 4 mm when moist) that has been soaked in sterile saline for injection on top of the opening. For more details on the materials and tools needed, please refer to Table 3.2.

2.4 Stabilization of Animal on Stereotactic Device

To ensure accurate insertion of the needle into the brain at the correct depth and a stable hold on the animal's head, a stereotactic device (Lab Standard Mouse Stereotaxic Instrument, Stoelting, IL) is used. Because the penetration of the needle into the brain has to be at a very precise depth (in this case, 2.5 mm) and then withdrawn again at a precise height (1 mm), the head of the animal needs to be extremely stable to prevent any sudden movement (especially from a loosening hold of the instrument on the head during the injection) that may result in an injury to the brain. To do this, the maxillary incisors of the mouse are hooked onto the tooth bar of the stereotaxic device, and then the nose "holder" bar is lowered onto the animal's snout firmly but gently to ensure the anterior portion of the animal's head stays in place. The posterior

Table 3.1
Tools and materials for animal preparations for the intracranial model of brain metastasis

	Item	Note	Amount	Sterility
Tools	Shaver	Wahl BravMini Cordless Trimmer, Wahl *Keep fur-free; Clean with brush after every use* *Keep well-oiled between use*	1	Wiped with 70% ethanol
	Heat pad	Foot Warmers, Heat Factory, Vista, CA *For maintenance of body temperature*	>2	Wiped with 70% ethanol or Clidox before use
	Cotton swab	Generic, wooden with cotton tip	1 packet (>20 counts)	Autoclaved
Materials	Eye ointment	Puralube Vet Ointment, Dechra Pharmaceuticals	1 tube	Sterile from manufacturer
	Carprofen	5 mg/kg	Calculated based on weight	Sterile from manufacturer
	Ketamine	100 mg/kg	Calculated based on weight	Sterile from manufacturer
	Xylazine	20 mg/kg	Calculated based on weight	Sterile from manufacturer

half of the head is stabilized by positioning the jaw cuffs or the ear bars into place either side of the mouse's head and then securing the final orientation firmly but gently so as not to injure the animal's head with too much pressure. For more details on the tools needed for this step, please refer to Table 3.3.

2.5 Stereotactic Injection

The preparation of the tumor cell suspension, needle and syringe setup, as well as additional postinjection solution(s) is vital to a successful injection to generate an intracranial brain metastasis. For an intracranial injection, it is necessary to ensure that the cell suspension "slurry" (resuspended in Hank's balanced salt solution, HBSS) has as little cell aggregates or "clumps" as possible. Due to the minute amount of cell slurry injected into the hole or pocket created in the brain, approximately 1–2 μL, a Hamilton syringe with a very fine barrel lumen fitted to an equally fine needle is used. Any clumps or cell aggregates may clog the needle, potentially increasing the risk of injecting too much cell suspension into the brain while trying to dislodge the clog, with the resulting pressure causing physical damage to the brain. Alternatively, creation of pockets of air that are difficult to remove from the needle and syringe setup also risks shearing the cells as the plunger is drawn and withdrawn repeatedly to remove any air bubbles. It is also

Table 3.2
Tools and materials for the surgical procedure to expose the brain for the intracranial model of brain metastasis

	Item	Note	Amount	Sterility
Tools	Electric/dental drill	Bone Micro Drill System, Harvard Apparatus *Clean well after use*	1	Sterilized with microbead sterilizer
	Drill burrs	0.9 mm and 1.4 mm drill burrs, Fine Science Tools *Clean well after use*	1	Sterilized with microbead sterilizer
	Heat pad	Foot Warmers, Heat Factory, Vista, CA *For maintenance of body temperature*	>2	Wiped with 70% ethanol or Clidox before use
	Scissors	Roboz, MD	1	Autoclaved
	Forceps	Roboz, MD	2	Autoclaved
	Cotton swab	Generic, wooden with cotton tip	1 packet (>20 counts)	Autoclaved
	Absorbable gelatin sponge	Size 100, Surgifoam, Ethicon 1974	1 large piece	Sterile from manufacturer
	Insulin syringe or 1 mL syringe with 27G needle	Terumo, MD	1	Sterile from manufacturer
Materials	Saline for injection/ sterile PBS	Gibco, Thermo Fisher Scientific	1 bottle, 100 mL	Sterile from manufacturer

Table 3.3
Tools for the stabilization of the animal on a stereotactic device for the intracranial model of brain metastasis

	Item	Note	Amount	Sterility
Tools	Mouse Stereotaxic Instrument	Mouse head holder Model 921, Kopf *Clean well after use*	1	Wiped with 70% ethanol or noncorrosive antiseptic agent before use
	Cotton swab	Generic, wooden with cotton tip	1 Packet (>20 counts)	Autoclaved

important to note that any cell slurry that is not immediately loaded into the syringe should be set aside on ice to maintain the viability of the cells while the initial cohort of mice are being injected.

In terms of cell numbers injected, the amount can range from a few cells (10–100) per injection for aggressive tumors or to test for "stemness" of a cell line using a limited dilution assay, to

100,000 cells per inoculation for an injection to establish tumors for efficacy studies. As with all tumor model generation protocols, researchers are strongly advised to optimize the cell numbers (titration of tumor cell numbers in a pilot model development study) for each different tumor cell line destined for a future study. Furthermore, it is also possible to "inject" a 1 mm x 1 mm or smaller tumor chunk (usually a patient-derived tumor xenograft, PDX) in the cortex to propagate the tumor, although this technique causes more damage to the brain and is quite superficial.

An additional 1 mL syringe filled with saline for injection or PBS should be prepared prior to the tumor cell injection for use during the withdrawal of the entire needle from the brain post-injection. During the withdrawal of the inoculation needle out of the brain, a constant flow of sterile PBS is applied on the area immediately where the needle is penetrating the brain to facilitate fluidic pressure on the opening as the needle is drawn out. This is intended to prevent any tumor cells sitting within the "pocket" or hole created by the needle from flowing out, particularly as the withdrawal of the needle may create a small "vacuum" that may wash some tumor cells out no matter how carefully and slowly it is withdrawn. This fluidic pressure created by the PBS stream is an additional precautionary step to ensure as many injected cells remain within the pocket made by the needle as possible. Finally, a saline-soaked piece of gelatin sponge is again placed on the brain opening to ensure the surface of the brain is kept moist while also keeping some degree of pressure on the tumor cell "pocket." For more details on the materials and tools needed, please refer to Table 3.4.

2.6 Surgical Procedures to Close the Brain

For the intracranial model of brain metastasis, there are three different ways to close the brain opening, complementary to the three types of brain openings created, namely, the (1) burr-drilled hole opening, (2) skull "flap" opening, or (3) cranial window. For the burr-drilled opening, the hole in the skull is sealed with a "plug" of dental bone wax, compacted gently but firmly into the hole with a spatula. For the skull "flap" opening, the "flap" is returned to its original position, covering the opening in the skull. The edges of the flap are sealed to the rest of the skull with glue, making sure not to apply excess amounts but a sufficient amount to seal the flap securely to the skull again. For the third method of the cranial window, the creation of the skull opening needs to be made at least 10 days (14 days is ideal) prior to the tumor cell inoculation to provide sufficient time for the animal to heal and any inflammation to regress. This includes proper placement of the round coverslip centered onto the skull opening, then secured in place with a dental acrylic and glue mix around the edges, creating an approximately 2-mm-thick "ring" to bond the coverslip to the skull. On the day of tumor inoculation, the cranial window is removed by

Table 3.4
Tools and materials for stereotactic injection of tumor cells into the brain in an intracranial model of brain metastasis

	Item	Note	Amount	Sterility
Tools	Hamilton microliter syringe	Model 701, 10 ul, removable needle, Hamilton *Rinse at least twice with PBS after use*	1	Autoclaved
	Needle	28G, 2″, point style 2 (cat 7803-02), Hamilton	1	Autoclaved
	Absorbable gelatin sponge	Size 100, Surgifoam, Ethicon 1974	1 large piece	Sterile from manufacturer
	Cotton swab	Generic, wooden with cotton tip	1 packet (>20 counts)	Autoclaved
	Insulin syringe or 1 mL syringe with 27G needle	Terumo, MD	1	Sterile from manufacturer
Materials	Saline for injection/ sterile PBS	Hospira, IL/Gibco, Thermo Fisher Scientific	1 bottle, 100 mL	Sterile from manufacturer
	Hank's Balanced Salt Solution (HBSS)	Gibco, Thermo Fisher Scientific	1 bottle, 100 mL	Sterile from manufacturer

prying the acrylic-glue "ring" away from the skull, usually with the coverslip intact within the ring frame, and therefore removed in the same instance. The brain surface is then gently cleaned with sterile saline for injection with a sterile cotton swab and kept moist with a saline-soaked gelatin sponge, before transfer to the stereotactic device for tumor cell inoculation. After the inoculation process, the cranial window is reassembled using a new sterile round coverslip and fresh acrylic-glue mixture to secure it back in place on the skull. Upon placing the coverslip onto the opening, if there is a layer of air between the coverslip and the surface of the brain, a drop of sterile saline can be added between the two layers to ensure a healthy level of moisture at the brain surface. However, it is essential to make sure that excess moisture around the edges of the coverslip is removed to ensure the bonding of the acrylic-glue material to the skull is secure, as moisture will negatively interfere with a proper bond formation. For more details on the materials and tools needed, please refer to Table 3.5.

Table 3.5
Tools and materials for the surgical closure of the brain for the intracranial model of brain metastasis

	Item	Note	Amount	Sterility
Tools	Kelly forceps	Roboz, MD	1	Autoclaved
	Forceps	Dumont #3 and #5, Roboz, MD	2	Autoclaved
	Scissors	Microscissors – Curved Vannas, RS-5611, Roboz, MD	1	Autoclaved
	Spatula	Roboz, MD	1	Sterile from manufacturer
	Ethilon Nylon Sutures	5-0 Ethilon, Ethicon X-698G	1	Sterile from manufacturer
	Absorbable gelatin sponge	Size 100, Surgifoam, Ethicon 1974	1 large piece	Sterile from manufacturer
	Cotton swab	Generic, wooden with cotton tip	1 packet (>20 counts)	Autoclaved
Materials	Surgical drape	3 M Steridrape Incise Drape, 3 M, MN	1 roll	Sterile from manufacturer
	Coverslip, round	Plastic 7 mm circle cut from Ted Pella square plastic coverslip (cat. 2225) or Glass 7 mm circle #1, Warner Instruments CS-7R	1 packet	Cleaned with 70% ethanol, UV irradiated
	Petri dish, 6 cm	Corning	1 package of 10	Sterile from manufacturer
	Saline for injection/ sterile PBS	Gibco, Thermo Fisher Scientific	1 bottle, 100 mL	Sterile from manufacturer
	Dental acrylic powder	Alternatively, very fine acrylic powder, nail salon grade	–	
	Krazy Glue	All purpose, precision tip	1 tube	
	Bone wax	W31 Ethicon	1 packet	

3 Methods

3.1 Intracranial Model of Brain Metastasis

To generate a model with singular brain metastasis (i.e., for treatment studies or PDX brain metastasis propagation), the intracranial model is the optimal choice. While this model skips all the steps of the metastasis process, most hypotheses interrogated require a model with only a singular metastasis per animal for practical reasons. This includes but is not limited to: high-resolution imaging where the exact position of the metastasis needs to be known and limited to a certain distance from the surface of the brain, harvest of brain metastasis tissue sample (without including surrounding normal brain in the sample) for sensitive biochemical or molecular assays, or the restrictive nature of the tumor sample for manipulation in vitro (fluorescent or bioluminescent labeling) before implantation into an in vivo host, to name a few. Furthermore, the readout of efficacy studies in the treatment of brain metastasis is vulnerable to errors with different numbers of metastases in different animals within or between the treatment cohorts, particularly for the initial testing of the treatment compound.

Preparation and stabilization of animal for surgery. The animal is anesthetized after at least 30 min post-administration of a suitable analgesic. After the animal reaches an appropriate level of nociception (no reaction with the pinch-reflex test on the foot), the surgical procedure is initiated while monitoring reflex at subsequent steps of the surgery. After the animals are anesthetized, particular care has to be taken to maintain the body temperature in a range between 25 and 38 °C, maintained by using heated pads (Foot Warmers, Heat Factory, Vista, CA). The two maxillary (upper) incisors of the mouse are separated from its tongue by inserting the wooden end of a clean and sterile cotton swab (or equivalent instrument) between them. The maxillary incisors are then hooked onto the "loop" or "tooth bar" of the animal stabilization apparatus, while the rest of the body of the mouse is gently but firmly pulled taut by the base of its tail. The skull is now stably in place for further surgical manipulation. Fur at the dorsal portion of the skull is shaved (skip this step for hairless mice, i.e., Nu/Nu mice). The area for incision is sterilized with 70% ethanol.

Surgical procedure to expose the brain. Using a pair of forceps to lift the skin, an incision is carefully made around the dorsal portion of the head of the animal, cutting a "flap" of skin-free, with one side untouched to form the "hinge" of the flap. The "flap" of skin is flipped to one side to expose the surface of the skull. Using a sterile cotton-tipped swab, moisture and connective tissue are carefully cleaned away from the surface of the skull with gentle but firm pressure. When suitably clean, the electric drill is prepared by inserting the desired drill bit and securely fastening it to the device. The drill setup is tested for power supply and rotation. The drill is used to make the relevant opening in the skull, taking care to put the correct amount of pressure on the skull surface with the drill so that the fragile brain tissue underneath is not injured in the drilling process. There are three different types of openings typically created for this procedure (refer to Fig. 3.1 for schematic representation of each version): (1) burr-drilled hole of ~1 mm, (2) skull "flap" opening made by drilling four critical points along the perimeter of the intended flap and then gradually thinning the connecting "lines" between these points predominantly at three sides to allow the piece of bone to flip open with one side held to

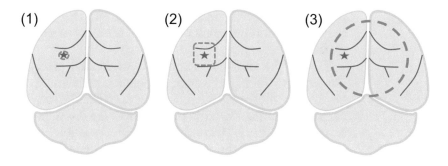

— Visible vessels on brain surface
-- Skull opening boundary
★ Point of injection

Fig. 3.1 Schematic of the brain from the top (plus large blood vessels), position of injection for each different variation of cranial openings

the skull by connective tissue/bone, or (3) skull opening to fit a brain "window" for high-resolution live imaging [5]. One of the reasons for choosing a "flap" over a "burr-drilled hole" opening is to improve the precision with which the stereotactic injection is executed, making it easier to avoid larger blood vessels due to an increased visible area of the surface of the brain. Regardless of which type of opening is made, the exposed surface of the brain is then kept moist with a piece of sponge pre-moistened with sterile saline for injection. The animal is then transferred to the stereotactic device for the next step.

Stereotactic injection. The two maxillary (upper) incisors of the mouse are separated from its tongue by inserting the wooden end of a clean and sterile cotton swab (or equivalent instrument) between them. The maxillary incisors are hooked onto the "loop" or "tooth bar" of the animal stabilization apparatus while placing a finger under the head and neck area of the animal to stabilize it in place. Using the other free hand, the left and right jaw holder cuffs are secured on both sides of the jaw, making sure that there is firm but not too much pressure to hold the jaw in place. The skull is now stably in place for further surgical manipulation.

Cells in suspension are centrifuged, washed once with PBS or Hanks balanced salt solution (HBSS), and then resuspended in HBSS at the relevant volume to yield the desired concentration of cells per injection. Sterile, autoclaved sets of Hamilton syringes, needles, and plungers (Cat. No. 80314, Model 701, Hamilton) are assembled carefully in a sterile hood. The cell solution is then aspirated into the syringe, taking care to avoid or remove any air bubbles trapped in any part of the syringe. The syringe setup is then attached to the syringe-needle holder of the stereotactic apparatus and securely fastened. The angle of the injection should be set between 0° and 10° away from the vertical tangent to the surface of the brain that is to be injected. The sterile sponge is removed from the surgical opening of the skull to expose the brain surface.

Using the knobs to control the horizontal plane, the area of the brain to be injected is positioned right below the tip of the needle. Slowly, the needle is brought downward toward the surface of the brain using the knob that controls the distance of the needle from the platform. The position of both the animal and the needle is adjusted so that the tip of the needle just touches the surface of the brain. The orientation is checked again to confirm that this is the correct location intended for the needle to penetrate the brain, avoiding any medium-large vasculature visible to the naked eye at the surface of the brain (Fig. 3.1). Figure 3.1 above denotes the approximate position where the tumor cells are injected into the brain from a top view, but this position can vary depending on each researcher's optimization of the technique and the requirements of different experimental conditions. Using the height adjustment knob and paying close attention to the notches of measurement on the apparatus, the needle is injected 2.5 mm into the brain and then withdrawn upward by 1 mm (Fig. 3.2). The cell slurry is injected slowly and carefully into the "pocket" created by the partial withdrawal of the needle, by depressing the syringe plunger gently but with sufficient pressure with a steady fingertip. A 1 mL syringe prefilled with sterile saline solution is prepared before withdrawing the needle. When ready, the entire needle is slowly withdrawn upward and away from the brain by rotating the height adjustment knob while simultaneously applying a steady stream of saline solution on the site of injection to maintain hydrostatic pressure on the opening to prevent cells from "leaking" out. Finally, when the needle has been entirely withdrawn, the saline stream is continued for a few more seconds before dabbing excess saline with a sterile piece of gauze around the opening but without touching the brain surface. The skull opening is covered with a piece of pre-moistened sterile sponge, and then the animal is transferred to the surgical hood for the closure process.

Surgical procedures to close the brain. The moistened sponge is carefully removed from skull opening. Next, a clean, sterile cotton tip is used to remove excess moisture from around the opening. To close the three different brain openings, the following procedures are followed accordingly: (a) for the drilled hole opening, a small piece of dental wax slightly larger than the hole diameter is placed over the opening and firm pressure applied with a spatula to "plug" the hole. (b) For the "flap" opening, the flap is replaced in its original position and dabbed with an adequate amount of glue along the edges of the flap to seal the flap to the skull. (c) For the brain "window" opening, a mixture of acrylic powder and glue is prepared in a sterile 6 cm dish, with the proportion of each adjusted to achieve a viscous but partially runny/easily moldable mixture. A

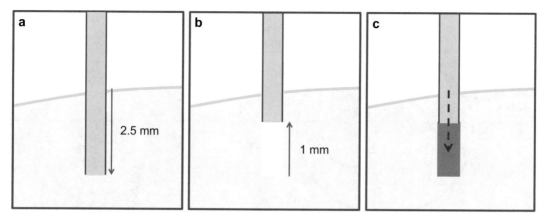

Fig. 3.2 Schematic of the brain from side, depth of injection for intracranial model

few drops of sterile saline solution in dripped onto the exposed brain surface. Any excess is removed with a clean and sterile cotton swab. A round coverslip is carefully placed onto the skull opening, making sure it is centered over the opening. Again, any excess liquid around the edges of the coverslip is carefully dabbed away. Using the wooden end of a sterile cotton-tipped swab, the acrylic-glue mixture is transferred onto the edge of the coverslip, making a 2-mm-thick ring around the coverslip to hold it in place (Fig. 3.3). Once the mixture is dry, the mouse can be placed gently into its cage to recover (animals with brain windows are housed in solidarity). This window is usually created 10–14 days before inoculation of tumor cells. On the day of inoculation, the window (coverslip and glue) is gently but firmly pried away from the skull of the animal with a forcep and any debris removed with fine forceps and a moistened cotton swab. The exposed brain surface is protected with a saline-moistened sponge placed over the opening. The animal is transferred to the stereotactic injection hood for stabilization on the stereotactic stage and subsequent injection of tumor cells as per the procedures above. The surgery is completed by covering the opening with a new (clean and sterile) coverslip and then sealing the edges to the skull with acrylic-glue mix.

For the hole and flap closure method, the piece of skin is replaced over the wound, and the edge of the flap of skin is sutured closed around the opening. Depending on the skill of the person suturing, continuous suturing or singular sutures (four to six individual sutures interspaced evenly around the cut edge of the skin flap) can be used to seal the skin opening. Finally, the animals are placed in their respective cages, and the cages are placed on a heat mat or close to a suitable heat source for optimal recovery. Once an entire cage of animals is fully recovered, the cage is returned to the housing rack.

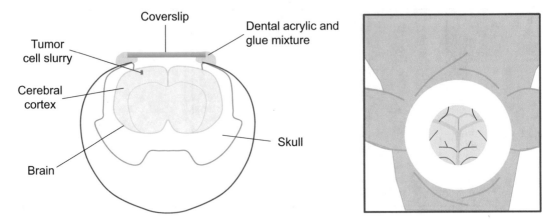

Fig. 3.3 Cross-sectional overview (left) and top view (right) of intracranial window setup with coverslip

3.2 Intracarotid Model of Brain Metastasis

To keep the BBB structurally intact (to a limited degree) and to mimic the final step of the metastasis process in vivo, the intracarotid model of brain metastasis is selected for experimental establishment of metastasis. Using this technique, the biological process of migrating cancer cells in the blood stream traveling in the direction of the brain (cell flow toward the heart is blocked in the surgical process) and then extravasating into the brain is captured. Although there is some degree of artificial manipulation of the amount of cells that migrate into the brain circulation, this technique does not manually disrupt the vasculature network in the brain as with an intracranial injection where the needle bores a hole into the brain and destroys any microvasculature and nervous tissues in its path. Additionally, this model also recapitulates multiple metastases typical in the clinical setting where patients usually present with three or more metastases. This technique requires a lot of experience, practice and a deft pair of hands to master the microsurgery technique.

Briefly, the animal is anesthetized with ketamine-xylazine and the left carotid artery surgically exposed with the aid of a dissection microscope (Fig. 3.4a). The external carotid artery (ECA) is ligated, and a 30G needle is inserted into the lumen of the common carotid artery (CCA), connected to a catheter constructed from fine tubing. At the other end of the catheter, a syringe (with a 30G needle) containing the tumor cell suspension at the desired concentration is inserted (Fig. 3.4b). A suitable amount of tumor cell suspension is slowly injected into the internal carotid artery (ICA), which supplies blood to the brain parenchyma. Usually, approximately 100,000 cells resuspended in 5 µL of PBS is injected and chased with 40 µL of PBS (alternatively, 100 µL containing the desired amount of cells is injected, without the subsequent PBS chasing). The hole in the artery is closed with a 10-0 suture under the dissection microscope to prevent bleeding and restore blood circulation. The surgical opening to expose the carotid artery is then sutured close, and the mouse is monitored

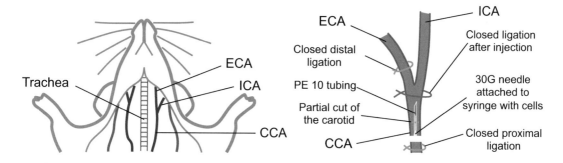

Fig. 3.4 Schematic of carotid artery exposure, injection entry point, sealing off of external artery

with frequent intervals during recovery to ensure circulation to the brain is sufficient.

3.3 Intracardiac Model of Brain Metastasis

To conserve even more of the biological process of metastasis progression, researchers favor the intracardiac model of brain metastasis as the more technically attainable and practical route of metastatic tumor cell inoculation. While there are publications in literature that utilize intravenous (tail vein) injection to establish models of brain metastasis [6], tail vein injections direct the inoculated tumor cells to pass through the circulation via the lungs before being distributed to the brain, causing many of the inoculated tumor cells to be "trapped" in the lung's microvasculature. The intravenous injection of metastatic tumor cells usually results in predominantly lung metastases rather than brain metastases, if any brain metastasis is successfully established in the first place (although some brain metastases have successfully been established this way). Thus, intracardiac injection of tumor cells into the left ventricle of the heart allows a larger proportion of inoculated tumor cells to filter through the circulation in the brain, accumulating or trapping "migrating" cells that then colonize the organ, conditions permitting.

Technically, there are a few variations of this method of tumor cell inoculation as detailed in the following section. With the advent of imaging technology, ultrasound-guided intracardiac injections have further improved the accuracy of injection directly into the left ventricle of the heart, reducing error and incidence of metastasis in irrelevant organs.

If using mice strains with fur, the fur around the chest (anterior) or on the left side of the animal under the arms until the end of the rib cage is shaved, depending on the direction of the intended injection. A hair removing cream (i.e., Nair) is applied to make sure the area is completely free of hair if intending to execute ultrasound-guided intracardiac injection. The animal is anesthetized with isoflurane, keeping a constant and stable supply of isoflurane while making sure that the nose and head of the animal are within appropriate orientation in the nose cone. The animal is ensured to be under the appropriate level of nociception by using

the pinch-reflex test on the foot. If there is no reaction, the procedure is continued while monitoring nociception at subsequent
steps of the surgical procedure. The intended area of injection is
sterilized with 70% ethanol.

Depending on the procedure of choice, the injection device is
prepared using one of the following two ways:

(a) Using an intracardiac injecting device. This device is made
    using 10 cm of clear polyethylene tubing [7], with a 27G
    needle (detached from plastic attachment) filed at one end and
    this blunted end inserted into one end of the tube (allows
    ~50% or 4 mm of the needle to project from the tube). An
    intact but blunted 27G needle is inserted into the other end of
    the tubing and attached to a 1 mL syringe filled with tumor
    cells in suspension. It is imperative to ensure that all air is
    removed from the entire length of the injection device.

(b) Using a 27G needle attached to a 1 mL tuberculin syringe.
    The syringe is filled with a suspension of tumor cells at the
    desired concentration in PBS. Again, meticulous care is taken
    to ensure that all air (or air bubbles) is removed from the entire
    length of the needle and syringe.

The animal is placed on its back (supine position). One of three
different points of entry is used:

(a) Anterior chest point of entry (Fig. 3.5a) (Optional: A midline
    skin incision is made over the upper chest cage to expose the
    clavicle and upper ribs.). The second intercostal space is identi-
    fied by directing the needle to the midline, starting 1–2 mm to
    the left of the sternum and angled 45° relative to the chest
    wall. The needle is inserted to its entire depth of 4 mm. A suc-
    cessful needle insertion will result in pulsatile flow of bright
    red blood into the transparent tubing of the injection device or
    into the syringe of a 27G needle. If there is no blood return or
    if dark blood with minimal pulsation is observed, the insertion
    of the needle is incorrect, and the needle is withdrawn. After
    assuring its patency, the needle can be reinserted. Once accu-
    rately positioned, the needle of an injection device is manually
    held stable while a second person slowly injects a total volume
    of 0.1–0.2 mL of tumor cell suspension. Injections are success-
    ful if the person executing the injection feels no resistance. If
    resistance is felt during injection, the animal should be eutha-
    nized. Any incision or opening is sutured close and the ani-
    mal's recovery is monitored.

(b) Lateral (left) chest point of entry (Fig. 3.5b). Using a sterilized
    gloved index or middle finger, the location of the heart is
    determined by feeling carefully along the left anterior portion
    of the chest for the point of the strongest heartbeat (approxi-
    mate location of where the heart should reside anatomically).
    From this point, an imaginary line is drawn toward the midline

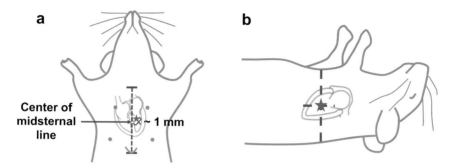

a

Center of
midsternal — ⊗ ~ 1 mm
line

b

Fig. 3.5 Schematic of the chest/heart from top, position of injection from top; chest/heart from side, position of injection from side

of the lateral left portion of the chest. The cross section where the two lines meet will be the point of injection (refer to Fig. 3.5b). The needle is inserted slowly into the chest cavity until a pulsatile flow of bright red blood into the syringe occurs. This indicates successful penetration of the needle into the left ventricle cavity of the heart. If there is no blood return or if dark blood with minimal pulsation is observed, the insertion of the needle is incorrect, and the needle is withdrawn. After assuring its patency, the needle can be reinserted. Once accurately positioned, the needle of an injection device is manually held stable while the other hand is used to slowly inject 0.1–0.2 mL of tumor cell suspension into the heart.

(c) Ultrasound-guided intracardiac injection [8]. With the anaesthetized animal placed on its back (supine) and facing slightly to the right, all four limbs are secured on a heated platform. A thick layer of ultrasonographic gel is applied onto the bare (fur-free) thorax above the heart. A stand-mounted 707B ultrasonographic probe (Visualsonics) is used to find the left ventricle of the heart. A syringe fitted to a 27G needle containing tumor cell suspension at the desired concentration is secured onto a needle holder and its position adjusted in place under the ultrasonographic probe until the needle can be visualized in the ultrasonographic image on the monitor. When the trajectory of the needle is deemed suitable, the needle is advanced through the intercostal space into the left ventricle. A pulse of bright red blood will flow into the syringe, indicating an accurate insertion into the left ventricle of the heart. 0.1–0.2 mL of tumor cell suspension is slowly injected into the heart (at an approximate rate of 20 μL over 30 s for a 100 μL injection). Carefully, the needle is withdrawn and the breathing and recovery condition of the animal is monitored.

It is vital to keep the animal in a cage that is placed on a heat pad or an alternative heat source that can keep the cage and animal warm during recovery. Once all animals in the cage have recovered, the cage can be returned to its original position on the rack.

3.4 Spontaneous (Resected) Model of Brain Metastasis

Ultimately, the ideal model for brain metastasis is one that will spontaneously metastasize from the inoculated primary tumor to the brain in due course (preferably following a timeline that facilitates efficacy studies and is reflective of metastatic progression in clinical settings). The approach to creating a model with these properties, however, necessitates (1) establishing a tumor cell line from a primary tumor that maintains its tumorigenic properties with sequential passages in vitro or in vivo, (2) genetically engineering the cell line to be fluorescent or bioluminescent to enable practical monitoring of tumor progression in vivo (particularly if there is more than one metastatic nodule and/or more than one site of metastasis), and (3) exhibition of brain "tropism", making it more likely than not this subclone of metastatic cells metastasizes to the brain in favor of other organs that are common sites of metastasis.

Researchers have developed a number of "brain-tropic" tumor cell lines to date, a number of which have been described in the literature including the 4T1 murine breast cancer cell line (aggressive, metastatic versions: 4T1.2, 4T1Br4), MDA-MB231-Br human breast cancer cell line, WM239-131/4-5B1 or WM239-131/4-5B2 human melanoma cell line [9], B16-B series of murine melanoma cancer cell line with brain-tropic properties, and A549 NSCLC human melanoma cell line, to name a few. Most of these cell lines have been genetically modified to express a fluorescent protein (i.e., GFP or mCherry) so that the tumor lesions that eventually form can be imaged with an in vivo imaging system like the IVIS Spectrum or quantitated postmortem with qPCR using primers for the fluorescent molecule. Alternatively, some of the tumor cell lines are engineered to express a bioluminescent protein (i.e., F-luciferase or G-luciferase) whose enzymatic activity upon metabolic processing of the relevant substrate (luciferin or coelenterazine) can be imaged with the IVIS Spectrum in vivo imaging system or using a plate reader that can measure bioluminescence from enzymatic activity of secreted G-luciferase present in the blood collected from tumor-bearing animals (and correlated with tumor burden based on a standard curve) [10, 11].

Firstly, either a parental tumor cell line of interest (particularly if there are genetic modifications to the cells that cannot be compromised) or a primary tumor cell line established from a brain metastatic lesion (while this is rare in study cohorts, there are incidences of spontaneous brain metastasis) is acquired and then genetically engineered to express a fluorescent protein (for metastatic cell sorting and enrichment) and, if desired, in combination with a bioluminescent tag, to enable subsequent monitoring of tumor progression using in vivo imaging modalities (i.e., IVIS Spectrum). The tumor cells are inoculated via intracardiac injection as per the protocol in Sect. 3.3 or orthotopic inoculation if the cell line yields at least a low incidence of brain metastasis spontaneously. The animal is then monitored for signs of CNS tumor progression via the relevant in vivo imaging modality of choice. Additionally, the physical condition and behavior of the animal may yield additional clues of the progression of brain metastases. For orthotopic primary tumor inoculation only: if an orthotopic primary tumor was established to generate spontaneous brain metastasis, the primary tumor's growth is monitored and the tumor is resected before it reaches ethical end stage, preferably at a size when the surgical resection of the tumor will minimize the surgical wound on the skin (~150 mm³ for a mammary tumor). The primary tumor-free animal is then allowed to recover and subsequently monitored for metastatic tumor progression. When an animal has a confirmed diagnosis of brain metastasis, the animal is euthanized and the brain is processed into single-cell suspension and sorted for metastatic tumor cells with the fluorescent marker using fluorescence-activated cell sorting (FACS). The yield for cell sorting is usually relatively low and may require a substantial amount of time to expand into a sufficient quantity for the next round of inoculation. Although the sorted metastatic tumor cells can grow with the standard culture media used for the parental cell line, various publications have suggested the use of brain-conditioned media to promote optimal growing conditions for brain-tropic cell lines [12]. When the brain metastatic subline reaches sufficient numbers, the next cycle of in vivo passaging can be done to enrich for brain-tropic cells from the current cycle. This cycle of in vivo passaging of each subsequent subline of brain-tropic metastatic cells is repeated for a total of three to five cycles (dependent on the degree of brain tropism of the enriched sublines). In parallel, advanced sublines can be orthotopically injected in a small cohort (five animals) to evaluate the rate of brain metastasis, and the decision for further enrichment can be made accordingly. Finally, the most enriched brain-tropic metastatic subline is then orthotopically injected into the animal and the cohort monitored for brain metastasis using the relevant imaging modalities or ex vivo blood assays as is suitable. The schematic in Fig. 3.6 depicts a summary of this process and the various options used to generate metastatic cell lines with increasing brain tropism.

4 Discussion

The selection of the type of brain metastatic model to answer the relevant scientific hypothesis is a crucial decision in the experimental design process. There are multiple parameters to take into consideration, including but not limited to: which step of the brain metastatic process needs to be interrogated, interaction of the brain metastatic lesions with the immune axis and the subsequent

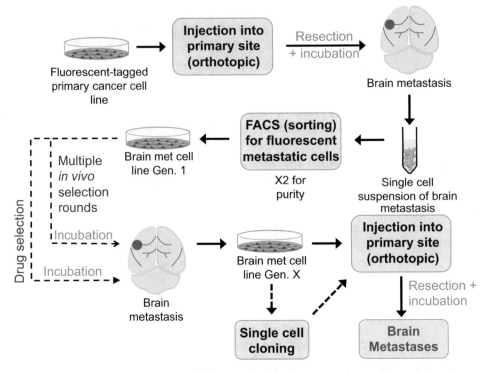

Fig. 3.6 Schematic of protocol utilizing multiple rounds of in vivo passaging or drug selection to generate brain-tropic, metastatic cancer cell lines

effect of the pharmacological agent used, the source of the tumor cells (mouse vs. human or untreated vs. drug-treated), the degree of brain tropism of the cell lines used, or the amenability of available imaging equipment to monitor tumor growth or regression, to name a few.

The main distinguishing features of each of the four models of brain metastasis described above are the steps of the metastatic cascade for which they each reflect, or attempt to mimic, in the clinical presentation of the disease in patients. As depicted in the schematic shown in Fig. 3.7, the intracranial model of brain metastasis is the most direct approach, which involves implantation of the tumor cells directly into the brain itself, allowing only the final step of colonization and growth in the brain microenvironment to form established "metastatic" malignant lesions followed by subsequent biological readout of tumor growth, survival, or response to pharmacological agents. This approach, while technically challenging and labor-intensive, has the highest take rate, making it ideal to test for downstream effects of genetically modifying the tumor or the host (e.g., gene knockout mouse models), as well as the efficacy of combinations of pharmacological agents targeting different cancer-dependent pathways. Furthermore, the intracranial brain metastasis model also allows the researcher to control the exact position and depth of tumor cell inoculation,

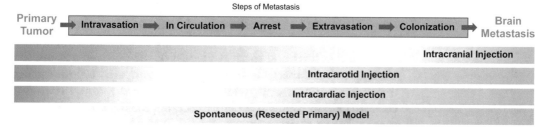

Fig. 3.7 Schematic showing the range of steps in the metastatic process each model of brain metastasis captures (shaded blue)

allowing for high-resolution imaging of a singular metastatic brain lesion (e.g., with multiphoton intravital microscopy), which typically requires the highest imaging plane of the lesion to be no deeper than 300 μm from the surface of the brain-coverslip interface. This is likely one of the few models that will enable in vivo longitudinal imaging of the interaction of the brain metastatic cells with other cells, architectural factors (e.g., tumor or tissue vasculature), and/or noncellular components of the brain microenvironment (i.e., collagen deposition).

Both the intracarotid and intracardiac route of injection generate brain metastasis models that mimic migration of "shed" tumor cells from the primary tumor in the circulation, before their subsequent arrest in the brain microenvironment's vasculature and extravasation out of the blood vessels, followed by colonization and growth in the brain as a metastatic lesion. Both these techniques have their advantages and disadvantages, with the intracarotid technique requiring highly skilled and trained personnel, delicate microsurgery skills to tease apart the carotid artery from surrounding tissue, injection of tumor cells at a steady rate to reduce the chances of stroke, and then execution of a minute suture to close the injection opening to prevent leakage of tumor cells and/or blood loss, followed by wound closure and ensuring subsequent survival of the animal. On the other hand, the intracardiac injection technique is straightforward, with a clearer readout of fresh blood return into the syringe when the left ventricle is successfully breached, as well as less chance of a stroke if the tumor slurry is carefully prepared and the animal's body temperature is carefully controlled. However, the intracardiac injection technique distributes the tumor cells to both the brain and the rest of the body at their first pass in the circulation, increasing the likelihood of metastasis forming in other organs other than the brain (there is a risk of developing bone metastases as the next predominant organ of colonization). As an advantage, the intracardiac introduction of tumor cells is superior to intravenous inoculation for establishing brain metastasis, as the intravenous route passes the cells through the lungs first, trapping most of the tumor cells in this organ causing predominantly lung metastasis, before subsequently circulating the cells in the blood to the heart and then out to the rest of the body. However, in comparison, the intracarotid injection approach filters the tumor cells in the circulation through the brain before entering the heart and then to the rest of the body, increasing the chances of brain metastasis establishment compared to intracardiac injection of tumor cells, while also reducing the time the tumor cells are held in circulation before entering the brain microenvironment, likely increasing their survival. Thus, there is a fine balance between needing a high success rate of brain metastasis establishment in a few animals using the time-consuming nature of the intracarotid technique, compared to a lower take rate and reduced brain specificity of the intracardiac technique, which on the contrary is less labor intense due to the ability to inoculate a significantly larger cohort in the same amount of time. Ultimately, the brain tropism of the tumor cell lines used will tip the balance, and both these techniques of brain metastasis establishment require the usage of fluorescent- or luminescent-tagged cells for subsequent monitoring of metastatic tumor take and growth.

Recently, single-cell suspensions prepared from patients' brain metastatic tissue (patient-derived xenografts, PDXs) were transduced with luciferase and then injected into SCID mice utilizing an improved intracarotid injection method [13]. The researchers claim that anterograde ligation of the external carotid artery and retrograde ligation of the common carotid artery promoted tumor cell migration toward the brain and prevented "backflow" into the heart or externally to other organs while protecting against inflammation or toxicity via the external carotid artery. Additionally, experimental data showed that these PDX brain metastasis models recapitulated the histological, molecular, and genomic patterns of their counterpart patient tumors [13]. It should be noted, however, that in comparison with the intracranial inoculation of tumor cells directly into the brain, the intracarotid route produces brain metastases that usually take a significantly longer amount of time to establish and grow compared to the same cells introduced intracranially.

Ultimately, a tumor model that can be inoculated orthotopically into the primary organ of origin, which then spontaneously metastasizes to the brain (with resection of the primary tumor before it reaches a certain size to allow for longer incubation time of circulating tumor cells), would be the ideal model of brain metastasis to interrogate a wide range of preclinical hypotheses that would be translationally relevant in the clinic. Again referring to Fig. 3.6, the spontaneous (resected) model of brain metastasis covers the entire metastatic cascade from escape from the primary tumor by intravasation into the circulation, traversing through the systemic circulation to the brain and arrest of the metastatic tumor cells in the brain, followed by extravasation, colonization of the brain TME, and progression of brain metastasis. In parallel with cutting-edge imaging technology to monitor brain metastatic burden and progression, this model would allow for interrogation of interventions at various steps of the metastatic cascade leading up to brain metastasis. Of clinical relevance, this model mimics the "natural" biological progression of brain metastasis, allowing the formation of a brain-tumor barrier that will be closer to that of brain metastasis patients than one in which the mechanical forces of a direct implantation into the brain could artificially alter or damage. However, one does have to keep in mind the significant challenge of timing the manifestation of brain metastases across different animals within the same study cohort and the relative average brain metastatic burden across the different treatment groups.

In regard to establishing a brain-tropic metastatic cell line, much effort has been invested into various strategies. An elegantly executed initiative to select for metastatically aggressive clones arising from parental tumor lines inoculated orthotopically into the primary sites of origin utilized pharmacological pressure with each in vivo "enrichment" cycle to eventually generate brain metastasis [9, 14]. Increasingly metastatic clones/clonal lines were harvested from the metastatic organs and further passaged in vivo, coupled with pharmacological selection to increase metastatic potential of the lines. In addition to the standard of care therapies and dosage regimen that endeavor to reflect clinical intervention of the disease, this strategy may also more faithfully mimic the genetic and phenotypic changes the cancer cells would encounter in a patient, potentially improving the metastatic model itself (and the brain-tropic cell lines derived from it) as a whole for further experimental interrogation [9, 14].

Other parameters for discussion in regard to the establishment and maintenance of robust brain-tropic metastatic "cell lines" include whether to culture them in media enriched with growth factors specific to the brain or brain-conditioned media (Neurocult or "mushed" brain-conditioned media) or in standard cell culture media with increased amounts of nutrients (i.e., 20% FBS or higher); two-dimensional (adherent cells in a flask) vs. three-dimensional cultures more reflective of the natural environment of these brain metastatic cells in the brain; duration of culture on plastic (too many in vitro passages seem to diminish luciferase-positive clones, *personal communication*); and the necessity of the immune axis to interrogate the experimental hypothesis, as this will determine the use of syngeneic or xenograft models, to name a few. Ultimately, it is up to the researcher to first determine the resources available, followed by the technical limitations and feasibility for specialist training to perform the intricate techniques needed to successfully establish the relevant brain metastasis models described above. A good review summarizing the tumor cells' sequential spread throughout different organs/parts of the body can be found in a review written by Daphu and colleagues [15]. Additionally, a deeper dive into literature to tease out the intricacies of the different variations (and more advanced versions) of the different brain metastasis models, their applicability for the particular cancer subtype they are studying, and the pros and cons for each model in respect to the hypothesis they wish to address can aid in determining the optimal model for the study.

5   Notes (Additional Troubleshooting, Tips, and Tricks)

Table 3.6 details additional troubleshooting notes, tips, and tricks for the respective brain metastasis models described above.

Acknowledgment

The author would like to thank Gino B. Ferraro and Sylvie Roberge for their invaluable input on the technical details of certain methods described above and for reviewing the manuscript in its final stages, as well as support from Rakesh K. Jain in facilitating their development of these expertise.

*The content, views, or opinions presented in this chapter in no way reflect the official policy or position of Merrimack Pharmaceuticals, Inc. or any of its affiliates.

Table 3.6
Glossary table of troubleshooting, tips, and tricks for generating the different animal models of brain metastasis

Model	Procedure	Notes
All	Anesthesia	• Additional anesthetic doses at one-third the initial dose can be administered if the anesthetic is observed to be wearing off during the surgical process.
Intracranial	Stereotactic injection—stabilization on device	• The first animal requires more adjustment trial and error; subsequent animals are easier to manipulate if one jaw cuff is left in place and the other adjusted to free the animal from the jaw hold after the injection. • If the process of stabilizing the head proves to be too challenging initially, a platform of sorts can be placed under the animal's head and torso to keep the animal at the desired height while adjusting the jaw holders. When the orientation is deemed acceptable, the platform can be removed carefully to observe if the head is held stably in place. If gentle pressure is applied on the skull and the head remains in place, then the animal is ready for the subsequent step in this procedure.

Table 3.6
(continued)

Model	Procedure	Notes
Intracranial	Stereotactic injection—cell slurry	• Given the minute volume feasible for injection into the brain (usually between 1 and 2 µL), the final preparation resembles a "slurry" of cells, with minimal addition of HBSS required.
Intracranial	Stereotactic injection—depth	• The depth of the needle suggested in the Methods section was 2.5 mm and then withdrawn 1 mm to create a 1.5 mm pocket. • It is suggested that a range of 1–3 mm can be used depending on preexisting known characteristics of the model of brain metastasis being studied (i.e., frequently observed in a particular section of the brain or a specific depth/layer) or results based on optimization/pilot experiments observing aggressiveness and growth rate of the particular tumor cell line/model.
Intracranial	Stereotactic injection	• Some researchers prefer timing the injection process to ensure steady and consistent flow of cell slurry into the pocket across 1–3 min, largely dependent on the cell lines they are using and their experience with these lines. • Some researchers necessitate timing the needle withdrawal process as well to ensure a steady and consistent force is applied, to prevent creating a vacuum above the injected cell slurry if a needle is withdrawn too quickly. The timed process can vary anywhere between 1 and 3 min.
Intracardiac	Cell slurry for injection	• As a reference, the number of cells injected via intracardiac inoculation range anywhere between 1×10^5 and 1×10^6 cells. However, depending on how likely the cells are to aggregate in the syringe, this number should be adjusted to prevent stroke or fatalities during the injection process. A pilot experiment should ideally be carried out to determine the maximum number of cells that can be delivered via intracardiac injection for a particular cell line or tumor slurry.
Intracardiac	Injection completion	• Upon completion of the injection, one may observe return of pulsatile blood flow into the tubing of the injection device or syringe, although this is not always the case.
Intracardiac	Survival rate of procedure	• It has been noted that keeping the animals suitably warm so heat loss does not occur during the recovery period can increase survival of animals by as much as 50% of the cohort.
Spontaneous	Symptoms of brain metastasis	• Physiological signs of CNS or brain metastasis include imbalanced posture or movement, movement of the animal favoring one side of the body—sometimes constant circular motion in one direction, abnormal and erratic behavior, paralysis of one side of the body, etc.
Spontaneous	No. of cells injected	• It has been observed that a lower cell number injected orthotopically in a breast cancer model increased the likelihood of brain metastases, partially by increasing the time to ethically permitted end-stage tumor size [11].

References

1. Valiente M, Ahluwalia MS, Boire A, Brastianos PK, Goldberg SB, Lee EQ, Le Rhun E, Preusser M, Winkler F, Soffietti R (2018) The evolving landscape of brain metastasis. Trends Cancer 4(3):176–196. https://doi.org/10.1016/j.trecan.2018.01.003

2. Askoxylakis V, Arvanitis CD, Wong CSF, Ferraro GB, Jain RK (2017) Emerging strategies for delivering antiangiogenic therapies to primary and metastatic brain tumors. Adv Drug Deliv Rev 119:159–174. https://doi.org/10.1016/j.addr.2017.06.011

3. Yang W, Warrington NM, Taylor SJ, Whitmire P, Carrasco E, Singleton KW, Wu N, Lathia JD, Berens ME, Kim AH, Barnholtz-Sloan JS, Swanson KR, Luo J, Rubin JB (2019) Sex differences in GBM revealed by analysis of patient imaging, transcriptome, and survival data. Sci Transl Med 11(473). https://doi.org/10.1126/scitranslmed.aao5253

4. Sun T, Plutynski A, Ward S, Rubin JB (2015) An integrative view on sex differences in brain tumors. Cell Mol Life Sci 72(17):3323–3342. https://doi.org/10.1007/s00018-015-1930-2

5. Askoxylakis V, Badeaux M, Roberge S, Batista A, Kirkpatrick N, Snuderl M, Amoozgar Z, Seano G, Ferraro GB, Chatterjee S, Xu L, Fukumura D, Duda DG, Jain RK (2017) A cerebellar window for intravital imaging of normal and disease states in mice. Nat Protoc 12(11):2251–2262. https://doi.org/10.1038/nprot.2017.101

6. Aprelikova O, Tomlinson CC, Hoenerhoff M, Hixon JA, Durum SK, Qiu T-h, et al. (2016) development and preclinical application of an immunocompetent transplant model of basal breast cancer with lung, liver and brain metastases. PLoS ONE 11(5):e0155262. https://doi.org/10.1371/journal.pone.0155262

7. Conley FK (1979) Development of a metastatic brain tumor model in mice. Cancer Res 39:1001–1007. https://doi.org/0008-5472/79/0039-0000502.00

8. Balathasan L, Beech JS, Muschel RJ (2013) Ultrasonography-guided intracardiac injection: an improvement for quantitative brain colonization assays. Am J Pathol 183(1):26–34. https://doi.org/10.1016/j.ajpath.2013.03.003

9. Cruz-Munoz W, Man S, Xu P, Kerbel RS (2008) Development of a preclinical model of spontaneous human melanoma central nervous system metastasis. Cancer Res 68(12):4500–4505. https://doi.org/10.1158/0008-5472.CAN-08-0041

10. Chung E, Yamashita H, Au P, Tannous BA, Fukumura D, Jain RK (2009) Secreted Gaussia luciferase as a biomarker for monitoring tumor progression and treatment response of systemic metastases. PLoS One 4(12):e8316. https://doi.org/10.1371/journal.pone.0008316

11. Kodack DP, Chung E, Yamashita H, Incio J, Duyverman AM, Song Y, Farrar CT, Huang Y, Ager E, Kamoun W, Goel S, Snuderl M, Lussiez A, Hiddingh L, Mahmood S, Tannous BA, Eichler AF, Fukumura D, Engelman JA, Jain RK (2012) Combined targeting of HER2 and VEGFR2 for effective treatment of HER2-amplified breast cancer brain metastases. Proc Natl Acad Sci U S A 109(45):E3119–E3127. https://doi.org/10.1073/pnas.1216078109

12. Kim SH, Redvers RP, Chi LH, Ling X, Lucke AJ, Reid RC, Fairlie DP, Martin A, Anderson RL, Denoyer D, Pouliot N (2018) Identification of brain metastasis genes and therapeutic evaluation of histone deacetylase inhibitors in a clinically relevant model of breast cancer brain metastasis. Dis Model Mech 11(7). https://doi.org/10.1242/dmm.034850

13. Liu Z, Wang Y, Kabraji S, Xie S, Pan P, Liu Z, Ni J, Zhao JJ (2019) Improving orthotopic mouse models of patient-derived breast cancer brain metastases by a modified intracarotid injection method. Sci Rep 9(1):622. https://doi.org/10.1038/s41598-018-36874-3

14. Francia G, Cruz-Munoz W, Man S, Xu P, Kerbel RS (2011) Mouse models of advanced spontaneous metastasis for experimental therapeutics. Nat Rev Cancer 11(2):135–141. https://doi.org/10.1038/nrc3001

15. Daphu I, Sundstrom T, Horn S, Huszthy PC, Niclou SP, Sakariassen PO, Immervoll H, Miletic H, Bjerkvig R, Thorsen F (2013) In vivo animal models for studying brain metastasis: value and limitations. Clin Exp Metastasis 30(5):695–710. https://doi.org/10.1007/s10585-013-9566-9

Chapter 4

Intravital Imaging of Brain Tumors

Cathy Pichol-Thievend, Boris Julien, Océane Anézo, Beatrice Philip, and Giorgio Seano

Abstract

Intravital imaging on live animals has provided new insights into the dynamics of tumor cells within their orthotopic microenvironment. In this chapter, we present a detailed method for intravital imaging of glioblastoma (GBM) cells in the mouse brain, with particular emphasis on the interactions of GBM cells with the surrounding vasculature.

This method involves the implantation of a cranial window and longitudinal intravital imaging as well as the analysis of tumor cells within their orthotopic microenvironment in vivo at a single-cell resolution using multiphoton imaging.

Key words Intravital microscopy, Glioblastoma, Tumor microenvironment, Longitudinal imaging

1 Introduction

Brain tumors refer to a collection of intracranial neoplasms of the central nervous system classified into primary and secondary brain tumors. Primary brain tumors can arise from cells of glial origin or non-glial origin, while secondary brain tumors are the distant metastasis of a primary tumor localized elsewhere in the body [1].

Tumors of glial origin, referred to as gliomas, are the most common type of primary brain tumors. These include glioblastoma (GBM), ependymoma, astrocytoma, oligodendroglioma, and oligoastrocytoma. GBM is the most aggressive of all and can originate from multiple cell types [2]. Standard treatment includes partial or radical tumor resection accompanied by chemo and radiotherapeutic interventions. Despite complete resection combined with aggressive treatments, the survival rate remains extremely low. Therapeutic success is compromised by the glioblastoma stem cells (GSCs) that escape treatment and, with cooperation from their niche components, regrow into a new and even

Giorgio Seano (ed.), *Brain Tumors*, Neuromethods, vol. 158,
https://doi.org/10.1007/978-1-0716-0856-2_4, © Springer Science+Business Media, LLC, part of Springer Nature 2021

more aggressive tumor. Overcoming these therapeutic challenges would require a thorough understanding of the mechanisms underlying therapeutic resistance, the journey this small population of GBM-initiating cells embarks on to develop into a full-fledged tumor, the process of vascular recruitment and nutrient supply to the developing tumor, and how signaling between the components of the tumor microenvironment cooperate in tumorigenesis.

Most of our understanding from imaging experiments are through stationary observations that have limited our ability to decipher the intratumoral properties including the dynamic interplay between the tumor, vasculature, and immune cells [3]. Real-time imaging techniques like magnetic resonance, computed tomography scan, and ultrasonography have markedly improved in vivo imaging with a resolution that reaches the organ level. However, they lack cellular and subcellular resolution for single-cell imaging and thus fail to address the biological complexity in a multicellular organism. Recent advancements in the field of microscopy have combined nonlinear microscopy with ultrafast scanning technology making intravital microscopy a powerful tool for in vivo imaging and visualization of complex biological mechanisms at a subcellular resolution [4]. Intravital microscopy (IVM) has since been used extensively in various branches of biology including neurobiology and immunology. It has also emerged as an invaluable microscopic imaging tool suitable for deciphering various aspects of tumor pathophysiology as this allows dissecting complex aspects of tumor heterogeneity and microenvironment interactions, including tissue metabolism, angiogenesis, tissue remodeling, immune response, and tumor cell invasion [5, 6].

The availability of numerous spectrally distinct fluorescent proteins, tracers, and markers has enabled visualization of tumor-host interaction, tumor cell invasion and homing the host tissue, tumor metastasis, and phenotypic changes within their natural environment at subcellular resolution using confocal or multiphoton imaging [7].

In this chapter, we describe the methods involved in intravital multiphoton imaging of a primary brain tumor, cranial window, and tumor implantation, followed by step-by-step description of long-term longitudinal multiphoton imaging and image analysis (Figs. 4.1 and 4.2).

2 Materials

2.1 Mice

All animal studies must be reviewed and approved by the relevant animal-care committees and must conform to the institutional and national ethics regulations. In the presented protocols, all animal procedures were approved by the Institut Curie's Ethics Committee

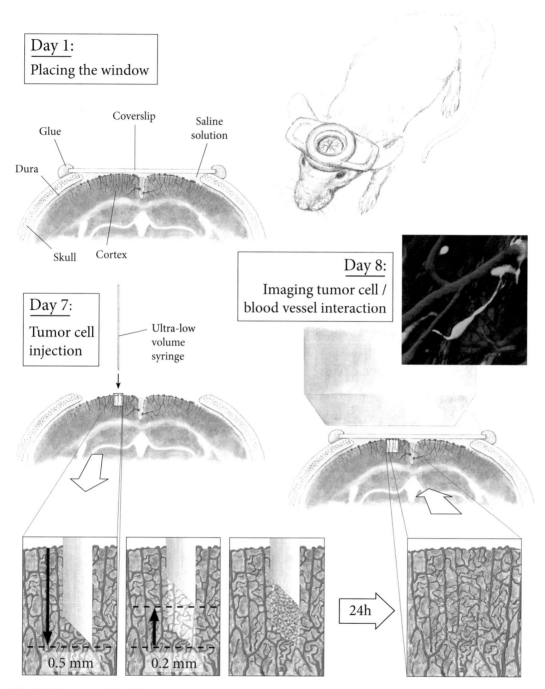

Fig. 4.1 Representation of surgical procedure and stage for intravital imaging of GBM cells. (Top right) Drawing of a mouse bearing a cranial window. For immobilization before imaging, the plastic wings (proximal end of a 3-mL syringe) are fixed on top of the window. (Top left) Cross-sectional rendering of the cranial window at day 1. (Bottom left) Injection site of tumor cells in the brain for intravital imaging 7 days after placing the window. (Bottom right) Multiphoton imaging of tumor cells (green)/blood vessel (dextran-TAMRA red) interaction. 24 h after injection, vessel co-option can be observed. (Reproduced from [12])

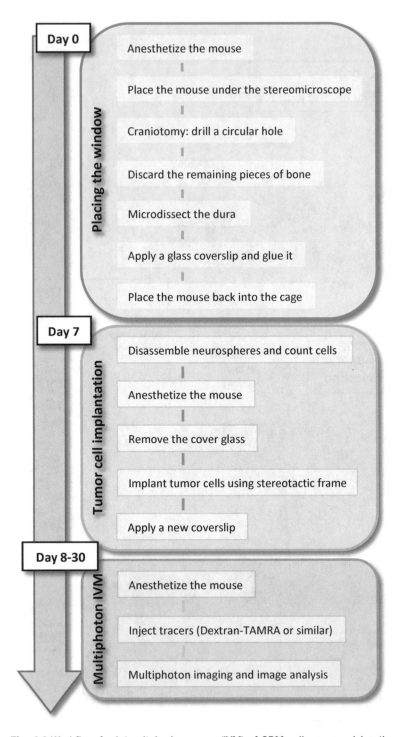

Fig. 4.2 Workflow for intravital microscopy (IVM) of GBM cells summarizing the main steps of the procedure. This method consists of three main steps: (1) placing a glass window on top of the brain, (2) Injection of tumor cells at day 7, and (3) multiphoton imaging of tumor cells and blood vessel interaction 24 h after injection

and the French Ministry of Higher Education, Research and Innovation as well as conducted in accordance with the criteria outlined in the *Guide for the Care and Use of Laboratory Animals* prepared by the National Academy of Sciences and published by the National Institutes of Health. The protocol was performed on Swiss nude female or male 7- to 9-week-old mice (Note 1). Mice were housed individually in an enriched environment (Note 2). Animal welfare was monitored daily and mice received the appropriate care. The animals were sacrificed if an endpoint was reached (Note 3).

2.2 Analgesia, Anesthesia, and Antibiotic Strategy

Buprenorphine (Buprecare®, Axience) at 0.1 mg/kg subcutaneous (s.c.); Bupivacaïne® (Mylan), 2.5 mg/mL, 3 × 30 μL (s.c. locally); carprofen, Rimadyl® (Zoetis), 5 mg/kg (i.p.); acetaminophen (Doliprane®, Sanofi) diluted at 2 mg/mL in water bottle (Note 4).

2.2.1 Analgesia

2.2.2 Anesthesia (Note 5)

Ketamine, Imalgene® (Merial)/xylazine, Rompun® (Bayer), 90 mg/9 mg/kg, (s.c.).

2.2.3 Antibiotics

Amoxicillin, 140 mg/mL; clavulanic acid, 35 mg/mL (Noroclav®, Bayer) 125 mg/kg/day diluted in NaCl 0.9% (s.c.).

2.3 Equipment and Reagents

– Laminar flow hood.

– Stereotaxic frame for mice with syringe holder.

2.3.1 Equipment

– Thermal pad.

– Bright-field stereomicroscope (SteREO Discovery V12, Zeiss).

– Micro driller (Harvard Apparatus, 75-1887) with 1.0 mm round Carbide drill bit.

– Surgical instrument sterilizer.

– 10 μL Hamilton syringe 701RN (VWR, 549-1137) with needle 28G, 51 mm, type 4 with an angle of 45°.

– Scissors A: Vannas microdissecting spring scissors curved 3 mm cutting edge 0.15 mm tip width 3IN overall length (Roboz Surgical, RS5611).

– Scissors B: Microdissecting scissors straight sharp points 23 mm blade length 3 1/2IN overall length (Roboz Surgical, RS5190).

– Scissors C: Bonn Scissors 3.5" straight 15 mm Sharp/Sharp (Roboz Surgical, RS5840).

– Forceps A: Dumont #1 forceps 0.20 × .12 mm tip size 120 mm inox. (Roboz Surgical, RS5040).

– Forceps B: Dumont #3C Forceps Dumostar tip size 0.17 X .10 mm. (Roboz Surgical, RS4968).

- Forceps C: Dumont #4 Forceps Dumostar tip size 0.06 X .02 mm Biologie tips (Roboz Surgical, RS4974).

- Round glass coverslip 7 mm (Harvard Apparatus 64-0723).

- Long cotton swabs.

- Sterile gauze (Laboratoire Tetra Medical).

- Sterile Gelfoam (Bloxang, Bausch & Lomb).

- 1 mL syringe (Terumo, SS + 01H1) with a 25G needle (Terumo, AN*2516R1) for product injections.

2.3.2   Reagents

- Sodium chloride, 0.9% (Braun).

- Chlorhexidine (Gilbert Healthcare).

- Histocompatible cyanoacrylate glue and white acrylic powder, dental or nail grade (Note 6).

- Ophthalmic lubricant (Farmila Thea Farmaceutici S.p.A.).

2.4   Cells and Cell Culture Reagents

GBM patient-derived cell lines were cultivated as neurospheres, in serum-free conditions to enrich for GSCs. These cells generate tumors that preserve patient-specific disease phenotypes. MGG8 cell line derived from GBM patient was previously established in the Department of Neurosurgery at the Massachusetts General Hospital (MGH) [8]. MGG8 cells were grown in NeuroCult™ NS-A Basal Medium (Human; #05750, STEMCELL Technologies) supplemented with NeuroCult NS-A proliferation kit (#05753, STEMCELL Technologies); human recombinant epidermal growth factor (EGF), 20 ng/mL (#78006.1, STEMCELL Technologies); human recombinant basic fibroblast growth factor (bFGF), 20 ng/mL (#78003.1, STEMCELL Technologies); and heparin, 5 μg/mL (#07980, STEMCELL Technologies). Cells screened for mycoplasma with MycoAlert Plus Mycoplasma Detection Kit (Lonza) tested negative.

Viable neurospheres appeared semitransparent and phase bright. Growth was monitored daily (morphology, size, media depletion, etc.). Half of the medium change was eventually performed prior to its acidification (when the medium starts to turn orange/yellow) and before neurospheres became too large (<100 μm). Neurospheres that reached around 100–150 μm in diameter are passaged by centrifugation at 192 g for 5 min [9]. For implantation, MGG8 cells were resuspended in NeuroCult™ complete culture medium at 20,000 cells/μL and kept on ice until injection (Note 7).

2.5   Imaging (Multiphoton Setup)

Multiphoton imaging uses nonlinear infrared excitation of samples with two or more photons and allows reaching fairly deep tissue [10]. Multiphoton microscopy (MPM) has become an indispensable imaging tool and a technology of choice for IVM of the brain and presents two main advantages compared to confocal micros-copy. MPM allows deeper tissue penetration (300–500 μm) while retaining single-cell resolution with less photobleaching and pho-totoxicity compared to confocal imaging [11].

IVM enables direct visualization of the dynamics of tumor cells and their microenvironment at single-cell resolution. The variables for intravital microscopy are multiple, and here we have listed the strictly necessary features. The multiphoton microscope used has to be as penetrant as possible, upright, and with enough space between objective and stage. We recommend a water-immersion objective with magnification between 10× and 40×. Gas anesthesia should be accessible and close to the microscope since the mouse under the microscope will be kept anesthetized with isoflurane.

3   Methods

3.1 Surgery for Cranial Window Setup

Chronic cranial window procedure may be used to study the behavior of brain tumors in the cortical area for the long term. It consists of removing a 7-mm-large circle of the cranium (craniot-omy) and the area below the dura (membrane of the brain) and replacing them with a glass coverslip (Figs. 4.1 and 4.4a).

1. 1 h prior to surgery, administer antibiotics (amoxicillin, clavu-lanic acid solution, 125 mg/kg/day, s.c.), and repeat the injection 24 h later.

2. 30 min prior to surgery, administer 0.1 mg/kg buprenorphine intraperitoneally (i.p.).

3. Anesthetize the mouse with ketamine/xylazine (90 mg/9 mg per kg body weight, i.p.), check the pedal reflex and monitor its respiration before starting and throughout the all procedure.

4. Place the mouse on a thermal pad (37 °C) to maintain its body temperature.

5. Apply ophthalmic lubricant on each eye with a cotton swab, and inject 0.5 mL of sodium chloride 0.9% (s.c.) to prevent dehydration.

6. 5 min before surgery, inject local anesthesia, Bupivacaïne® (Mylan), 2.5 mg/mL, subcutaneously in three different points of the head (3 × 30 μL s.c. locally).

7. Using scissors C and forceps A, remove a circle of 10 mm of the skin and connective tissue covering the skull. Expose and clean the periosteum to the temporal crests with a sterile cot-ton swab (Note 8).

8. Place the mouse on the immobilization stage under the stereomicroscope, and adjust the focus in order to obtain the best imaging quality (Note 9).

9. Draw a 6-mm circle over the frontal and parietal region using a pen.

10. Use high-speed bone micro drill with a 1.4-mm-diameter burr tip, and make a groove around the margins of the circle. Slowly drill by circular movement in the groove, and apply cold saline regularly to avoid heating, which can damage the below brain tissue. Bleeding often occurs at this stage (Note 10).

11. When the bone becomes loose and the circle formed by the drilling starts moving, stop drilling.

12. Use forceps A to separate the bone flap from the *dura mater* underneath very carefully, and at the same time apply sterile saline and Gelfoam between the detaching bone and the dura. The dura capillaries may tear, resulting in bleeding (Note 10).

13. Apply Gelfoam soaked with saline on the *dura mater* to keep the brain moist, leaving it on until the bleeding stops.

14. Use forceps B to remove small pieces of the bone attached to the edge of the circle to avoid the regrowth of the bone (Note 11).

15. Using forceps C make a small opening in the dura near the edge and then cut the dura with the scissor A all along the circle through the bone and along the sagittal sinus, without damaging it.

16. Place a round glass coverslip, 7 mm in diameter, on top of the bone, and put saline solution drop by drop under the window to keep the exposed part of the brain wet (Note 12).

17. Glue the coverslip to the bone (1:1 mixture of histocompatible cyanoacrylate glue and fine acrylic powder). Be careful of not applying glue in direct contact with the brain (Note 13).

18. Allow the mouse to recover on the heating pad until it awakens and put it in its cage for recovery (one mouse per cage).

19. 6 h after surgery, inject i.p. carprofen, at 5 mg/kg, and repeat the procedure every 18 h if the animal shows any sign of pain. Moreover, acetaminophen diluted at 2 mg/mL should be added in the water bottle.

20. Monitor the mouse post-surgery twice a day during the first 72 h and daily afterward.

3.2 Tumor Cell Implantation

Here, to monitor cells in vivo in their orthotopic environment, MGG8 cells were infected with lentiviruses containing a vector expressing EGFP under the control of the CMV promoter. Cells were implanted 7 days after craniotomy to bypass potential inflammation.

3.2.1 Cell Preparation

1. Collect cells at early passage (less than 10), grown as neurospheres in serum-free medium, and centrifuge at room temperature for 5 min at 192 g.

2. Resuspend cell pellet in 1 mL of complete medium to obtain single-cell suspension, and count them with a hemocytometer.

3. Adjust the cell concentration to 20,000 cells/μL suspended in NeuroCult complete media.

4. Keep cells on ice until injection.

3.2.2 Tumor Cell Implantation

In order to perform IVM, cells should be implanted within the brain tissue but relatively superficial (0.1–0.3 mm deep). Here, we injected 20,000 MGG8 cells at multiple sites of injection in order to increase tumor cell implantation efficiency and imaging chances (three injection sites per brain hemisphere) (Note 14).

1. 30 min before surgery, inject 0.1 mg/kg buprenorphine intraperitoneally (i.p.).

2. Anesthetize the mouse with an injection (i.p.) of ketamine and xylazine mixture (90 mg/9 mg per kg body weight), test its pedal reflex, and monitor its respiration before starting.

3. Place the mouse on a thermal pad (37 °C) to maintain its body temperature and in the stereotactic frame under a stereomicroscope, and immobilize its head with three fixation points: teeth bar in the front and ear bars inserted in ear canal on each side.

4. Apply an ophthalmic lubricant on each eye to prevent dehydration.

5. Use the forceps B to remove the cover glass and the glue.

6. Bone regrowth from the skull can occur that may obscure the imaging window. If this is the case, remove the new bone delicately with fine forceps.

7. Clean the bone area using a dry cotton swab and forceps and rinse the surface of the brain with sterile saline (Note 8).

8. Use Gelfoam to absorb saline.

9. Fill a Hamilton syringe with the cells (20,000 cells/μL) and place a 28G-gauge needle.

10. Fix the syringe on the stereotactic injection device and move it on top of the brain.

11. Insert the needle to a depth of 0.5 mm into the brain and wait 30 s.

12. Raise the needle of 0.2 mm and very slowly inject 1 μL of cell suspension (20,000 cells) for 30 s.

13. Wait 1 min before removing the needle from the brain and slowly raise it in around 30 s.

14. If necessary, repeat the injection in three different sites per brain hemisphere.

15. Implant a new coverslip gluing it with a cyanoacrylate glue/acrylic powder mixture.

3.3 Imaging Tumor Cell-Blood Vessel Interaction Using Multiphoton Microscopy

Blood vessels form a very important component of tumor microenvironment. Most of the tumors are highly dependent on tumor vasculature, and its amount and function may radically influence tumor progression and fate. Tumor can induce blood vessel formation through the mechanism of angiogenesis or use the pre-existing vasculature (Fig. 4.3) [12–14]. Few functional blood vessels within the tumor induce tumor hypoxia and acidic environment caused by the reduction of oxygen and nutrients, thus stimulating tumor aggressiveness, invasion, and immunosuppression [15]. Moreover, the phenomenon of vessel co-option, i.e., the movement of tumor cells toward and then along the pre-existing vasculature, has been recently described to be an important strategy to reach vasculature in glioblastoma [12–14].

For these reasons and many others, the study of the dynamic relationship between tumor cells and blood vessels is of vital importance in order to understand tumor progression and resistance to therapy. Blood vessels and tumor cells can be labeled and visualized

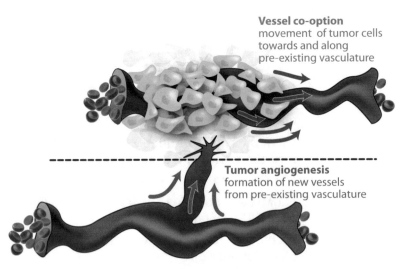

Fig. 4.3 How to reach the perivascular niches? Vessel co-option and tumor angiogenesis. Schematic on how tumors get vasculature: distinction between vessel co-option (movement of tumor cells toward and along pre-existing surrounding vasculature) and tumor angiogenesis (formation of new vessels from preexisting vasculature). The two processes can be coexistent in tumors. (Reproduced from [13])

with multiple different methods. Here below we describe some of them:

<table>
<tr><td>3.3.1 Blood Vessel Imaging</td><td>

1. Retro-orbital injection of dextran-TAMRA.

   Large dextrans (100 k to 2 M Da) transiently circulate in the bloodstream without extravasation through the blood-brain barrier (BBB). Thus, they can be used in order to visualize the bloodstream in a range not larger than 2 h after intravenous injection. In order to visualize blood vessels, 0.1 mL of 2 MDa dextran-TAMRA (10 mg/mL) is injected retro-orbitally before imaging (Note 15). Images are usually very bright and the 3D structure of the bloodstream within the vasculature is evident. However, fluorescent dextran accumulates in the brain tumor environment, and after 3–4 injections, image quality is strongly reduced.

2. Tracking of fluorescently labeled red blood cells (RBC).

   RBCs can be collected through cardiac puncture from old donor mice of the same strain used in the experiment. RBCs need then to be ex vivo labeled with the far-red lipophilic fluorescent dye (1,1-dioctadecyl-3,3,3,3-tetramethylindodicarbocyanine perchlorate (DID), Invitrogen) [16]. DID-labeled red blood cells are finally inoculated in the interested mouse by tail vein injection. This will allow long-term imaging since DID fluorescence is stable and red blood cell half-life is approximately 10–14 days. This method is theoretically very powerful since it allows for very refined real-time quantifications of red blood cell velocity, trajectory, and planar versus turbulent flow (tumor hemodynamics). However, if used just for tumor vessel labeling, it could be misleading since some small or abnormal vessels are not well perfused by red blood cells and so invisible to intravital microscopy.

3. Endothelial cell reporter mice.

</td></tr>
</table>

Blood vessels can be visualized using fluorescent reporter transgenic mouse strains [17, 18]. In these transgenic mice, a fluorescent reporter is placed under the control of an endothelial-specific promoter, thus allowing direct visualization of the blood vasculature without the need for antibody staining. Many endothelial-specific fluorescent reporters have been developed, and some are also used in combination with the Cre/lox site-specific recombination system allowing the control of the temporal expression of the fluorescent reporter.

3.3.2 Tumor Cell Labeling

Tumor cells can be permanently labeled with a multitude of fluorescent proteins, such as GFP, DsRed, and tdTomato, or transient fluorescent chemical probes, such as carbocyanine dyes. This is easier and cleaner than blood vessel labeling since it can be done in vitro before tumor cell implantation.

3.3.3 Multiphoton Imaging

In order to perform IVM, mice need to be anesthetized with isoflurane gas and placed on a stage that would allow the constant gas anesthesia and a proper immobilization of the head and the cranial window (see below). The water-immersion objective should be a long working distance. The imaging session cannot be too long since isoflurane anesthesia cannot be delivered to the mouse for more than 3 h. Inject before and every hour 0.5 mL of sodium chloride 0.9% (s.c.) to prevent dehydration, and apply ophthalmic lubricant on each eye with a cotton swab. Heating pad and physiological monitoring are highly recommended during every imaging session.

A general consideration on conventional multiphoton imaging that must be taken in account is that this type of live imaging is not able to reach deep tissue; thus just the first 300–500 μm of brain tissue will be accessible for imaging. This may be very relevant in the studies of cellular or tissue processes that necessarily happen in the core of tumor tissues, i.e., deep for definition. Other types of IVM may be performed if the equipment is available, such as optical frequency dynamic imaging (OFDI), optical coherent tomography (OCT), ultrasonography, and stereomicroscopy (Fig. 4.4). Moreover, it is highly recommended to combine IVM with traditional histological imaging in order to confirm the crucial discoveries from IVM.

A serious issue in IVM is represented by the motion artifacts due to respiration and heartbeat. In the IVM of brain tissue, reducing these pitfalls is potentially easier, by properly immobilizing the head of the mouse. We developed a simple method to physically stabilize the cranial window. To do so, a proximal end of a 3-mL syringe was cut with a scalpel in order to form a ring with two plastic holders (Figs. 4.1 and 4.4a). This ring was then glued on the top of the cranial window using biocompatible glue. An in-house produced holder was fabricated and placed on the top of the microscope stage (Fig. 4.4a).

The objective magnification should be selected in accordance with the scientific IVM needs. For tissue imaging, stereomicroscopes or 5–10× magnification objectives can be used. If the scientific interest is in single-cell and cell-to-cell imaging, at least 20× magnification should be employed. Otherwise, 40–60× magnifications may be necessary for intracellular imaging (Note 16).

In order to perform longitudinal IVM, finding the same imaging spot over different imaging sessions is highly important. For this, motorized stages and references within the brain tissue and/ or the cranial window may be necessary (Note 17).

Importantly, multiphoton IVM can cause phototoxicity and photobleaching if the imaging scanning velocity or the power of the laser is inappropriate. It is also true that imaging with too low laser power and with a too fast scanning velocity may strongly reduce the image resolution. Therefore, it will be necessary to find a proper balance between these imaging features that allows the best imaging quality with no toxicity or bleaching.

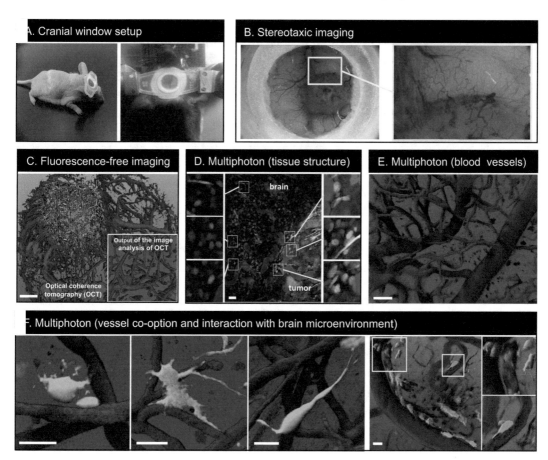

Fig. 4.4 Examples for intravital microscopy. (**a**) Pictures of the cranial window placed on nude or C57BL/6 7- to 9-week-old mice. Example of mouse immobilization by using the stereotactic frame before imaging thanks to the plastic wings (proximal end of a 3-mL syringe glued on the skull and hold by an in-house manufactured holder). (**b**) Stereomicroscopy of GBM GL261-bearing mouse with cranial window at 1 month after tumor implantation. The right panel is the magnification of the yellow square. (**c**) In vivo imaging using optical frequency dynamic imaging (also called optical coherent tomography) of the whole brain under the cranial window. Scale bar, 1 mm. (Reproduced from the Nikon Small Word Competition and from [28]). (**d**) Intravital multiphoton imaging of brain nuclei around the nodular Gl261-DsRed GBM implanted in H2B-EGFP mouse (CAG::H2B-EGFP C57/BL6). GFP (*green*, nuclei within the brain tissue), DsRed (*red*, tumor cells), and BlueCascade (*blue*, 2 MDa-dextran, blood flow). Insets are magnifications of the yellow squares. Scale bar, 20 μm. (Reproduced from [28]). (**e**) Multiphoton imaging of the cortical blood flow using 2 MDa-dextran-TAMRA. Scale bar, 20 μm. (**f**) Examples of vessel co-option of GBM MGG8-EGFP cells implanted in nude mice. MGG8-GFP (tumor cells), Red (2 MDa-dextran-TAMRA, blood flow). Scale bar, 20 μm. (Reproduced from [12])

Scanning multiphoton imaging allows acquiring 3D images. This provides very powerful data collection that can be properly managed and interpreted uniquely with good image analysis tools.

3.4 Image Processing and Analysis

Although complex, image analysis of 3D multiphoton images can provide the researcher with very large amount of interesting results. Multiple tools can be of help for the analysis and processing of 3D images, such as ImageJ/Fiji, Bitplane Imaris, MathWorks MATLAB, Amira-Avizo, deconvolution software, and many others. The analysis and quantification of 3D and 4D images may require specific training and ad hoc setup, but it is strictly necessary in order to be able to get the maximum of results from IVM (Reviewed in [19, 20]).

4   Notes

Note 1: Age, sex, and strain of the mice are adjustable to the needs of the specific experiment. Nevertheless, multiple parameters have to be considered:

- Cranial window implants lead to the settlement of a rigid structure on the skull bone for several weeks; thus the use of postnatal or young adult mice with strong growth potential is not recommended: the cranial window structure may interfere with the normal growth of the skull or maybe lost leading to neurodamage and/or death of the animal.

- Sex has to be chosen in accordance with the tumor type and may interfere with your choice of housing (see Note 2).

Note 2: It is preferable to individually house mice since there is the eventuality of cranial window loss in case of the housing of multiple mice per cage. Avoid placing objects with angles, like metallic or plastic houses, that the mice could use to trap/stick their device with the high risk to lose it.

Note 3: In practice, around 80% of cranial window would succeed if the technologist is skilled. In some case, animals can accidentally lose their window, this constitutes an endpoint except if the technologist can replace the window with a new one immediately after this happens. In that case, cranial window can be replaced following the procedure in Sect. 3.2.2.

Note 4: If anesthesia is not properly working, cranial window implantation is a potentially painful procedure. Animal welfare should be periodically monitored during and after surgery.

Note 5: Isoflurane gas anesthesia may be employed, but with a modified mask that does not obstruct the cranial window implantation (Fig. 4.4a).

Note 6: Histocompatible cyanoacrylate glue and white acrylic powder can be replaced with appropriate dental cement.

Note 7: Concentration and resuspension medium, as well as time on ice should be adapted to the cell type. We strongly recommend to implant only cells at low passage, since the tumor initiation ability is reduced during the time of cell culture. It is also important to be sure that cells are well dissociated, to avoid occlusion of the needle during tumor implantations. Our experience suggested to not keep cells for more than 3–4 h since the cell survival radically drops down after this time.

Note 8: The connective tissue attached to the bone must be accurately removed by applying the proper pressure with a cotton stick.

Note 9: The head can be immobilized by using surgical tape on the skin of the mouse in order to apply a gentle pressure on the mouse head.

Note 10: Excessive speed or force applied with the drill might damage the underlying brain tissue. Bleeding is not uncommon; to stop it use abundant saline and Gelfoam. The skull needs to be dry when drilled; therefore, we recommend to dry it with a cotton swab before restarting drilling.

Note 11: Carefully removing all fragments of the bone from the edge of the cut is essential to avoid regrowth of the bone. This can be done properly only by using a stereomicroscope with good light.

Note 12: Adding saline between glass and the brain is essential to avoid glue coming in contact with the brain and to keep the brain moist. During this step, drops should not spread out the space between glass and the brain and the bone must necessarily be dry; otherwise the glue will not stick on it and glass.

Note 13: Be particularly careful to avoid touching the surface of the brain. Since the glue is hydrophobic, maintain the bone free area of the skull filled up with saline to repel any glue that may escape. The glue needs to be produced by mixing the right amount of glue and acrylic powder. To 3–4 drops of glue, add progressively the powder in order to obtain the proper viscosity. This step must be done shortly; otherwise the glue will solidify before being applied. Alternatively, dental cement can also be used to fix the coverslip to the bone.

Note 14: The area of implantation may be crucial in order to find the cells by intravital microscopy after cranial window reimplantation. We recommend taking notes on the implantation coordinates.

Note 15: Retro-orbital injection of the venous sinus must be performed by skilled technologists. This procedure has been previously described [21].

Note 16: The breathing movement and the physiological tissue shaking highly reduce the imaging resolution. Therefore, although possible [6], we discourage subcellular investigations by IVM if one does not employ refined setup to digitally or physically reduce the motion artifacts due to the respiration and the heartbeat.

Note 17: A useful spatial reference may be provided by the brain vasculature itself that is also visible without fluorescent dyes. Otherwise, if a motorized stage is available, a reference on the cranial window may be used to set the ($x = 0$; $y = 0$) position at the beginning of each imaging session.

5   Conclusions

IVM has provided the opportunity to study the dynamics of biological processes within a tissue at the single-cell level and in real time, thus addressing biological questions that could not be answered by in vitro studies. This powerful technique has been instrumental in the assessment of tissue physiology (e.g., blood flow variation under physiopathological conditions) [22, 23], as well as the dynamics of individual cells both in developmental conditions and during cancer progression (reviewed in [24]).

In recent years, the use of optical windows and the development of imaging tools at the subcellular level have provided unique insights in our current understanding of cancer progression. IVM has been used to track the dynamic of cancer cell invasion and dissemination revealing diverse modes of cell motility and migration in melanoma and breast cancer [25, 26]. IVM has also been instrumental to study the role of the microenvironment during cancer progression, in particular, the interaction of cancer cells with the immune system but also with the surrounding tumor vasculature (reviewed in [5]). With this technique, the cellular dynamics of tumor angiogenesis, blood vessel architecture, and physical parameters have been characterized [15, 16, 27–29]. IVM has also allowed visualizing the mechanism of vessel co-option employed by tumor cells to reach the pre-existing vasculature and invade the brain during GBM progression [12].

This imaging technique has also been useful in assessing drug delivery within the vasculature providing new insights into the mechanism of drug resistance [30, 31]. Finally, subcellular imaging techniques such as photobleaching have provided interesting findings into the molecular dynamics and activity of key proteins involved in cancer dissemination (reviewed in [32]).

In recent years, astonishing progress has been made in the development of IVM tools and applications; however researchers are still facing unique limitations imposed by IVM. Methods to improve the temporal resolution (by increasing scanning speed and image collection rate) [33, 34] and tissue penetration (using adaptive optics [35, 36] and/or three-photon microscopy [37, 38]), as well as long-term imaging techniques, are currently developed and constitute the main challenges IVM will have to solve in order to make new exciting discoveries.

Acknowledgments

We thank Pauline Deshors, Aafrin Pettiwala, and Guillaume Bourmeau (Institut Curie Research Center, Orsay-Paris) for critical reading and discussion. A special thank goes to Renaud Chabrier for illustration at Fig. 4.1. This work was supported by the

Fondation ARC pour la recherche sur le cancer, the Inserm-CNRS ATIP-Avenir grant, the European Research Council (ERC) under the European Union's Horizon 2020 (grant agreement no. 805225), and the NanoTheRad grant from Paris-Saclay University.

References

1. Krammer MJ et al (2011) Modern management of rare brain metastases in adults. J Neuro-Oncol 105(1):9–25

2. Alcantara Llaguno SR, Parada LF (2016) Cell of origin of glioma: biological and clinical implications. Br J Cancer 115(12):1445–1450

3. Prager BC et al (2019) Cancer stem cells: the architects of the tumor ecosystem. Cell Stem Cell 24(1):41–53

4. Kirui DK, Ferrari M (2015) Intravital microscopy imaging approaches for image-guided drug delivery systems. Curr Drug Targets 16(6):528–541

5. Gabriel EM et al (2018) Intravital microscopy in the study of the tumor microenvironment: from bench to human application. Oncotarget 9(28):20165–20178

6. Weigert R et al (2010) Intravital microscopy: a novel tool to study cell biology in living animals. Histochem Cell Biol 133(5):481–491

7. Suetsugu A et al (2018) Visualizing the tumor microenvironment by color-coded imaging in orthotopic mouse models of cancer. Anticancer Res 38(4):1847–1857

8. Wakimoto H et al (2012) Maintenance of primary tumor phenotype and genotype in glioblastoma stem cells. Neuro-Oncology 14(2):132–144

9. Kloepper J et al (2016) Ang-2/VEGF bispecific antibody reprograms macrophages and resident microglia to anti-tumor phenotype and prolongs glioblastoma survival. Proc Natl Acad Sci U S A 113(16):4476–4481

10. Denk W, Strickler JH, Webb WW (1990) Two-photon laser scanning fluorescence microscopy. Science 248(4951):73–76

11. Helmchen F, Denk W (2005) Deep tissue two-photon microscopy. Nat Methods 2(12):932–940

12. Griveau A et al (2018) A glial signature and Wnt7 signaling regulate glioma-vascular interactions and tumor microenvironment. Cancer Cell 33(5):874–889. e7

13. Seano G (2018) Targeting the perivascular niche in brain tumors. Curr Opin Oncol 30(1):54–60

14. Seano G, Jain RK (2019) Vessel co-option in glioblastoma: emerging insights and opportunities. Angiogenesis 23(1):9–16

15. Martin JD, Seano G, Jain RK (2019) Normalizing function of tumor vessels: progress, opportunities, and challenges. Annu Rev Physiol 81:505–534

16. Kamoun WS et al (2010) Simultaneous measurement of RBC velocity, flux, hematocrit and shear rate in vascular networks. Nat Methods 7(8):655–660

17. Monvoisin A et al (2006) VE-cadherin-CreERT2 transgenic mouse: a model for inducible recombination in the endothelium. Dev Dyn 235(12):3413–3422

18. Motoike T et al (2000) Universal GFP reporter for the study of vascular development. Genesis 28(2):75–81

19. Wiesmann V et al (2015) Review of free software tools for image analysis of fluorescence cell micrographs. J Microsc 257(1):39–53

20. Kherlopian AR et al (2008) A review of imaging techniques for systems biology. BMC Syst Biol 2:74

21. Yardeni T et al (2011) Retro-orbital injections in mice. Lab Anim (NY) 40(5):155–160

22. Masamoto K et al (2012) Repeated longitudinal in vivo imaging of neuro-glio-vascular unit at the peripheral boundary of ischemia in mouse cerebral cortex. Neuroscience 212:190–200

23. Zhang S, Murphy TH (2007) Imaging the impact of cortical microcirculation on synaptic structure and sensory-evoked hemodynamic responses in vivo. PLoS Biol 5(5):e119

24. Weigert R, Porat-Shliom N, Amornphimoltham P (2013) Imaging cell biology in live animals: ready for prime time. J Cell Biol 201(7):969–979

25. Manning CS, Hooper S, Sahai EA (2015) Intravital imaging of SRF and notch signalling identifies a key role for EZH2 in invasive melanoma cells. Oncogene 34(33):4320–4332

26. Prunier C et al (2016) LIM kinase inhibitor Pyr1 reduces the growth and metastatic load of breast cancers. Cancer Res 76(12):3541–3552

27. Peterson TE et al (2016) Dual inhibition of Ang-2 and VEGF receptors normalizes tumor vasculature and prolongs survival in glioblastoma by altering macrophages. Proc Natl Acad Sci U S A 113(16):4470–4475

28. Seano G et al (2019) Solid stress in brain tumours causes neuronal loss and neurologi-

cal dysfunction and can be reversed by lithium. Nat Biomed Eng 3(3):230–245

29. Uhl C et al (2018) EphB4 mediates resistance to antiangiogenic therapy in experimental glioma. Angiogenesis 21(4):873–881

30. Laughney AM et al (2014) Single-cell pharmacokinetic imaging reveals a therapeutic strategy to overcome drug resistance to the microtubule inhibitor eribulin. Sci Transl Med 6(261):261ra152

31. Orth JD et al (2011) Analysis of mitosis and antimitotic drug responses in tumors by in vivo microscopy and single-cell pharmacodynamics. Cancer Res 71(13):4608–4616

32. Nobis M et al (2018) Molecular mobility and activity in an intravital imaging setting—implications for cancer progression and targeting. J Cell Sci 131(5):jcs206995

33. Akemann W et al (2015) Fast spatial beam shaping by acousto-optic diffraction for 3D non-linear microscopy. Opt Express 23(22):28191–28205

34. Li B et al (2019) An adaptive excitation source for high-speed multiphoton microscopy. Nat Methods 17(2):163–166

35. Ji N et al (2012) Characterization and adaptive optical correction of aberrations during in vivo imaging in the mouse cortex. Proc Natl Acad Sci U S A 109(1):22–27

36. Zheng W et al (2017) Adaptive optics improves multiphoton super-resolution imaging. Nat Methods 14(9):869–872

37. Horton NG et al (2013) In vivo three-photon microscopy of subcortical structures within an intact mouse brain. Nat Photonics 7(3):205

38. Ouzounov DG et al (2017) In vivo three-photon imaging of activity of GCaMP6-labeled neurons deep in intact mouse brain. Nat Methods 14(4):388–390

Part II

Ex Vivo Models

Chapter 5

Glioblastoma Patient-Derived Cell Lines: Generation of Nonadherent Cellular Models from Brain Tumors

Joris Guyon, Tiffanie Chouleur, Andreas Bikfalvi, and Thomas Daubon

Abstract

Glioblastomas are brain tumors derived from astrocytes or oligodendrocytes. These tumors have a heterogeneous structure composed of a necrotic and vascularized center and an invasive periphery. The rapid growth of glioblastoma and its ability to invade surrounding tissues make this cancer difficult to treat. The median survival of patients afflicted with the disease ranges from 12 to 15 months. The standard treatment, based on the Stupp protocol, consisting of a resection of the tumor mass and adjuvant temozolomide treatment and/or radiotherapy, allows delaying recurrence for only a few months. Since tumor cell invasion is the central element in tumor recurrence, it is important to understand the mechanisms of tumor invasion and the interactions between tumor cells and their microenvironment. To this aim, suitable models have been developed and are constantly updated. In this article, we describe in detail experimental procedures based on the spheroid technology which allow mechanistical studies related to proliferation, survival, migration, and invasion.

Keywords Glioblastoma, Patient-derived cell, Spheroids, Stem-like cells, Invasion, In vivo model

1 Introduction

1.1 Glioblastoma: Basic Considerations

Glioblastomas (GBMs, WHO grade IV gliomas) are brain tumors derived from oligodendrocytes or astrocytes [1]. GBMs are very aggressive and are characterized by a high proliferation rate, a highly vascularized and necrotic core, and infiltrative cells. The median survival of patients after diagnosis is very low, around 14 months, despite optimal treatments. The Stupp protocol is commonly used as a first-line treatment for GBMs. It consists of maximal tumor resection, followed by radiotherapy and temozolomide chemotherapy [2]. In the last decade, antiangiogenic therapies appeared promising, such as the use of humanized anti-VEGF antibodies (bevacizumab, Roche) [3]. Antiangiogenic therapy in experimental models leads to a decrease in vascularization and tumor mass. Many clinical studies have been conducted in GBM

Giorgio Seano (ed.), *Brain Tumors*, Neuromethods, vol. 158,
https://doi.org/10.1007/978-1-0716-0856-2_5, © Springer Science+Business Media, LLC, part of Springer Nature 2021

patients with this treatment; however, little progress has been made in terms of overall survival [4]. Indeed, these tumors were considered mainly clonal masses with identical molecular characteristics. Recent studies, including single-cell RNA sequencing, identified different clusters of cells within the tumor [5]. Inter-patient and intra-tumor heterogeneity is an issue for research and clinical application. Moreover, current treatments cannot overcome the presence of invasive cells that cause tumor recurrence. It is therefore necessary to better understand the mechanisms of tumor cell invasion that lead to the resistance of GBMs to current therapies. To gain better insights into GBM development and invasion, suitable experimental models are needed.

1.2 Comparison Between Glioblastoma Cell Models

During decades, tumor cells have been grown as adherent cultures in serum-complemented medium (Fig. 5.1a). These cell models, such as the well-known U87MG cells, have been useful in some experimental settings. However, monolayer cultures do not accurately recapitulate the in vivo conditions [6]. Furthermore, GBM cells grown in monolayers may drift away from their original phenotype and may not recapitulate classical GBM in vivo [7]. Also, classical U87MG cells when implanted in vivo do not form invasive strands [8]. In contrast, patient-derived xenograft (PDX) models better reproduce GBM features when implanted in vivo. These models are based on small pieces of human tumor material, growing in agar-overlaid flask, that are sequentially transplanted into rodent brains [9, 10] (Fig. 5.1a). Genetic mutations are well conserved between patient and rodent tumors [11]. The limitation of this tumor model is the short time of culture between two implantations, restraining the possibilities of experimental procedures to be conducted. A good compromise between the monolayer culture cells and patient-derived tumor fragments is the use of nonadherent GBM cells in the absence of serum (Fig. 5.1a). GBM patient-derived cells when grown in neurobasal medium spontaneously have a low adhesion and mainly grow as spheroids [12]. Unlike monolayer cultures, spheroids grown in 3D partially mimic the conditions of a tumor in vivo, including availability of oxygen and nutrients. The spheroid core is under hypoxic and low-glucose conditions, while the periphery is well oxygenized, and metabolites and growth factors are directly available [6]. This culture model preserves exchanges between cells and even maintains symbiotic interactions between cells normally found in a tumor in vivo. Furthermore, human GBM cells from suspension culture, such as P3, NCH421K, and BL13, implanted into immunodeficient mice, closely recapitulate features of patient tumors (neo-angiogenesis, necrosis, invasive strands) [13, 14].

In this chapter we describe a procedure for isolating, culturing, and genetically modifying GBM cells (overexpression, shRNA, CRISPR-Cas9), for further experimental studies.

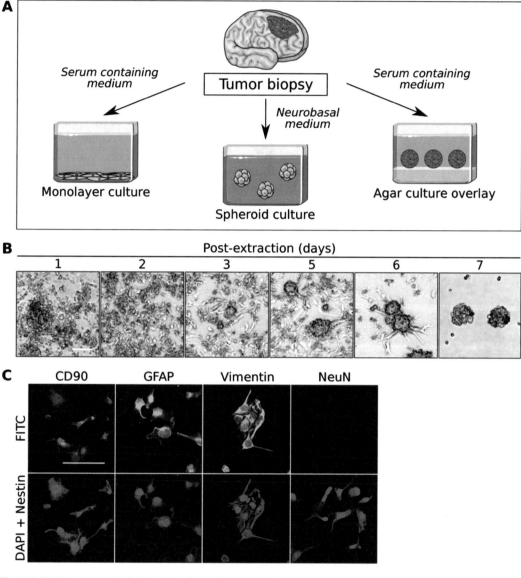

Fig. 5.1 Glioblastoma cell-derived models. (**a**) Overview of the three experimental GBM models, modified from ref. [11]. (**b**) Establishment of a GBM patient-derived cell line. Cells are extracted from patient tumor samples and cultured in complete neurobasal medium (cNBM). Cell clusters are observed at 1 week. Scale 100 μm. (**c**) Cells from spheroids are plated and stained for nestin, GFAP, vimentin, Thy-1, or NeuN. Scale 50 μm

2   Materials

2.1   Dissociation of Fresh Tumor Sample

The following reagents are used: phosphate buffered saline (PBS) 1X prepared from sterile water with DPBS 10X (Pan-Biotech, P04–53-500, stored at 4 °C), complete neurobasal medium (cNBM) elaborated from NBM (Gibco, 21103–049, stored at 4 °C) supplemented by B27® (Gibco, 12587), 10 ng/mL basic fibroblast growth factor (bFGF, PeproTech, 100-18B), 500 U/μL heparin, 100 U/mL penicillin, and 50 mg/mL streptomycin (Gibco, 15140–122). Cells are grown at 37 °C, 5% CO_2, and 95% humidity.

For the use of fresh GBM tumor tissue, written permission according to the local guidelines is obtained. Anonymized tumor samples are put into sterile flasks (Falcon 75 cm², 10497302) containing complete Dulbecco's Modified Eagle Medium elaborated (cDMEM) from DMEM (4.5 g/L glucose/pyruvate; Gibco™ 41966052) supplemented by 100 U/mL penicillin, 50 mg/mL streptomycin, and 20% of fetal bovine serum (FBS) and transferred rapidly to the laboratory (<6 h, longer time is possible but not recommended). Tumor pieces are then cut with a Swann-Morton surgical scalpel (Fisher, 12397999) and dissociated with Trypsin-EDTA 1X at 0.05% final concentration (Gibco, 25300–054, stored at 4 °C) or 2 mg/mL collagenase D (Roche, stored at 4 °C). Pieces are mechanically flushed and crushed through a 40 μm pore strainer (Fisher Scientific, 07–201-430) with the back of the plunger of a syringe (Terumo, SS10ES1). Longer-term culture is done in cNBM.

2.2 Processing Spheroids for Experiments

Spheroids are dissociated with Accutase (Gibco, ref.: A11105–01, stored at 4 °C), which provides gentle cell dissociation. Cell suspensions stained with trypan blue 0.4% (ThermoFisher, T10282) are automatically counted with a cell counting chamber slide (Countess®, Invitrogen, C10283). Cells are then cultured in a 96-well round-bottom plate (Falcon, 08–772-212) containing 0.4% methylcellulose (Sigma). Methycellulose stock solution is diluted in NBM for a 2% final concentration. The solution is stirred overnight at 25 °C and then filtered before use. Spheroids spontaneously assemble within a day.

2.3 Infection and Stabilization with Fluorescent Probe, shRNA, or CRISPR-Cas9

Lentiviral particles are produced in HEK293T cells with polylysine (Sigma, P4832–50, stored at 4 °C) and Opti-MEM, CaCl₂ 2.5 M (Sigma, C5080, stored at 4 °C), vectors pSD11 (VSV-G, viral envelope) and pSD16 (packaging construct), filtered HEPES 2 M (Sigma, H3784-100G, stored at 4 °C), and HEPES buffered saline (HeBS, Sigma, 51,558–50, stored at 4 °C). Matrigel (Corning™ 354,230) is diluted to a final concentration of 0.2 mg/mL in cold cNBM. It is used for coating coverslips, plate wells, or flasks to obtain adherent cells for infection and then incubated for 30 min at 37 °C. Polybrene (optional but will improve plasmid uptake) is used for infection steps. One percent Virkon solution is required for decontamination of cell culture material and medium that is in contact with virus. Complete neurobasal medium cNBM is used as culture medium.

2.4 Spheroid Invasion in Collagen I Matrix

To prepare the collagen gel, type I collagen (Corning™ 354,236), to a final concentration of 1 mg/mL, 1X PBS, 0.023xV$_{collagen}$-NaOH 1 M, and sterile H_2O sufficient to 100 μl are used for one spheroid. To avoid spheroid damage by pipetting, 200 μl tips are cut, and 5 mm of each tip are removed by scissors.

2.5 Immunofluore-scence	Spheroids or dissociated cells are fixed by a solution of 4% paraformaldehyde (PFA) in PBS (Santa Cruz, sc-281692). Blocking solution is prepared with PBS/2% fetal calf serum/1% bovine serum albumin (Euromedex, 04–100-812-C). Primary antibodies include anti-cortactin (Millipore 05–180), anti-Thy1 (Santa Cruz, sc-53,116), anti-GFAP (DAKO Z0334), anti-vimentin (Abcam ab16700), anti-nestin (mouse origin, Sigma 388 M-14, and rabbit origin, Invitrogen PA511887), and anti-NeuN (Millipore, MAB377). Suitable secondary anti-rabbit or anti-mouse fluorescent antibodies and ProLong reagent are purchased from Invitrogen.
2.6 In Vivo Experiments	Male RAGγ2C$^{-/-}$ mice are housed in an accredited animal facility. All animal procedures are performed according to the institutional guidelines and approved by the local ethics committee (C2EA −50; Comité d'éthique pour l'Expérimentation Animale Bordeaux). A stereotaxic apparatus (KOPF) is used for spheroid implantation. Anesthesia consists of a mixture of xylazine and ketamine.
2.7 Tools and Procedures for Image Acquisition and Processing	A video microscope with controlled chamber (Nikon) is used with appropriate software (Nikon NIS Element software) for taking pictures of proliferation or invasion and a confocal microscope for the analysis of spheroid sections. A NanoZoomer device (2.0 HT, Hamamatsu) and a confocal microscope (Eclipse Ti, Nikon) are used for the acquisition of brain tumor sections. Pictures are then quantified with Fiji software (ImageJ).

3 Methods

3.1 Processing of Brain Tumor Cells	Tumor pieces are taken by the neurosurgeon from two different areas, which include the dense/angiogenic core and distant invasive areas [14]. Tumor pieces are put in a culture flask containing 50 mL of cDMEM. Tumor samples are cut under sterile conditions into small pieces of a few squared millimeters with two sterile scalpels to facilitate enzymatic dissociation. Samples are washed twice in a 50 mL conical-bottomed tube containing 25 mL of PBS and then centrifuged (250 g, 5 min). Supernatants are removed and tumor pieces are incubated at 37 °C with trypsin or collagenase D, in a water bath for 30 min. Samples are regularly agitated to avoid pelleting. Digested tumor pieces are washed twice in PBS and further mechanically dissociated using a syringe and by filtering through a 40 μm strainer. After washing and centrifugation, the flow-through is placed in a culture flask containing 10 mL of cNBM (Fig. 5.1b).

3.2 Spheroid Culture and Identification of Stem Cell Markers

First, cells are washed with PBS for debris removal. The medium is replaced every other day, and small cell clusters are observed within a few days (Fig. 5.1b). After 1 week, cell clusters detach from the flask bottom and float in cNBM. Spheroids are then centrifuged and washed in PBS, and Accutase is added for 5 min in a 37 °C incubator. After mechanical dissociation by up-and-down pipetting, cells are diluted with PBS, counted, placed in fresh cNBM, and cultured for 1 week.

To identify stem cell markers, dissociated spheroids are plated on Matrigel-coated coverslips and stained for nestin, glial fibrillary astrocytic protein (GFAP), vimentin, or CD90 (Thy-1). As negative controls, cells are stained with the neuronal marker NeuN. Cells stain positively for stem cell markers, but not for NeuN, as shown in Fig. 5.1c.

Cells are then used in different experiments as outlined in Sects. 3.3–3.7.

3.3 Lentiviral Infection of Cells

3.3.1 Preparation of Lentiviral Particles

Lentiviral particles are prepared in HEK293T cells, which are transfected with the following mix: 4 μg of pSD11, 10 μg of pSD16, and 20 μg of the lentiviral vectors (pXL317-GFP). 50 μL CaCl$_2$ at 2.5 M is then added to the mix. The mixture of CaCl$_2$/vectors is added drop by drop to 500 μL HeBS in a Falcon tube while shaking on a vortex at low speed. The transfection mix is then added to the cells by pipetting drop by drop in the culture medium to homogenously cover the whole plate. Six hours after transfection, culture medium is removed and replaced with 8 mL fresh Opti-MEM supplemented with 20 mM HEPES. Supernatants are harvested, filtered, and either directly used on cells or stored at −80 °C for later use.

3.3.2 Infection of Cells

Cells are cultured on plates coated with Matrigel (0.2 mg/mL), incubated overnight, and infected with lentiviral particles. In case of low infection, dissociated cells are incubated with lentiviral particles, 0.4% methylcellulose, and cNBM medium. Cells are then centrifuged in U-bottom plates for 2 h at 300 g and incubated at 37 °C. When fluorescent lentiviral vectors are used, spheroids can be visualized after 2 days as a control for infection efficiency.

3.4 Spheroids in 2D Migration Assays

Spheroids are prepared in a 96-well round-bottom plate. 10^4 cells in 200 μL medium containing 0.4% methylcellulose are distributed in each well. Spheroids can be used 3 days after preparation but after no more than 1 week. To improve the process, cells are centrifuged at 37 °C for 2 h at 300 g, and spheroids are observed directly after centrifugation. Cells can then be used in a spheroid migration assay (Fig. 5.2a). Matrigel-coated coverslips are freshly prepared and incubated with cNBM for 30 min prior to the experiment. Spheroids are then washed twice with PBS and deposited onto the coverslips. Spheroids rapidly adhere on the coverslips and cell migration is observed directly after attachment (Fig. 5.2a). The areas of migration are then measured and compared to control (Fig. 5.2a).

3.5 Spheroids in 3D Assays: Quantification of Invasion and Proliferation

Collagen matrix is prepared in a tube on ice with type I collagen at 1 mg/mL final concentration, PBS 1X, $0.023xV_{Collagen}$ NaOH 1 M, and sterile H_2O. The solution is incubated on ice for 30 min. In the meantime, spheroids are collected in 500 μL tubes and washed twice with PBS 1X. Spheroids are pipetted with 100 μL of the collagen solution and inserted in the center of a well of a 96-well plate. The gel is polymerized for 30 min at 37 °C, and 100 μL of medium is then added on top of the gel. Inhibitors, activators, or other molecules are added in the solutions. Pictures are taken 24 h after collagen inclusion using a video microscope in brightfield and fluorescence (Fig. 5.2b). Fiji software is used for image analysis. As spheroids are GFP positive, a threshold is applied to obtain a black and white mask (Image ➔ Adjust ➔ Threshold) which then automatically creates a selection corresponding to GFP-positive areas (Edit ➔ Selection ➔ Create Selection). At this step, it is recommended to verify that the selected area fits the total area of brightfield images. Delimitating the spheroid core at 24 h is done manually. These different areas are measured (Analyze ➔ Measure), and each of the measurements is saved in the region of interest (ROI) manager. To calculate the invasive area, the core area is subtracted from the total area. Normalization is done according to the spheroid core area or its circumference (Fig. 5.2b).

Alternative quantification with image processing for no GFP cells can be used. Unsharp mask (Process ➔ Filters ➔ Unsharp mask) is applied and background is subtracted (Process ➔ Substract background). Picture is blurring (Process ➔ Filters ➔ Gaussian Blur), and a binary picture is created (Process ➔ Binary ➔ Make Binary) to also obtain a black and white mask. The mask fields are closed and filled (Process ➔ Binary ➔ Close- + Fill Holes). Mask is smoothed with a median filter (Process ➔ Filters ➔ Median), and the following steps are the same from the previous Create Selection.

Cilostamide, a cyclic nucleotide phosphodiesterase inhibitor, is known to decrease cell invasion [15] and is used as an example herein. Using the second quantification technique, we compared the invasion rate of spheroids, in the presence or absence of cilostamide. The images and results are available in Fig. 5.2c.

Spheroids can also be directly used in the 96-well round-bottom plates for proliferation experiments. As seen in Fig. 5.2d, spheroids are incubated for 7 days, and pictures are taken at different time points. Spheroid size is quantified automatically using Fiji software by applying a threshold for the GFP signal as described above (Fig. 5.2d). Spheroids can be monitored for more than 2 weeks.

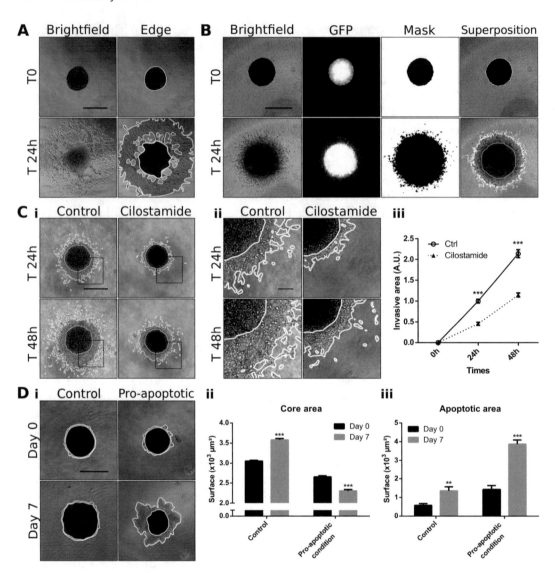

Fig. 5.2 In vitro experiments based on spheroids. (**a**) Visualization of spheroid migration at 0 and 24 h. Scale 500 μm. (**b**) Quantification of spheroid invasion: brightfield images, fluorescent GFP images, corresponding masks from Fiji, and a merged image between brightfield and the selection created from the mask. Invasive area = (total area − core area)/core area. Scale 500 μm. (**c**) Effect of cilostamide on P3 spheroids. (**i**) Images of the spheroids at 24 and 48 h in brightfield. Scale 500 μm (**ii**) magnified images from (**i**) (scale 100 μm) and (**iii**) quantification of invasive areas (******$P < 0.01$ and *******$P < 0.001$, Student's t-test). (**d**) Effect of hypoxia on cell apoptosis. (**i**) Images of spheroids at day 0 and day 7 and quantification of core (**ii**) and apoptotic areas (**iii**) (******$P < 0.01$ and *******$P < 0.001$, Student's t-test). Scale 500 μm

3.6   Immunofluore-scence Staining on Spheroid Sections or in 3D

3.6.1   Spheroid Sections

Spheroids are washed twice in PBS and then included in OCT for quick freezing in 2-methylbutane in a small freezing container. Using a cryostat, blocks are cut into 10 μm-thick sections and placed onto gelatin-coated slides. Classical immunofluorescence is then performed on spheroid sections. Briefly, sections are fixed with 4% paraformaldehyde (PFA) for 10 min, washed three times with PBS, and permeabilized with PBS/0.1% Triton X100 for 15 min. After blocking and circling with a hydrophobic solution (PAP pen, Dako), sections are incubated overnight at 4 °C with primary antibodies diluted in the blocking solution. After three washes in PBS, secondary fluorescent antibodies diluted in blocking solution are added on the sections and incubated for 1 h at room temperature. Sections are mounted after three PBS washes with ProLong Gold reagent. Examples of stained spheroid sections are shown in Fig. 5.3a.

3.6.2 Immunostaining on 3D Spheroids

Spheroids are washed with PBS and fixed with 4% PFA for 30 min, washed three times in PBS, and permeabilized with PBS/0.1% Triton X100 for 1 h. After blocking for 1 h with the solution described above, spheroids are incubated with primary antibodies on a rotator during 24 h. After five PBS washes, samples are incubated with secondary fluorescent antibodies overnight. Spheroids are washed again extensively in PBS, then transferred to a glass bottom dish, and covered with ProLong Gold reagent and a coverslip. Spheroids are then imaged by confocal microscopy in the following days (Fig. 5.3b), and 3D reconstruction is performed by using NIS software (Fig. 5.3b).

3.7 Orthotopic Patient-Derived Xenograft (PDX) and Tumor Phenotyping

Human spheroids are implanted into the brain of immunodeficient rodents using a stereotaxic apparatus [14]. Mesenchymal GBMs are particularly adapted for this assay. In brief, mice are placed on the stereotaxic table after anesthesia, the skin is opened with a scalpel, and a small hole is drilled in the skull to facilitate the insertion of a Hamilton syringe (Sigma-Aldrich, 20980-U). Five spheroids are injected in the mouse brain, 2.2 mm left of the bregma and 3.4 mm deep. Tumor growth is observed after 2–3 weeks by MRI or by luciferase imaging. Depending on the tumor type, the maximum volume is reached between 30 and 60 days. At the end of the experiment, mice are euthanized and brains are collected for immunohistology. Sections of 10 µm thickness are incubated with anti-nestin and anti-CD31 antibodies to detect human tumor cells and mouse blood vessels, respectively (Fig. 5.3c).

Most of the glioblastoma features are observed several weeks after spheroid implantation. A large contralateral invasion, mainly via the corpus callosum, is present (Fig. 5.3c, top zoom image). By magnifying the images, many isolated single cells are also detected around the tumor core (Fig. 5.3c, bottom zoom image). Blood vessel density in the tumor may be increased when compared to the normal brain (Fig. 5.3c, middle zoom image). Necrotic areas can also be detected by hematoxylin/eosin staining and hypoxic regions by pimonidazole or by immunofluorescence using anti-carboxic anhydrase IX (CAIX) antibodies [14, 16].

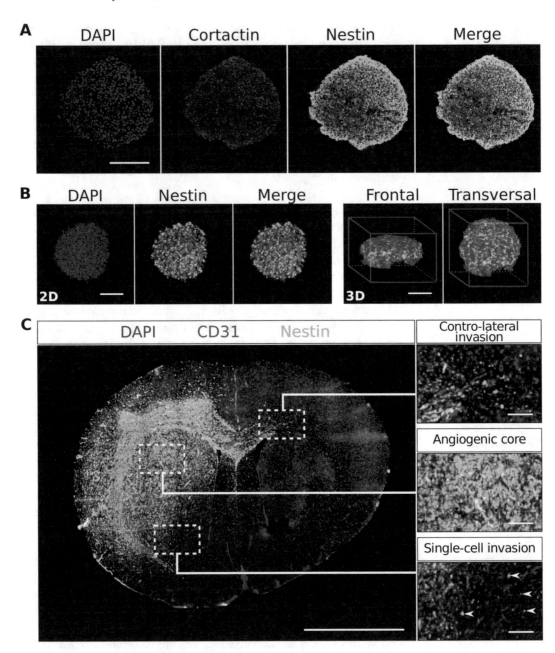

Fig. 5.3 Histological features of GBM spheroids and of tumors after spheroid implantation in the mouse brain. (**a**) Immunofluorescence of spheroid sections. Spheroids were included in OCT, frozen to be cut in 10 μm sections, and stained with anti-cortactin (red) and anti-nestin (green) antibodies and DAPI (blue) for nucleus staining. Scale: 250 μm. (**b**) Spheroids are stained with anti-nestin antibodies and DAPI and imaged by confocal microscopy. Scale 250 μm. 3D views of z-stacks are created by using NIS software (Nikon) (**c**) Immunohistofluorescence of a NCH421K tumor section showing tumor cells and blood vessels stained with anti-nestin (green) and anti-CD31 (red) antibodies, respectively. Scale 2.5 mm. Magnified views of contralateral invasion, angiogenic core, and single-cell invasion areas are represented in the right panels. Scale 250 μm. Arrows indicate single-cell invading stroma

4 Conclusion

GBM patient-derived cell lines are useful models to study GBM development. These models have several advantages: (1) in vitro 3D organization, (2) maintenance of a correct genetic profile, and (3) histological features of the human disease in orthotopic xenografts. Tumor heterogeneity observed between patients and within a tumor is recapitulated to some extent by these cell lines after intracranial implantation in mice. The four GBM subtypes (mesenchymal, neural, pro-neural, and classical) can be implanted in vivo, but mesenchymal subtype tumors have a more efficient growth. Typical histological features of glioblastoma are observed: necrosis, pseudopalisading cells, large and leaky vessels, hemorrhage, single and contralateral invasions, and the latter mainly via the corpus callosum. Nevertheless, some cell lines give rise to more angiogenic and necrotic tumors, while others are more invasive [9]. The inconvenience of xenograft models is the immunodeficient background which does not reflect the immune-competent environment found in human disease. Syngeneic tumors such as derived from CT2A or GL261, or genetically engineered mouse models (GEMMs), have the advantage of a full immunocompetent background, but lack the genetic background found in human tumors [17]. Thus, all these models have advantages and inconveniences and must be used in a specific scientific context in which the experiments done in the selected model will lead to meaningful results.

To our opinion, GBM patient-derived cell lines can be used for many experiments related to glioblastoma development, especially tumor cell invasion. Furthermore, they are also useful for testing the efficacy of antitumor therapies or of other pharmacological agents.

5 Notes and Troubleshooting

- Always use cut tips when manipulating spheroids in order to keep their integrity. Tips are cut with scissors and autoclaved before use.

- Spheroids assemble spontaneously in the flask. For optimization, U-bottom well plates should be used (10,000 cells per well) with methylcellulose.

- To wash spheroids, centrifugation is not necessary as they drop spontaneously by gravity.

- For the invasion assay, spheroids should be place in the center of the well to let cells invade from the spheroid core. If the spheroids are not positioned in the center of the well, they should be repositioned by slightly tapping the side of the plate.

- For spheroid inclusion in OCT, they should be properly embedded by avoiding bubbles for facilitating cryo-sectioning.

- Different patient-derived cell lines may be used for xenograft assays. Mesenchymal subtype tumors develop faster than proneural tumors. Some cell lines form angiogenic tumor in vivo, while others are more invasive.

Acknowledgments

Joris Guyon is a recipient of Toulouse Hospital scholarship, and Tiffanie Chouleur is a PhD student supported by IDEX Bordeaux-McGill University. Andreas Bikfalvi got financial support from Ligue Contre le Cancer, ARC, ANR, and TRANSCAN-ERA-NET. Thomas Daubon is supported by ARC and ERA-NET grants.

Author Contributions
JG, TC, AB, and TD wrote the manuscript. JG designed all Figs. TD and AB discussed the manuscript.

References

1. Kotliarova S, Fine HA (2012) SnapShot: glioblastoma multiforme. Cancer Cell 21(5):710–710. e711. https://doi.org/10.1016/j.ccr.2012.04.031

2. Stupp R, Hegi ME, Gilbert MR, Chakravarti A (2007) Chemoradiotherapy in malignant glioma: standard of care and future directions. J Clin Oncol Off J Am Soc Clin Oncol 25(26):4127–4136. https://doi.org/10.1200/JCO.2007.11.8554

3. Gatson NN, Chiocca EA, Kaur B (2012) Anti-angiogenic gene therapy in the treatment of malignant gliomas. Neurosci Lett 527(2):62–70. https://doi.org/10.1016/j.neulet.2012.08.001

4. Weller M, Yung WK (2013) Angiogenesis inhibition for glioblastoma at the edge: beyond AVAGlio and RTOG 0825. Neuro-Oncology 15(8):971. https://doi.org/10.1093/neuonc/not106

5. Patel AP, Tirosh I, Trombetta JJ, Shalek AK, Gillespie SM, Wakimoto H, Cahill DP, Nahed BV, Curry WT, Martuza RL, Louis DN, Rozenblatt-Rosen O, Suva ML, Regev A, Bernstein BE (2014) Single-cell RNA-seq highlights intratumoral heterogeneity in primary glioblastoma. Science 344(6190):1396–1401. https://doi.org/10.1126/science.1254257

6. Lenting K, Verhaak R, Ter Laan M, Wesseling P, Leenders W (2017) Glioma: experimental models and reality. Acta Neuropathol 133(2):263–282. https://doi.org/10.1007/s00401-017-1671-4

7. Torsvik A, Stieber D, Enger PO, Golebiewska A, Molven A, Svendsen A, Westermark B, Niclou SP, Olsen TK, Chekenya Enger M, Bjerkvig R (2014) U-251 revisited: genetic drift and phenotypic consequences of long-term cultures of glioblastoma cells. Cancer Med 3(4):812–824. https://doi.org/10.1002/cam4.219

8. Newcomb EW, Lukyanov Y, Alonso-Basanta M, Esencay M, Smirnova I, Schnee T, Shao Y, Devitt ML, Zagzag D, McBride W, Formenti SC (2008) Antiangiogenic effects of noscapine enhance radioresponse for GL261 tumors. Int J Radiat Oncol Biol Phys 71(5):1477–1484. https://doi.org/10.1016/j.ijrobp.2008.04.020

9. Bougnaud S, Golebiewska A, Oudin A, Keunen O, Harter PN, Mader L, Azuaje F, Fritah S, Stieber D, Kaoma T, Vallar L, Brons NH, Daubon T, Miletic H, Sundstrom T, Herold-Mende C, Mittelbronn M, Bjerkvig R, Niclou SP (2016) Molecular crosstalk between tumour and brain parenchyma instructs histopathological features in glioblastoma. Oncotarget 7(22):31955–31971. https://doi.org/10.18632/oncotarget.7454

10. Wang J, Miletic H, Sakariassen PO, Huszthy PC, Jacobsen H, Brekka N, Li X, Zhao P, Mork S, Chekenya M, Bjerkvig R, Enger PO (2009) A reproducible brain tumour

model established from human glioblastoma biopsies. BMC Cancer 9:465. https://doi.org/10.1186/1471-2407-9-465

11. Huszthy PC, Daphu I, Niclou SP, Stieber D, Nigro JM, Sakariassen PO, Miletic H, Thorsen F, Bjerkvig R (2012) In vivo models of primary brain tumors: pitfalls and perspectives. Neuro-Oncology 14(8):979–993. https://doi.org/10.1093/neuonc/nos135

12. Kim SS, Pirollo KF, Chang EH (2015) Isolation and culturing of glioma cancer stem cells. Curr Protoc Cell Biol 67:23.10.1–21.10.10. https://doi.org/10.1002/0471143030.cb2310s67

13. Boye K, Pujol N, Isabel DA, Chen YP, Daubon T, Lee YZ, Dedieu S, Constantin M, Bello L, Rossi M, Bjerkvig R, Sue SC, Bikfalvi A, Billottet C (2017) The role of CXCR3/LRP1 cross-talk in the invasion of primary brain tumors. Nat Commun 8(1):1571. https://doi.org/10.1038/s41467-017-01686-y

14. Daubon T, Leon C, Clarke K, Andrique L, Salabert L, Darbo E, Pineau R, Guerit S, Maitre M, Dedieu S, Jeanne A, Bailly S, Feige JJ, Miletic H, Rossi M, Bello L, Falciani F, Bjerkvig R, Bikfalvi A (2019) Deciphering the complex role of thrombospondin-1 in glioblastoma development. Nat Commun 10(1):1146. https://doi.org/10.1038/s41467-019-08480-y

15. Zimmerman NP, Roy I, Hauser AD, Wilson JM, Williams CL, Dwinell MB (2015) Cyclic AMP regulates the migration and invasion potential of human pancreatic cancer cells. Mol Carcinog 54(3):203–215. https://doi.org/10.1002/mc.22091

16. Allen E, Jabouille A, Rivera LB, Lodewijckx I, Missiaen R, Steri V, Feyen K, Tawney J, Hanahan D, Michael IP, Bergers G (2017) Combined antiangiogenic and anti-PD-L1 therapy stimulates tumor immunity through HEV formation. Sci Transl Med 9(385). https://doi.org/10.1126/scitranslmed.aak9679

17. Huse JT, Holland EC (2009) Genetically engineered mouse models of brain cancer and the promise of preclinical testing. Brain Pathol 19(1):132–143. https://doi.org/10.1111/j.1750-3639.2008.00234.x

Chapter 6

Organotypic Brain Cultures for Metastasis Research

Lucía Zhu and Manuel Valiente

Abstract

Despite the emergence of brain metastasis as an unmet clinical need, our knowledge of the underlying biology is limited. Current treatments provide limited positive responses in most patients, which have generally poor prognosis upon diagnosis. Modeling of brain metastasis has been based on brain metastatic cell lines that are injected in mice. This has remained the only relevant preclinical platform for the validation of mechanistic findings obtained in vitro. Here, we describe the use of organotypic brain cultures for brain metastasis research, which recapitulate the in vivo phenotype in an ex vivo procedure, and discuss their multiple applications for basic and clinical research, thus providing a useful tool for improving the current landscape of brain metastasis preclinical models.

Key words Organotypic brain cultures, Metastasis, Therapy, Tumor microenvironment, Patient-derived organotypic culture (PDOC)

1 Introduction: Experimental Models for Brain Metastasis Research

Brain metastasis constitutes a major clinical problem due to its increasing incidence, poor prognosis, and limited available therapies that improve survival of most patients [1]. Modeling of this advanced progression of cancer has been challenging limiting major advances on the knowledge of the underlying biology. Human and mouse brain metastatic (BrM) derivatives have been previously described as valuable resources for brain metastasis research [2–6]. However, in vitro cell lines do not fully recapitulate the complex multistep process of the disease. Thus, different experimental mouse models are used for mimicking the human course of the disease. These include human and mouse BrM cell lines from most common sources of brain metastasis (lung and breast cancer and melanoma) that are inoculated into immunosuppressed or immunocompetent hosts, respectively, to induce metastasis [2–6]; genetically engineered mouse models (GEMMs) that generate spontaneous brain metastases [7–9]; and patient-derived xenografts (PDXs) that maintain pathological features of the

Giorgio Seano (ed.), *Brain Tumors*, Neuromethods, vol. 158,
https://doi.org/10.1007/978-1-0716-0856-2_6, © Springer Science+Business Media, LLC, part of Springer Nature 2021

human brain metastases of origin [10–13]. The study of brain metastasis using these preclinical models has allowed to mechanistically dissect several molecular processes involved at different stages of the metastatic disease [4, 14], including the important contribution of the tumor-associated microenvironment (i.e., reactive astrocytes) [5, 15]. Although mice are the gold-standard experimental model for brain metastasis research, they are costly and thus impose obvious economical limitations for extensive validation of preliminary results obtained in vitro.

Organotypic brain cultures ex vivo constitute a novel experimental platform for brain metastasis research since they faithfully recapitulate phenotypically and functionally metastatic cells in vivo [4, 14] and allow rapid evaluation of multiple experimental conditions at the same time at affordable economical costs. Organotypic brain cultures that we describe here are short-term assays that can be maintained up to 7 days with intact cellular architecture and functionality of different resident cell types present in the brain tissue as well as metastatic cells able to colonize the brain (Fig. 6.2d–g). Common assays performed for interrogating functional and mechanistic insights of brain metastasis using organotypic brain cultures include evaluation of the efficacy of monotherapies or combined therapies (targeted or immunotherapy, including blocking antibodies), cellular response to radiotherapy, physical interactions between cancer cells and the brain microenvironment (i.e., vascular co-option) [4], and the contribution of the tumor-associated microenvironment to the metastatic disease [5, 15] (Fig. 6.1a–c).

Organotypic cultures could be also used to mimic different stages of the colonization process including early versus late stages. Early stages are reproduced by plating cancer cells on the brain slice, while advanced stages involve a preliminary step in which brain metastases are induced in vivo to later process the brain ex vivo (Fig. 6.2e). This allows interrogating pharmacologically and biologically clinically relevant stages of brain metastasis. Additionally, this approach also evaluates tumor-associated stromal cells that are only present when metastases are already fully established (i.e., pSTAT3+ reactive astrocytes) [5]. This tumor-associated microenvironment also includes non-brain resident recruited immune cells that contribute to the biology of the tumor-stroma entity (Fig. 6.1c).

2 Materials

2.1 Cell Lines

Human and mouse BrM cell lines from melanoma (B16/F10-BrM), lung (H2030-BrM, PC9-BrM, 482N1, 393N1), and breast adenocarcinoma (MDA231-BrM, CN34-BrM, ErbB2-BrM) have been previously described [2–5]. These BrM cell lines can be used

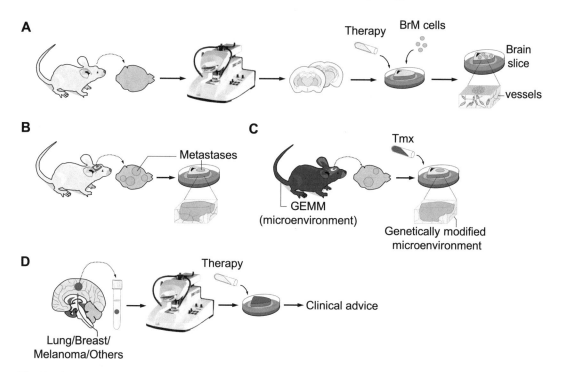

Fig. 6.1 Organotypic cultures for brain metastasis research. (*a*) Schema representing organotypic brain cultures mimicking early stages of brain colonization. (*b*) Organotypic brain cultures from established brain metastases. (*c*) Organotypic brain cultures using genetically engineered mouse models (GEMM) to interrogate brain metastasis in the context of genetic modifications induced in specific components of the microenvironment in a conditional manner (i.e., tamoxifen will activate Cre recombinase and delete the floxed gene). (*d*) Patient-derived organotypic culture (PDOC) as a platform to test personalized therapies

adding them directly to brain slices in culture obtained from tumor-free mice or inoculating them in mice to obtain brain slices with established metastasis generated in vivo.

2.2 Reporter Lentiviral Constructs

BrM cell lines are engineered with lentiviral constructs expressing firefly luciferase (Addgene ref.#19166) and the fluorescent protein GFP (Addgene ref.#36083), which allow noninvasive in vivo imaging in mice or ex vivo analysis of brain slice cultures and cancer cell tracking in histological sections, respectively.

2.3 Inoculation of BrM Cells to Generate Brain Metastasis

Hundred microliter of a BrM cell suspension in PBS containing 10^5 cells of a BrM cell line is injected intracardiacally in anesthetized 6–10-week-old athymic nu/nu (Envigo) or C57BL/6 mice (or the specific strain corresponding to the cancer cells), for human and mouse cell lines respectively, as described elsewhere [2, 3]. Metastases-bearing brains obtained at the endpoint of the disease (5–7 weeks after intracardiac inoculation of human BrM cell lines or 2 weeks in syngeneic mouse BrM cell lines) are used for preparing organotypic brain cultures. BrM cells could be also inoculated

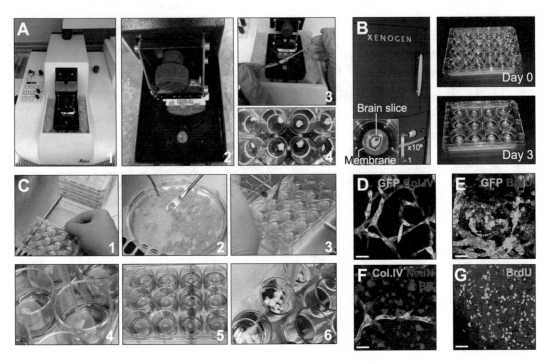

Fig. 6.2 Preparation and processing of organotypic brain cultures. (*a*) Brain sectioning at the vibratome and collection of sectioned brain slices in 24-well plates with supplemented HBSS. (*b*) Bioluminescence imaging of brain slices with an IVIS Xenogen machine. Imaging is performed when brain slices have established metastasis or, at a later time point, if cancer cells are plated on top (C4). (*c*) Preparation of organotypic brain cultures: (1) placing of membranes, (2) selection of brain slices, (3) plating brain slices on membranes, (4) seeding cancer cells on brain slices, (5) organotypic brain cultures ready for culturing, and (6) immunofluorescence staining of fixed organotypic brain cultures. (*d*) Organotypic brain culture mimicking early stages of brain metastasis showing cancer cells (GFP) co-opting the vessels (Col.IV). Scale bar 50 μm. (*e*) Organotypic brain culture with established brain metastasis showing proliferating (BrdU) cancer cells (GFP). Scale bar 75 μm. (*f*) Capillaries (Col.IV) and neurons (NeuN) in an organotypic brain culture without cancer cells. Nuclei are stained with bis-benzamide (BB). Scale bar 20 μm. (*g*) Patient-derived organotypic culture (PDOC) with proliferating cancer cells (BrdU). Blue corresponds to autofluorescence. Scale bar 100 μm

in mice in which certain components of the microenvironment have been genetically engineered (Fig. 6.1c). Thus, the behavior of inoculated syngeneic mouse BrM cell lines could be studied in the context of a modified brain microenvironment.

2.4 Agarose and Molds

Low melting point agarose (Lonza, ref.#50080) is added to Peel Away Disposable Embedding Molds (Electron Microscopy Sciences, ref.#70182) and brains embedded in it.

2.5 Double-Edge Razors

Double-edge stainless razor blades (Electron Microscopy Sciences, ref.#72000) are divided in two and used to section brains in the vibratome.

2.6 Vibratome

A vibratome (Leica VT1000 S) is used for slicing dissected brains with or without brain metastases (Fig. 6.1a–b) (see below) into 250 μm brain slices at speed = 2 and frequency = 5 (Fig. 6.2a).

2.7 Slice Culture Media

Brains are dissected on ice-cold Hank's balanced salt solution (HBSS) supplemented with HEPES (pH 7.4, 2.5 mM), D-glucose (30 mM), $CaCl_2$ (1 mM), $MgCl_2$ (1 mM), and $NaHCO_3$ (4 mM). Brains or brain slices are maintained on ice-cold supplemented HBSS during slicing at the vibratome and before culturing (Fig. 6.2a).

Brain slices are cultured in 12- or 24-well plates with 1 ml or 0.5 ml of slice culture media (DMEM, supplemented HBSS, fetal bovine serum 5%, L-glutamine [1 mM], 100 IU/ml penicillin, and 100 mg/ml streptomycin) per well at 37 °C, 5% CO_2 (Fig. 6.2c).

2.8 Membranes

0.8 μm Whatman® Nuclepore™ Track-Etched Membranes (Sigma-Aldrich) are used for culturing brain slices (Fig. 6.2b, c).

2.9 IVIS

An IVIS Xenogen machine (Caliper Life Sciences) is used for bio-luminescence imaging (BLI) of metastases-bearing mice, brains, and brain slices (Fig. 6.2b).

2.10 BrdU Pulse and Fixation

A bromodeoxyuridine (BrdU) (Sigma-Aldrich, ref. B9285, 10 mg/ml) pulse is given 2–4 h before fixing brain slices adding 20 μl per ml of media. Brain slices are fixed in 4% paraformaldehyde before (PFA 4%) performing immunofluorescence.

2.11 Immunofluorescence

Primary antibodies for immunofluorescence: GFP (Aves Labs, ref. GFP-1020, 1:1000), GFAP (Millipore, ref. MAB360, 1:1000), Iba1 (Wako, ref. 019–19,741, 1:500), NeuN (Millipore, ref. MAB377, 1:500), cleaved caspase-3 (1:500, ref. 9661; Cell Signaling), BrdU (Abcam, ref. ab6326, 1:500), collagen IV (Millipore, ref. AB756P, 1:500), and Olig2 (Millipore, ref. AB9610, 1:500). Secondary antibodies: Alexa Fluor anti-chicken 488, anti-rabbit 555, anti-mouse 555, anti-rat 555, anti-rabbit 633, and anti-mouse 647 (Invitrogen, dilution 1:300) (Fig. 6.2d–g).

2.12 Mowiol-NPG Anti-fade Reagent

Mowiol is solved in K_2HPO_4 buffer (1.36 g/L, pH 7.2) at a concentration of 25% weight/volume and mixed overnight at room temperature (RT). Glycerol is added 1:2 and mixed for 16 h at RT, followed by 15 min centrifugation at 12,000 rpm. NPG is solved 2.5% weight/volume in glycerol and mixed overnight at RT. NPG and Mowiol solutions (1:10) are mixed well by vortexing vigorously, followed by centrifugation at maximum speed for 2 min to avoid collecting bubbles.

2.13 Molecular Biology	RNA is extracted using RNeasy Mini Kit (Qiagen, ref. 74,106) and QIAshredder columns (Qiagen, ref. 79,654). Protein is extracted using RIPA buffer 1× (Cell Signaling ref. 9806).

3   Methods

By adapting previously reported methods [16], organotypic brain cultures ex vivo have been shown to recapitulate phenotypically and functionally brain colonization by metastatic cells [4, 14] (Figs. 6.1a and 6.2d). Thus, molecular mediators of this process can be assessed genetically and pharmacologically.

3.1   Brain Preparation

Tumor-free brains from C57BL/6 (or the corresponding strain of BrM cells) or athymic nu/nu mice, according to the mouse or human origin of BrM cells respectively, of 4–10 weeks old are dissected in supplemented ice-cold HBSS and embedded in 42 °C preheated 4% low melting agarose dissolved in supplemented HBSS. Agarose is added to square molds, and the brain is then placed inside and moved into all directions with a 20–200 μl tip to facilitate the embedding process. The solid agarose block is obtained by sectioning the plastic walls, trimmed to decrease the amount of agarose and glued into the sample holding disc. The brain is placed in coronal position with the cerebellum closer to the sample holding disc. If properly trimmed, up to 4–5 brains can be sectioned simultaneously.

If the intention is to section brains that have already established metastasis developed in vivo, mice at the endpoint of the metastatic disease (5–7 weeks after intracardiac inoculation of human BrM cell lines (i.e., H2030-BrM) or 2 weeks in syngeneic mouse BrM cell lines (i.e., B16/F10-BrM)) (Fig. 6.1b) are anesthetized (isoflurane) and injected retro-orbitally with D-luciferin (150 mg/kg) and subjected to BLI for confirming tumor development. Brains are finally dissected in supplemented HBSS and imaged by ex vivo BLI for localizing the metastases.

3.2   Vibratome Sectioning and Generation of Slices

The vibratome should be placed within a hood to minimize contamination risk. Half of a double-edge razor is placed in the razor-holding arm. Ice-cold HBSS is added to the vibratome cuvette. The cuvette should be always surrounded with ice. The sample holding disc with glued agarose-embedded brains is fixed in the cuvette using the fixing nail. Using the control panel to move the sectioning arm, the initial and endpoints are fixed together with the speed, frequency, and thickness. The automatic mode is selected, and embedded brains are sectioned into 250 μm slices. Upon sectioning slices are obtained from the cuvette using flat spatulas and transferred to a 10 cm petri dish with supplemented HBSS, which is kept cold on ice, where agarose could be easily removed and each slice divided at the hemisphere into two symmetrical pieces (Fig. 6.2a).

3.3 Adding BrM Cells to Brain Slices and Culture Tissue

The petri dish with the slices is transferred to the cell culture room. In a hood, 500–1000 µl of slice culture media is added to a 24–12-well plate where the membrane is carefully deposited on top of the liquid. Brain slices are placed with flat spatulas on the top of 0.8 µm pore membranes floating on slice culture media (one slice per membrane). If inhibitors are being tested, they are added to the media at the desired concentration. Brain slices are incubated at 37 °C and 5% CO_2 for 1–2 h so they can consolidate on the membrane. Then, if tumor-free brains are used, 3×10^4 BrM cells suspended in 2 µl of slice culture media are placed on the surface of the slice with a 1–10 µl micropipette and incubated for 16–24 h (Fig. 6.2c, panel 4). During this time cancer cells will infiltrate the brain tissue and interact with the capillaries—a process known as vascular co-option [4] (Fig. 6.2d). BrM cells could had been engineered before by lentiviral vectors (i.e., shRNA, siRNA, CRISPR/Cas9-induced *knockout*, etc.) to target genes of interest and to evaluate their contribution to the colonization of the brain [4, 5]. If conditional lentiviral vectors are used, the induction should be initiated after the 16–24 h period once the BrM cells are within the tissue. One possibility is to use conditional lentiviral vectors that are doxycycline dependent. The induction occurs after the addition of doxycycline (1 µg/ml) to the slice culture media. This antibiotic must be replaced every 48 h. After an incubation period of 3–7 days, qRT-PCR or western blotting of brain slices can confirm successful induction of *knockdown* or *knockout* in the cancer cells. Additionally, genetically engineered mouse models (GEMM) could be used to score candidate genes in the brain microenvironment [5] (see 5.2).

3.4 Bioluminescence Imaging (BLI) in Brain Slices with BrM (Day 0)

At this moment, 20 µl of D-luciferin (15 mg/ml) is added per ml of slice culture media and brain slices are imaged (Day 0) to confirm the presence of BrM cells using BLI with an IVIS Xenogen machine (Fig. 6.2b). If tumor-bearing brain slices are used, imaging is done right after obtaining the slices at the vibratome. Slices are placed in a 24-well plate with supplemented HBSS and D-luciferin is added as described. Image acquisition involves removing the lid of the plate within the IVIS to then detect bioluminescence by acquiring the signal for 1 s to 2 min, going through intervals of 10 s, 30 s, and 1 min, in order to avoid saturated BLI signal. From this point on, brain slices could be incubated under described culture conditions for up to 7 days. Brain slices must be always on the membrane; if they are found floating in the media, the specific slice must be discarded. Thus, it is very convenient to check the slices after the plate is transported. Within a 7-day cultured period, the cellular architecture is intact as confirmed by histological analysis of specific cellular markers that identify various brain cell types (i.e., endothelial cells (collagen IV), reactive astrocytes (GFAP), neurons (NeuN), microglia (Iba1), and oligodendrocytes (Olig2)).

3.5 BLI at the Experimental Endpoint

Brain slices are reimaged by BLI at the end of the experiment (Fig. 6.2b). Usually we do 3 days, which is the most versatile time point for evaluating readouts regarding proliferation and death once genes are targeted or inhibitors are tested. Cancer cell growth as measured by BLI is obtained by normalization of BLI at this time point over Day 0. Bioluminescence analysis is performed using Living Image software, version 4.5, by selecting each slice individually as a region of interest (ROI) and measuring the photons/s/cm²/steradian. For cultures over 3 days, slice culture media is replaced every 3 days. Consequently, inhibitors will be also replaced.

3.6 Fixation of Brain Cultures

At the end of the experiment, if proliferation is one of the readouts, a BrdU pulse could be given 2–4 h before fixation. For fixing the sample, the membrane with the brain slice is submerged in the media and carefully separated from it. Then slices are transferred with a brush from the culture plate to another 12- or 24-well plate for fixation with PFA 4%. Slices must be submerged within the PFA. The plate is incubated overnight at 4 °C in an orbital shaker.

3.7 Free-Floating Immunofluorescence

Free-floating immunofluorescence is performed for histological analysis (Fig. 6.2c, panel 6) and for evaluation of cellular death and/or proliferation by staining against anti-cleaved caspase 3 (CC3) and BrdU markers, respectively. Brain tissue architecture is visualized using respective markers for different cell types from the brain as previously described (Fig. 6.2d–g). The immunofluorescence is performed in the 24-well plate, adding a minimum volume of 350 μl/well. Brain slices are blocked in 10% NGS, 2% BSA, and 0.25% Triton X-100 in PBS for 2 h at RT shaking. Media must be removed carefully with a plastic Pasteur pipette avoiding drying of the tissue. Primary antibodies are incubated overnight at 4 °C in the blocking solution in an orbital shaker. The day after an incubation of 30 min at RT will conclude this step. After at least six washings in PBS-Triton 0.25%, secondary antibodies are added in the blocking solution and incubated for 2 h at RT. The plate should be shaking both during washings and incubation with the secondary antibody. After at least three washings in PBS-Triton 0.25%, nuclei are stained with bisbenzimide (1 mg/ml) for 7 min at RT shaking. This protocol of immunofluorescence applies to most of the primary antibodies, such as anti-GFP which is used to enhance the labeling of cancer cells and facilitate the quantification or components of the microenvironment to either score potential toxicities or the inhibitors tested or study their interactions with cancer cells

(Fig. 6.2f). However, the immunofluorescence against BrdU requires a special protocol. After finishing the regular protocol, for instance, to label GFP+ BrM cells, brain slices require to be treated with HCl 2 N 30 min at 37 °C in a water bath, where the plate floats, or in an oven, followed by 0.1 M borate buffer (pH 8.5) 10 min at RT shaking. After at least six washings with TBS, brain slices are blocked in 3% NGS in TBS-Triton 0.25% for 1 h, and primary antibody anti-BrdU is incubated for 72 h at 4 °C shaking. After at least six washings with TBS-Triton 0.25%, the secondary antibody is incubated in blocking solution for 2 h at RT shaking followed by at least three washings with TBS.

3.8 Mounting Brain Cultured Slices

Stained slices are transferred to a 10 cm plate with PBS with a brush and carefully mounted pushing them carefully through the slide that is partly submerged. Once all brain slices are positioned and the surrounding area of each slice is dry, Mowiol-NPG antifade reagent is added and the slide mounted.

3.9 Imaging of Stained Brain Organotypic Cultures

Immunofluorescence images of brain slices are acquired with a Leica SP5 upright confocal microscope ×10, ×20, ×40, and ×63 objectives and analyzed with ImageJ software (Fig. 6.2d–g).

3.10 qRT-PCR and WB on Organotypic Cultures

After whole RNA isolation of brain slices at the endpoint of the culture assays, the qRT-PCR analysis of the genes of interest is performed. Brain slices are mechanically lysed in RLT buffer with β-mercaptoethanol (1:100) and centrifuged at full speed for 3 min in QIAshredder columns. Homogenates are processed following manufacturer's indications for whole RNA isolation. Three brain slices in 500 μl of RLT buffer yield between 300 and 400 ng of RNA. Human versus mouse cells can be differentiated using specific primers.

Protein lysates of brain slices are obtained by mechanical homogenization of brain tissue in commercial RIPA buffer + protease inhibitors (PMSF 1 mM, NaF 1 mM, and Na_3VO_4 1 mM). Samples are maintained on ice for 30 min, vortexing every 10 min. Samples are centrifuged at 14,000 g at 4 °C for 10 min, and supernatants are collected for protein quantification. Three brain slices in 500 μl of lysis buffer yield between 500 and 1000 μg of protein.

4 Troubleshooting and Frequent Tips

- To solve the agarose in HBSS, heat the solution until boiling in a microwave. Do these three times until you observe that the solution is transparent. Do not start with preheated HBSS as this will facilitate the formation of aggregates. Place the agarose at 42 °C for at least 30 min before adding the brain.

- Brains must be carefully moved in all directions with a 20 µl tip when embedded in agarose for full covering of brain surface with the dense solution. This step is critical since it will avoid detaching of the brain from the agarose during sectioning. However, in case the embedded brain detaches from the agarose during sectioning, it is better to stop and start over.

- Maintain the brain in the desired position with the tip until agarose becomes slightly solid. This will avoid movement of the brain and will allow sectioning of the desired area.

- Cells must be carefully placed on top of brain slices. Plates must be manipulated and transported very smoothly, avoiding sudden movements, to maintain cell drop on the brain slice surface until cells penetrate the brain slice 16–24 h later.

- Minimizing manipulation and transportation of plates is recommended. Membranes and/or brain slices sink into the media upon frequent movement, and viability of brain slices is compromised.

- Replacement of media after BLI is recommended. Subsequent acquisition values of the same samples might be altered by the presence of residual D-luciferin.

- Brain slices must be completely submerged in PFA 4% for correct fixation.

- Avoid touching brain slices with tips during free-floating immunofluorescence. Tissue may easily break down.

- Mount brain slices with a brush by careful manipulation. Avoid spreading of creased tissue.

- Apply enough mounting media and let coverslip attach by capillarity. Avoid pressing the tissue to remove bubbles.

5   Applications of Organotypic Cultures in Brain Metastasis Research

5.1 Organotypic Cultures to Study Response to Therapy

The use of organotypic brain cultures for evaluating the efficacy of potential therapeutic agents for brain metastasis has been previously reported [4, 5]. We reported that silibinin, an inhibitor of activation of the transcription factor STAT3, efficiently reduces cancer cell growth in organotypic brain cultures with established macrometastases previously grown in vivo [5]. These findings were then validated in vivo, proving organotypic brain cultures as a reliable tool to evaluate efficacy of potential therapeutic agents and to generate relevant preliminary results that can be later validated in vivo. This approach is not only restricted to the use of chemical compounds but can be also extended to other therapeutic agents such as blocking antibodies [4, 5]. In this regard, anti-CD8 blocking antibody in organotypic brain cultures has been used for the mechanistic dissection of the immunosuppressive and pro-metastatic role of pSTAT3+ reactive astrocytes in brain metastasis [5]. Another example would be the use of anti-FasL blocking antibody in brain slices for obtaining mechanistic insights of the role of neuroserpin in early stages of brain colonization by disseminated cancer cells [4]. Altogether, evidence supports the use of organotypic brain cultures as a valid approach for both drug efficacy assays and functional studies, providing a potent complementary approach to those methodologies already available for drug discovery and basic research.

5.2 Organotypic Cultures to Study the Tumor-Associated Microenvironment

Among many advantages of using organotypic brain cultures for brain metastasis research, the possibility of evaluating the interaction of the tumor-associated brain microenvironment with the metastatic cells represents the major difference compared to in vitro experimental models.

Organotypic brain cultures outperform in vitro models when used for drug efficacy assessment. The ex vivo approach permits simultaneous evaluation of the contribution of each entity (cancer cells or microenvironment) to the pharmacological phenotype of interest, allowing at the same time mechanistic dissection of the observed phenotype. This can be scored by combining genetic silencing of the targeted gene together with the effect of the inhibitor. In this regard, the study by Priego and colleagues has nicely shown that organotypic brain cultures allow evaluation of drugs that target the pro-metastatic microenvironment in brain metastasis, specifically pSTAT3+ reactive astrocytes, instead of focusing only cancer cells [5]. For this purpose, a GEMM with a tamoxifen-inducible knockout of *Stat3* in reactive astrocytes and cancer cells that incorporate shRNA against the *STAT3* gene were used. Comparing silencing of the gene in cancer cells or the astrocytes versus the effect of the inhibitor (which targets both cancer cells and the microenvironment), authors were able to define a novel pro-metastatic role for pSTAT3+ reactive astrocytes in brain metastasis [5].

Increasing the knowledge about the mechanisms that metastatic cells use for interacting with and thus adapting to the brain microenvironment allows designing novel therapeutic options for patients with brain metastasis. In this regard, organotypic brain cultures can be also used to study physical interactions between metastatic cells and different components of the brain [4, 14, 17]. For this purpose, brain slices are cultured either with tumor plugs [17] or seeding cancer cells on top of them [4, 14] in order to study the brain parenchyma-metastasis interface. In this sense, organotypic brain cultures have provided a useful tool to visualize physical interactions between brain metastatic cells and endothelial cells that occur at early stages of brain colonization (Fig. 6.2d) by cancer cells, a process named vascular co-option. More importantly, organotypic brain cultures have also allowed functional validation of key molecules required for this process, such as the axon pathfinding molecule L1CAM [4, 14]. Since studying early stages of brain metastasis at the single cell level is complex and time-consuming, the fact that organotypic brain cultures recapitulate faithfully both at the phenotypic and at the transcriptomic level these processes [14] strongly supports this ex vivo approach as a valid and reliable tool for basic research on brain metastasis. Hence, organotypic brain cultures are able to provide results that could help to better design a subsequent necessary in vivo validation using mouse models.

5.3 Patient-Derived Organotypic Cultures

The use of organotypic cultures is not limited to research with experimental models, but they can be also utilized for performing patient-derived organotypic cultures (PDOCs) using surgically resected fresh human samples from a variety of cancers such as breast, lung, prostate, colon, head and neck, gastric, and esophageal [18–25] (Fig. 6.1d). PDOCs derived from brain tumors have previously been reported for exploring susceptibilities of glioblastoma tumors to different therapies [26]. More recently, the potential of PDOCs in brain metastasis has been demonstrated [27]. PDOCs allow rapid drug efficacy assessment of multiple therapies directly on the human sample, which maintains intact its pathological features and intra-tumor heterogeneity. Real-time results obtained from these assays might be used for clinical advice, which will validate PDOCs as valuable therapeutic tool for cancer precision medicine [28].

6 Conclusions

Experimental models extensively used for brain metastasis research include in vitro and in vivo models. However, in vitro models do not recapitulate entirely the complexity of the metastatic disease, and in vivo models present economical limitations for medium- and high-throughput-based applications. We describe here the use of organotypic brain cultures ex vivo for brain metastasis research, as they overcome main limitations imposed by the two previous models. Applications of this approach include drug screening, mechanistic dissection of different stages of brain colonization by metastatic cells, evaluation of different components of the tumor-associated brain microenvironment, and patient-derived organotypic cultures for precision medicine. Ex vivo cultures allow studying relevant molecular mediators and signaling pathways by both pharmacological and genetic approaches, and results have been reported to be validated in vivo and in human samples.

However, organotypic brain cultures do present limitations as an experimental model, since they do not consider steps of the metastatic disease that occur before extravasation of cancer cells from the vessels. In this regard, they do not allow interrogating the ability of metastatic cells to intravasate into capillaries from the primary tumor site, to survive in circulation, and, more importantly, to cross the blood-brain barrier (BBB). In addition, these cultures cannot be maintained over long periods of time in comparison with other 3D cultures such organoids. In conclusion, organotypic brain cultures constitute a useful, convenient, and advantageous experimental model for brain metastasis research and provide a rapid, easy, and cost-effective tool for basic and clinical applications.

Acknowledgments

We thank members of Brain Metastasis Group for critical discussion. Research in the Brain Metastasis Group is supported by MINECO-Retos SAF2017-89643-R (M.V.), Cancer Research Institute CLIP Award 2018 (M.V.), AECC (GCTRA16015SEOA) (M.V.), Bristol-Myers Squibb Melanoma Research Alliance Young Investigator Award 2017 (M.V.), Beug Foundation's Prize for Metastasis Research 2017 (M.V.), Worldwide Cancer Research (19-0177) (M.V.), H2020-FETOPEN (828972) (M.V.), Fundación Ramón Areces (CIVP19S8163), and La Caixa-Severo Ochoa International PhD Program Fellowship (L.Z.). M.V. is a Ramón y Cajal Investigator (RYC-2013-13365) and an EMBO YIP investigator.

References

1. Valiente M, Ahluwalia MS, Boire A et al (2018) The evolving landscape of brain metastasis. Trends Cancer 4:176–196. https://doi.org/10.1016/j.trecan.2018.01.003

2. Bos PD, Zhang XH-F, Nadal C et al (2009) Genes that mediate breast cancer metastasis to the brain. Nature 459:1005–1009. https://doi.org/10.1038/nature08021

3. Nguyen DX, Chiang AC, Zhang XH-F et al (2009) WNT/TCF signaling through LEF1 and HOXB9 mediates lung adenocarcinoma metastasis. Cell 138:51–62. https://doi.org/10.1016/j.cell.2009.04.030

4. Valiente M, Obenauf AC, Jin X et al (2014) Serpins promote cancer cell survival and vascular co-option in brain metastasis. Cell 156:1002–1016. https://doi.org/10.1016/j.cell.2014.01.040

5. Priego N, Zhu L, Monteiro C et al (2018) STAT3 labels a subpopulation of reactive astrocytes required for brain metastasis. Nat Med 24:1024–1035. https://doi.org/10.1038/s41591-018-0044-4

6. Malladi S, Macalinao DG, Jin X et al (2016) Metastatic latency and immune evasion through Autocrine inhibition of WNT. Cell 165:45–60. https://doi.org/10.1016/j.cell.2016.02.025

7. Kato M, Takahashi M, Akhand AA et al (1998) Transgenic mouse model for skin malignant melanoma. Oncogene 17:1885–1888. https://doi.org/10.1038/sj.onc.1202077

8. Meuwissen R, Linn SC, Linnoila RI et al (2003) Induction of small cell lung cancer by somatic inactivation of both Trp53 and Rb1 in a conditional mouse model. Cancer Cell 4:181–189. https://doi.org/10.1016/S1535-6108(03)00220-4

9. Cho JH, Robinson JP, Arave RA et al (2015) AKT1 activation promotes development of melanoma metastases. Cell Rep 13:898–905. https://doi.org/10.1016/j.celrep.2015.09.057

10. Lee HW, Lee J-I, Lee SJ et al (2015) Patient-derived xenografts from non-small cell lung cancer brain metastases are valuable translational platforms for the development of personalized targeted therapy. Clin Cancer Res 21:1172–1182. https://doi.org/10.1158/1078-0432.CCR-14-1589

11. Ni J, Ramkissoon SH, Xie S et al (2016) Combination inhibition of PI3K and mTORC1 yields durable remissions in mice bearing orthotopic patient-derived xenografts of HER2-positive breast cancer brain metastases. Nat Med 22:723–726. https://doi.org/10.1038/nm.4120

12. Contreras-Zárate MJ, Ormond DR, Gillen AE et al (2017) Development of novel patient-derived xenografts from breast cancer brain metastases. Front Oncol 7:252. https://doi.org/10.3389/fonc.2017.00252

13. Krepler C, Sproesser K, Brafford P et al (2017) A comprehensive patient-derived xenograft collection representing the heterogeneity of melanoma. Cell Rep 21:1953–1967. https://doi.org/10.1016/j.celrep.2017.10.021

14. Er EE, Valiente M, Ganesh K et al (2018) Pericyte-like spreading by disseminated cancer cells activates YAP and MRTF for metastatic colonization. Nat Cell Biol 20:966–978. https://doi.org/10.1038/s41556-018-0138-8

15. Chen Q, Boire A, Jin X et al (2016) Carcinoma-astrocyte gap junctions promote brain metastasis by cGAMP transfer. Nature 533:493–498. https://doi.org/10.1038/nature18268

16. Polleux F, Ghosh A (2002) The slice overlay assay: a versatile tool to study the influence of extracellular signals on neuronal development. Sci STKE 2002:pl9. https://doi.org/10.1126/stke.2002.136.pl9

17. Blazquez R, Pukrop T (2017) 3D Coculture model of the brain parenchyma-metastasis Interface of brain metastasis. Methods Mol Biol 1612:213–222. https://doi.org/10.1007/978-1-4939-7021-6_16

18. van der Kuip H, Mürdter TE, Sonnenberg M et al (2006) Short term culture of breast cancer tissues to study the activity of the anti-cancer drug taxol in an intact tumor environment. BMC Cancer 6:86. https://doi.org/10.1186/1471-2407-6-86

19. Davies EJ, Dong M, Gutekunst M et al (2015) Capturing complex tumour biology in vitro: histological and molecular characterisation of precision cut slices. Sci Rep 5:17187. https://doi.org/10.1038/srep17187

20. Vaira V, Fedele G, Pyne S et al (2010) Preclinical model of organotypic culture for pharmacodynamic profiling of human tumors. Proc Natl Acad Sci U S A 107:8352–8356. https://doi.org/10.1073/pnas.0907676107

21. Holliday DL, Moss MA, Pollock S et al (2013) The practicalities of using tissue slices as preclinical organotypic breast cancer models. J Clin Pathol 66:253–255. https://doi.org/10.1136/jclinpath-2012-201147

22. Gerlach MM, Merz F, Wichmann G et al (2014) Slice cultures from head and neck squamous cell carcinoma: a novel test system for drug susceptibility and mechanisms of resistance. Br J Cancer 110:479–488. https://doi.org/10.1038/bjc.2013.700

23. Carranza-Torres IE, Guzmán-Delgado NE, Coronado-Martínez C et al (2015) Organotypic culture of breast tumor explants as a multicellular system for the screening of natural compounds with antineoplastic potential. Biomed Res Int 2015:618021. https://doi.org/10.1155/2015/618021

24. Koerfer J, Kallendrusch S, Merz F et al (2016) Organotypic slice cultures of human gastric and esophagogastric junction cancer. Cancer Med 5:1444–1453. https://doi.org/10.1002/cam4.720

25. Naipal KAT, Verkaik NS, Sánchez H et al (2015) Tumor slice culture system to assess drug response of primary breast cancer. BMC Cancer 16:78. https://doi.org/10.1186/s12885-016-2119-2

26. Merz F, Gaunitz F, Dehghani F et al (2013) Organotypic slice cultures of human glioblastoma reveal different susceptibilities to treatments. Neuro-Oncology 15:670–681. https://doi.org/10.1093/neuonc/not003

27. Varešlija D, Priedigkeit N, Fagan A et al (2018) Transcriptome characterization of matched primary breast and brain metastatic tumors to detect novel actionable targets. J Natl Cancer Inst. https://doi.org/10.1093/jnci/djy110

28. Meijer TG, Naipal KA, Jager A, van Gent DC (2017) Ex vivo tumor culture systems for functional drug testing and therapy response prediction. Future Sci OA 3:FSO190. https://doi.org/10.4155/fsoa-2017-0003

Chapter 7

Human Glioblastoma Organoids to Model Brain Tumor Heterogeneity Ex Vivo

W. Dean Pontius, Lisa C. Wallace, Katrina Fife, and Christopher G. Hubert

Abstract

Intratumoral cellular heterogeneity is a clinically important aspect of most cancers. Diffusion limitations in solid tumors create oxygen and nutrient gradients that support different cellular phenotypes within distinct tumor regions. Until recently, methods for culturing glioblastoma (GBM) cells have largely ignored this aspect of the disease biology. We have developed a method for three-dimensional culture of GBM that recapitulates regional microenvironmental gradients ex vivo. These GBM organoids mimic the architecture of patient tumors and contain phenotypically diverse cell populations that respond differently to perturbation. The goal of this chapter is to provide readers with detailed methods for generating and applying GBM organoids. In addition, we review the past, present, and future of GBM research to frame the utility of this technology in the broader context of the field.

Key words Brain tumors, Organoids, Glioblastoma, GBM, Ex vivo models, Cancer stem cells, Heterogeneity, Microenvironment, Hypoxia

1 Introduction

1.1 Glioblastoma and Its Cellular Composition

Glioblastoma (GBM) is the most common type of primary brain cancer, with a dismal patient prognosis of 14–16 months [1]. Despite the pressing clinical problem GBM poses, the standard of care has marginal impact on patient survival. Poor patient response can be attributed to a number of factors, including cellular diversity and microenvironmental heterogeneity in GBM. The existence of tumor-initiating cancer stem cells (CSCs) that can self-renew and differentiate into all cancer cell types within a GBM tumor is becoming increasingly supported, despite controversy surrounding their exact molecular markers.

The GBM microenvironment is composed of a variety of non-tumor immune, neural, and glial cells as well as other tumor cells. These cell types create different stem-cell niches critical to both neural stem cell and CSC behavior. Patient tumors contain CSCs

Giorgio Seano (ed.), *Brain Tumors*, Neuromethods, vol. 158,
https://doi.org/10.1007/978-1-0716-0856-2_7, © Springer Science+Business Media, LLC, part of Springer Nature 2021

with varying phenotypes, dictated by the specific niche in which they are located. One population of highly proliferative CSCs resides in the oxygen- and nutrient-rich perivascular niche [2], while a second population of CSCs is present in the hypoxic niche located in the necrotic and avascular tumor core [3]. This hypoxic niche, while depleted of oxygen and nutrients, contains a population of quiescent CSCs with a slow turnover rate and different biological function from that found in the perivascular niche [4, 5].

While CSCs are heavily impacted by microenvironmental cues present in each niche, they are also able to modify their respective niches in a reciprocal fashion. CSCs secrete factors that induce angiogenesis [6] and immune suppression [7]. They also signal to neighboring differentiated GBM cells to promote tumor growth through mutual paracrine signaling via the BDNF-NRTK2 axis [8]. CSCs can even transdifferentiate into non-cancer-like cellular constituents of the tumor microenvironment, such as vascular pericytes that maintain tumor vessels and support tumor growth [9].

1.2 Methods for Maintaining Glioblastoma Stem Cells for Laboratory Investigation

Traditionally, the majority of cancer cell cultures, including long-time established GBM cell lines, have been grown in vitro with standard cell culture media such as DMEM containing 10% serum. Serum containing media, however, has shown to promote differentiation of GBM cells isolated from patients, causing phenotypic and genotypic divergence from the tumor of origin [10]. In fact, GBM cells cultured in serum containing media more closely resemble other serum cultured cancer cell lines at the transcriptional level than they do patient GBM tumors [11]. These findings have led to the adaption of methods from developmental neurobiology to study glioblastoma stem cells and underscore the profound influence that culture methods may have upon our understanding or misunderstanding of disease biology.

The current standard for culturing GBM CSCs is using neural stem cell media containing epidermal growth factor (EGF) and basic fibroblast growth factor (bFGF). CSCs can be grown in this media either as tumor spheres [10] or adherently on laminin coated plates [12], without any three-dimensional matrix structure. In these conditions, the multi-lineage potential and self-renewal of CSCs is preserved, as seen by phenotypic assays and marked by expression of neural stem cell markers such as SOX2 and nestin. These systems have been used to elucidate many important and clinically relevant aspects of glioblastoma stem cell biology. Such phenomena include the contribution of CSCs to tumorigenesis, resistance to standard of care therapy, tumor microenvironment remodeling, and more.

Due to their ease of culture and relative cellular homogeneity, CSCs enable high-throughput studies and accurate interpretation of bulk molecular and genetic analysis. Additionally, these cultures can be dissociated and orthotopically injected into immunocompromised mice to evaluate glioblastoma biology in the context of the cerebral microenvironment. While these widespread experimental systems have many advantages, they enrich primarily for rapidly growing CSCs and do not recapitulate the cellular and microenvironmental heterogeneity crucial to GBM biology within patient tumors.

1.3 Impact of Heterogeneity in Response to Treatment

The standard of care for GBM is surgical resection followed by concurrent temozolomide (TMZ) and radiotherapy [1]. Both TMZ and radiotherapy are treatment modalities with little targeted mechanism; they work on proliferative cells by inducing DNA damage in normal or cancerous tissue. However, there are a few problems with this approach that limit maximal efficacy in the clinic. First, many cancer cells within the tumor may not be actively proliferating. Second, CSCs have upregulated DNA damage response machinery [6]. Approaches targeting distinct subpopulations of CSCs residing in different niches have shown success in preclinical models; however these combination therapies have yet to be clinically translated [5]. GBM can acquire a phenotype of therapy resistance through cell-intrinsic mechanisms, but also through cell-extrinsic mechanisms where resistance is conferred by normal cells within the microenvironment. For example, paracrine signaling between reactive astrocytes and GBM through endothelin receptor promotes resistance to TMZ [13]. Similarly, the microenvironment itself can confer protection against therapeutic modalities. For example, radiation is less effective in hypoxic tumor regions that are unable to produce cell-damaging reactive oxygen species [14]. In addition to the *intra*tumoral heterogeneity present within GBM tumors, there is also *inter*tumoral heterogeneity that results in markedly different patient responses to treatment [15]. The intertumoral heterogeneity is a result of different genetic and epigenetic drivers of disease progression. Variation between tumors is exacerbated by each patient's unique genetic and physiologic background, treatment experience, and tumor evolution.

1.4 Organoids as "Mini-Tumor" Models of Cancer

The popularity of three-dimensional cell culture models has increased in recent years. There are many methods for growing cells and cancer cells in three dimensions, each with their own advantages and disadvantages. Organoids are a promising method for incorporating microenvironmental factors into in vitro cell culture in a variety of different cancers (e.g., glioblastoma, colorectal cancer, bladder cancer, breast cancer) and developmental contexts [16–19]. Organoids permit long-term co-culture of heterogeneous populations of cells, some of which would normally be out competed by faster growing populations after multiple passages in vitro. This co-culture maintains physiologically relevant cell-cell contacts that are absent from conventional 2D-culture methods, but present in vivo. Patient-derived organoids also mimic the morphological, genetic, and transcriptomic heterogeneity present within primary tumors [16]. Preservation of tumor complexity could open the door for researchers to explore new avenues of precision medicine.

To permit long-term propagation of patient samples, researchers have turned to organoids to generate "Living Biobanks" that retain tumor heterogeneity ex vivo better than establishing cell lines [16, 18]. There is increasing interest in capturing the spectrum of genetic and phenotypic heterogeneity present in patients who present with specific cancers. However, the majority of biobanking initiatives rely on cryopreserving patient samples for later characterization and contribution of clinical information to laboratory findings on a disease. Recent work has focused on the use of patient-derived organoids as an in vitro system that recapitulates characteristics of patient tumors. This allows for both preservation and perturbation of patient samples in vitro, as well as the ability to draw correlations between response and genetic and transcriptomic information.

Organoid banks can be used to mitigate certain challenges that can arise from patient-derived cell lines or xenografts. Despite their comparative ease of propagation, patient-derived cell lines quickly select for the populations of cells from the original tumor that are most fit to survive in vitro [10, 20]. While patient-derived xenografts capture GBM cellular heterogeneity and incorporate aspects of the original tumor microenvironment, they are laborious and expensive to generate and maintain and have been shown to suffer the same clonal selection after serial implantation [21]. Patient-derived organoids have been used in colon cancer research to characterize potential biological drivers of disease progression [16]. In addition, the introduction of known driver mutations or tumorigenic genetic aberrations has been used in normal developmental organoids to track the earliest stages of malignant transformation [22, 23]. However, until recently there was no way to generate organoids from GBM patient tissue, preventing the benefits of organoid research from being translated to the study of GBM.

1.5 Glioblastoma Organoids as a Complementary Model to Current In Vitro and In Vivo Systems

Parallel generation of diseased brain organoids and normal developmental cerebral organoids from the same patient have provided insight into the etiology of various neurodevelopmental disorders [24–26]. We developed a method for generating GBM organoids from primary patient samples or patient-derived cell cultures based on previous cerebral organoid protocols (Fig. 7.1) [17, 27]. These organoids, once mature and populated to high cell density, are large enough to compel cellular competition for resources and subsequent cellular adaptation, allowing spatial heterogeneity and physiologic niches to develop over time within the three-

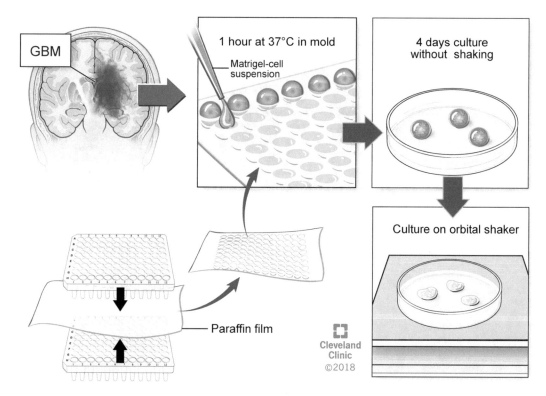

Fig. 7.1 Workflow for generating glioblastoma organoids from patient specimens. Dissociated or finely minced glioblastoma tissue, GBM tumor sphere cultures, or dissociated single GBM cells are suspended in 80% Matrigel at day 0. The Matrigel-cell suspension is added to pressed parafilm molds to form pearl-shaped structures. These structures polymerize at 37 °C for 1 h, after which they are unmolded and cultured in Neurobasal complete medium for 4 days without shaking. After 4 days, the nascent organoids are placed on an orbital shaker in a 37 °C incubator for the remainder of their time in period

dimensional structures. The two main niches present within GBM organoids are a hypoxic niche at the core of the structure and a perivascular-like proliferative niche at the outer rim (Fig. 7.2). These niches contain populations of cells with phenotypes similar to cells of analogous regions in patient tumors. Radiation-resistant, quiescent stem cells reside at the core of the organoid, while highly proliferative, comparatively radiation-sensitive cells can be found at the highly oxygenated perimeter (Fig. 7.2) [17, 28].

We demonstrated that the diffuse and invasive morphology of a patient GBM was maintained when orthotopically xenografted in mice after that sample was grown as an organoid, but the same patient sample grown in parallel as sphere culture formed nodular noninvasive masses upon xenograft. Lastly, the same GBM organoid protocol described previously can be used to study brain metastatic lesions of other cancer types that are also biologically influenced by the cerebral microenvironment [17].

Although more laborious, time-consuming to expand, and lower throughput than monolayer or sphere culture, GBM organ-

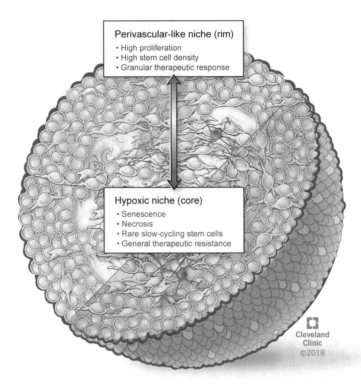

Fig. 7.2 Cross section of a glioblastoma organoid demonstrating regional micro-environmental and cellular heterogeneity. The three-dimensional structure and large size of glioblastoma organoids allow for the formation of oxygen and nutrient gradients. Over time, these microenvironmental gradients create niches that promote the growth of glioblastoma cells with diverse phenotypes. In a high oxygen/nutrient environment, such as the perivascular-like niche at the rim of the organoid, glioblastoma stem cells (blue) are highly proliferative and sensitive to standard therapeutic intervention. In low oxygen/nutrient environments, such as the hypoxic niche at the core of the organoid, cells are generally senescent and necrotic (yellow). A slow cycling, therapy-resistant stem cell population distinct from those in the perivascular-like niche is supported by the hypoxic period

oids may serve to bridge the gap between current in vitro and in vivo systems. While each system has its own benefits and drawbacks depending on the biological question at hand (Fig. 7.3), cancer organoids have established themselves as invaluable tools in the arsenal of cancer biologists, with the list of their applications continuing to grow.

2   Materials

All equipment and reagents should be handled in sterile conditions. Reagents should be stored at 4 °C unless noted otherwise.

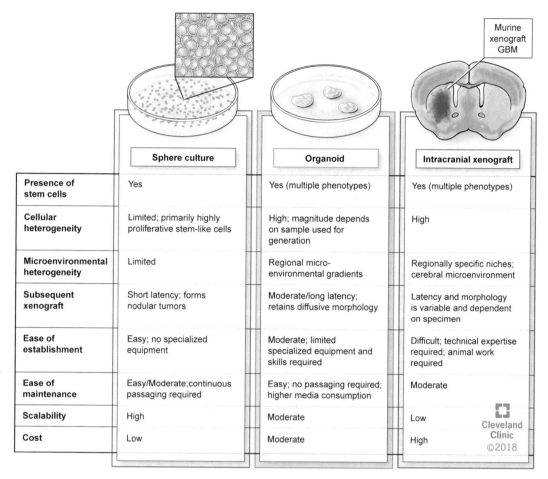

	Sphere culture	Organoid	Intracranial xenograft
Presence of stem cells	Yes	Yes (multiple phenotypes)	Yes (multiple phenotypes)
Cellular heterogeneity	Limited; primarily highly proliferative stem-like cells	High; magnitude depends on sample used for generation	High
Microenvironmental heterogeneity	Limited	Regional micro-environmental gradients	Regionally specific niches; cerebral microenvironment
Subsequent xenograft	Short latency; forms nodular tumors	Moderate/long latency; retains diffusive morphology	Latency and morphology is variable and dependent on specimen
Ease of establishment	Easy; no specialized equipment	Moderate; limited specialized equipment and skills required	Difficult; technical expertise required; animal work required
Ease of maintenance	Easy/Moderate;continuous passaging required	Easy; no passaging required; higher media consumption	Moderate
Scalability	High	Moderate	Low
Cost	Low	Moderate	High

Fig. 7.3 Comparison of systems for propagating glioblastoma specimens in a laboratory setting. Standard methods for studying glioblastoma in a laboratory setting include sphere culture and intracranial xenograft. Depending on the investigator's question, each system has advantages and disadvantages. Organoids may be a valuable model to bridge the gap between sphere culture and intracranial xenografts, since they retain aspects of the patient tumor lost in standard in vitro models while remaining simpler, more controllable, and less expensive to generate and maintain than in vivo models

Freeze-thaws of frozen reagents should be minimized by making aliquots of smaller volume.

To move GBM organoids, create wide-orifice p1000 tips by using a sterile razor blade to cut off the end of standard tips. If the orifice is too large, the surface tension created will be insufficient to hold the fluid and organoid. If the orifice is too small, there is a risk of breaking the organoid as it is aspirated.

All GBM specimens should be cultured in GBM stem cell conditions using Neurobasal complete medium as described below:

1. Neurobasal medium minus phenol red (ThermoFisher 12349015).

2. EGF (store at −20 °C) (final conc. 10 ng/mL) (R&D 236-EG).

3. bFGF (store at −20 °C) (final conc. 10 ng/mL) (R&D 4114-TC),

4. B27 minus vitamin A (final conc. 1X) (ThermoFisher 12587001).

5. Glutamine (final conc. 2 mM) (ThermoFisher 35050061).

6. Sodium pyruvate (final conc. 1 mM) (ThermoFisher 11360070).

7. Antibiotic-antimycotic (final conc. 1X) (Gibco 15240062).

This protocol describes methods for tumor sphere culture of GBM CSCs.

2.1 Macrodissection of Patient Tissue

1. Razor blade (sterile).

2. Red blood cell (RBC) lysis buffer (sterile filtered or autoclaved).

   (a) Distilled H_2O.

   (b) NH_4Cl (final concentration 155 mM).

   (c) $KHCO_3$ (final concentration 10 mM).

   (d) EDTA (final concentration 0.1 mM) (Note 1).

2.2 Stem Cell Culture of GBM Cell Lines and Primary Specimens

1. Neurobasal complete medium.

2. Petri dish/nonadherent cell culture plate.

2.3 Generation of Organoids from GBM CSCs and Patient Samples

1. Accutase (store at −20 °C).

2. Parafilm.

3. Matrigel.

4. Neurobasal complete medium.

5. Blunt forceps.

6. 96-well PCR plate

7. Cell culture/petri dish.

8. Dry ice.

9. Wet Ice.

2.4 Growth of GBM Organoids

1. Neurobasal complete medium.

2. Cell culture plate or dish (10 cm dish, 15 cm dish, 6-well plate, or 12-well plate).

3. Orbital shaker in culture environment (Note 2).

2.5 Dissociation of GBM Organoids for Subsequent Xenograft

1. Miltenyi Tumor Dissociation Kit and optional gentleMACS™ Dissociator.

2. Accutase (Millipore).

3. Razor blade.

2.6 Organoid Fixation

1. 4% paraformaldehyde.

2.7 Cryosectioning

1. 1XPBS.

2. 30% sucrose in distilled H_2O.

3. Tissue-Plus™ OCT compound (Fisher).

4. 100% isopentane.

5. Dry ice.

6. Peel-A-Way® disposable histology molds (Thermo).

2.8 Paraffin Embedding

1. Paraffin wax.

2. Base molds.

3. Histology cassettes.

4. 100% ethanol.

5. Clear-Rite 3™ clearing reagent.

2.9 Immunofluorescence

1. Blocking solution (1XPBS, 10% serum according to antibody host, and 10% Triton X-100).

2. 1XPBS.

3. PAP pen.

4. Primary antibody.

5. Secondary antibody.

6. DNA stain (DAPI, DRAQ5).

7. FluorSave™ reagent.

8. Microscope slides.

9. Coverslips.

2.10 Immunohisto-chemistry

1. Xylene.

2. Ethanol.
   (a) 50%.
   (b) 75%.
   (c) 95%.
   (d) 100%.

3. 1X Tris Buffered Saline with Tween (TBST).

4. Slide staining dish.

5. CST-SignalStain® citrate unmasking solution.

6. Hydrogen peroxide.

7. Blocking solution.

(a) 1XPBS, 10% serum according to animal host of secondary antibody.

(b) Commercial blocking solution (i.e., Background Sniper).

8. PAP pen.

9. Primary antibody.

10. Secondary antibody.

11. VECTASTAIN ABC reagent.

12. SignalStain® DAB Diluent.

13. SignalStain® DAB Chromogen Concentrate.

14. Gills 2 Hematoxylin.

15. Signature Series Clarifier™ 2.

16. Shandon™ Bluing Reagent.

17. Clear-Rite 3™ clearing reagent (Note 3).

18. Mounting medium.

19. Microscope slides.

20. Coverslips.

3   Methods

See Fig. 7.1 for a schematic diagram of the organoid generation process. All steps should be performed in sterile conditions in a sterile biosafety cabinet. Cells are cultured in a cell culture incubator at 37 °C and 5% CO_2. All media should be warmed to 37 °C before use.

3.1   Formation of Organoid Molds

1. Form organoid molds by taking wax paper off of a clean sheet of parafilm and placing it between two 96-well PCR plates.

   (a) Note 4.

2. Against a flat solid surface, apply even pressure to the top 96-well PCR plate to stretch parafilm and form dimples.

3. Pick up the top 96-well PCR plate with dimpled parafilm stuck to the bottom, and place in the container of dry ice for >30 s.

4. Once parafilm has been placed on dry ice for 30 s, lift the 96-well PCR plate, and immediately pull the parafilm off the bottom with sterile blunt forceps.

   (a) Note 5.

5. Place formed parafilm mold in cell culture/petri dish.

   (a) Note 6.

3.2   Macrodissection of Patient Tissue

1. In a sterile dish (i.e., 10 cm culture plate), finely mince pieces using a razor blade and applying even pressure (Note 7).

2. Transfer finely minced tumor chunks to a 15 mL conical tube, and spin at 200 g for 3 min to pellet tissue.

3. Aspirate supernatant and resuspend tissue in 10 mL RBC lysis buffer (optional but recommended for bloody tissues).

4. Incubate suspension at room temperature for 10 min with gentle shaking every 2–3 min (Notes 8 and 9).

5. Spin down suspension at 200 g for 3 min and aspirate RBC lysis buffer. Resuspend pellet in Neurobasal medium or PBS. Spin down again and aspirate prior to dissociation steps.

6. Dissociate sample using the Miltenyi Tumor Dissociation Kit according to manufacturer's instructions or by suspending in Accutase for 5–10 min with occasional gentle trituration.

7. Spin down suspension at 200 g for 3 min and resuspend pellet in Neurobasal medium.

8. *Optional*: A single-cell suspension can be created before formation of organoids by passing the cell suspension through a 70 μm filter (Notes 10 and 11).

9. Proceed to Sect. 3.4, step 8.

3.3 Stem Cell Culture of GBM Cell Lines and Primary Specimens

The success of organoid formation from patient specimens varies with the specific sample and institutional tissue procurement procedures. Although not all samples will form organoids, ones that are unable to grow as patient-derived xenografts have been used to form organoids. Additionally, specimens that do establish organoid models reproducibly grow in other standard in vitro and in vivo models.

1. CSCs should be cultured as tumor spheres in Neurobasal complete medium in a nonadherent cell culture/petri dish according to standard procedures [10] (Note 12).

3.4 Generation of Organoids from GBM CSC Tumor Spheres or Patient Tissue

1. Transfer GBM tumor spheres to a 15 mL conical tube, spin at 200 g for 5 min, and aspirate supernatant (Note 13).

2. Resuspend GBM tumor spheres in 1 mL of Accutase, and place Accutase-cell suspension in cell culture incubator for 3–5 min. Gently agitate or triturate to assist generation of a single-cell suspension.

3. Spin down Accutase-cell suspension at 200 g for 5 min and aspirate supernatant.

4. Wash cells with 1XPBS. Spin to pellet cells.

5. Resuspend cells in Neurobasal complete medium, and triturate to ensure a single-cell suspension.

6. Count cells and dilute to a concentration of 1000–20,000 cells per 4 μL of Neurobasal complete medium (Note 14).

7. Place cells on ice to cool.

8. Once cell suspension is cold, add 16 µL of ice cold Matrigel for every 4 µL of cell suspension or minced patient tissue, and pipette up and down to mix thoroughly (Notes 15–17).

9. Add 20 µL droplets of cell-Matrigel suspension to the parafilm mold (Note 18).

10. Place the droplets within the parafilm mold in a sterile cell culture/petri dish in a cell culture incubator for 1 h.

11. After organoids have solidified, they can be unmolded and added to a 10 cm cell culture dish with 20 mL Neurobasal complete medium (Note 19).

12. Place 10 cm cell culture dish in incubator without shaking for 4 days and without replacing the medium.

3.5 Growth of GBM Organoids

1. After 4 days of growth without shaking, replace medium with fresh Neurobasal complete medium (Note 20).

2. Place the cell culture dish with the organoids on an orbital shaker and shake at 80 rpm.

3. Organoid media should be changed every 2–3 days.

4. As organoids grow, they can be transferred to separate plates using a cut p1000 tip to create a wide orifice (Note 21).

5. Organoid health can be roughly inferred using a light microscope (Fig. 7.4). Disparate cells embedded within the Matrigel will migrate and proliferate, eventually causing the transparent organoid to become opaque. Dead cells will appear rounded and may be dark (less refractile) and will not show evidence of directionality or migration (Notes 22 and 23).

6. Organoids can be grown under these conditions for extended periods (>1 year). The maximum diameter of an organoid depends on the specific specimen and seems to remain relatively constant after the first few months (Notes 24 and 25).

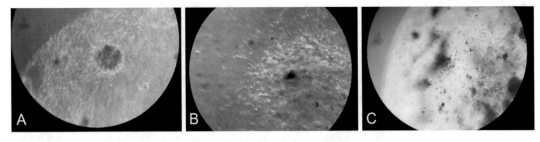

Fig. 7.4 Light microscope images of early organoid growth. Researcher's view through the eyepiece of a light microscope showing (**a, b**) multiclonal outgrowth and (**c**) migration/invasion of single cells throughout Matrigel (center of view) during the early development of successfully established patient organoids. The cellular density of the organoids at these stages is sparse. The cells will continue to expand and colonize the Matrigel as the organoids mature, eventually making it impossible to observe individual cells by light microscopy

3.6 Dissociation of Organoids for Subsequent Xenograft or Organoid Formation

1. Transfer organoids to a clean 10 cm cell culture dish without cell culture medium using a clipped, wide-orifice p1000 tip.

2. Mince organoids using a sterile razor blade, applying even pressure during the chopping motion.

3. Use Neurobasal complete medium to transfer organoid pieces and single cells to a 15 mL conical tube.

4. Spin down suspension at 200 g for 5 min and aspirate supernatant.

5. Resuspend cells in 1 mL of cold Accutase and place on ice for 15 min.

6. After 15 min on ice, place Accutase-cell suspension in 37 °C water for approximately 5–15 min. Cell suspension can be gently triturated or mixed to aid dissociation (Note 26).

7. Dilute Accutase-cell suspension with Neurobasal medium, and spin down at 200 g for 5 min and aspirate supernatant.

8. Wash cells with 1XPBS.

9. Resuspend cells in 1XPBS or Neurobasal (base) medium at desired concentration for orthotopic injection of cells as previously described [29] (Note 27).

10. Organoids can also be dissociated using the Miltenyi Tumor Dissociation Kit according to the manufacturer's protocol for glioblastoma xenografts (Note 28).

3.7 Cryoembedding

1. Transfer organoids to a 1.5 mL tube containing 1 mL 4% PFA using a wide-orifice p1000 tip.

2. Store organoids at 4 °C overnight. If embedding organoids in paraffin, proceed to Sect. 3.8.

3. After overnight fixation in 4% PFA, wash the organoid with 1XPBS three times.

4. Transfer the organoid from 1XPBS to a new 1.5 mL tube with 1 mL 30% sucrose in water. Store at 4 °C overnight.

5. Chill a beaker of 100% isopentane on dry ice.

6. Add a small amount of OCT to a cryomold, covering the bottom, and transfer the organoid to the cryomold (Note 29).

7. Add OCT to cover the organoid and fill the rest of the volume of the cryomold (Note 30).

8. Using long forceps, transfer the cryomold to the chilled isopentane. The OCT will start to freeze, becoming opaque in the process.

9. The blocks can be kept at −20 °C or −80 °C for short- or long-term storage, respectively.

3.8 Paraffin Embedding

1. Sort organoids by size.

    (a) Small organoids are roughly under 3 mm.

    (b) Large organoids are roughly over 3 mm.

2. Using a wide-orifice p1000 tip, transfer each organoid to a histology cassette, and process into paraffin wax according to the following processing schedules.

1.5 h processing schedule (small organoids)	
50% ethanol	3 min
75% ethanol	3 min
95% ethanol	3 min
95% ethanol	4 min
100% ethanol	2 min
100% ethanol	3 min
100% ethanol	4 min
Clear-Rite	2 min
Clear-Rite	3 min
Clear-Rite	4 min
Paraffin wax	15 min
Paraffin wax	15 min

2 h processing schedule (large organoids)	
50% ethanol	6 min
75% ethanol	6 min
95% ethanol	5 min
95% ethanol	8 min
100% ethanol	5 min
100% ethanol	5 min
100% ethanol	8 min
Clear-Rite	5 min
Clear-Rite	5 min
Clear-Rite	8 min
Paraffin wax	30 min
Paraffin wax	30 min

3. Transfer the organoids from the cassettes to metal embedding molds, and chill the bottom until the wax turns semisolid.

4. After the wax has partially frozen, add additional to the top of the mold.

5. Place the cassette top corresponding to each sample over the metal mold, and transfer to a cold plate to continue freezing until the wax is completely solid.

6. The block is finished and ready to be sectioned when the wax has completely solidified and is able to be removed from the mold. The embedded tissue can also be stored at room temperature or 4 °C.

3.9 Immunofluorescence (IF)

Commonly used markers for IF include SOX2 and OLIG2 for monitoring CSC localization and cleaved caspase 3 for measuring apoptosis after therapeutic treatment. See Table 7.1 for information on specific antibodies.

1. Unmask OCT sections by gently shaking in 1XPBS for 30 min at room temperature.

2. Remove slides from 1XPBS, and dry around the tissue sections using a folded Kimwipe (Note 31).

3. When the slide is sufficiently dry, circle the tissue sections with a PAP pen (Note 32).

4. Add enough blocking solution to cover the tissue section, and incubate at room temperature for 90 min.

5. Dilute primary antibody to desired concentration in blocking solution.

6. Once the tissue section has been blocked for 90 min, gently aspirate the blocking solution using a pipette.

7. Add 300–500 μL primary antibody solution to tissue section (Notes 33 and 34).

8. Incubate for 1 h at room temperature.

9. After 1 h incubation, wash the slide in 1XPBS for 10 min at room temperature.

10. Dilute secondary antibody (1:1000) in blocking solution to desired concentration.

11. Add 300–500 μL of secondary antibody mix to the tissue section, and incubate for 45 min at room temperature.

12. After 45 min incubation, wash the slide in 1XPBS for 10 min.

13. Stain the tissue for DNA by incubating for 10 min in:

    (a) 1XPBS + DAPI (1:10,000).

    (b) 1XPBS + DRAQ5 (1:1000).

14. Wash for 10 min with 1XPBS.

Table 7.1
Table of antibodies for glioblastoma organoid staining applications

Antigen	Application	Host	Supplier	Catalog no.	Dilution
Cleaved caspase 3	IHC/IF	Rabbit	Cell Signaling Technologies	9664S	1:500/1:400
Human nestin	IHC	Mouse	Millipore	ABD69	1:1000
Ki67	IHC	Rabbit	Abcam	Ab15580	1:200
Living Colors mCherry	IHC	Mouse	Clontech	632543	1:500
Olig2	IF	Rabbit	Millipore	AB9610	1:500
Phospho-histone H3	IHC	Rabbit	Cell Signaling Technologies	9701S	1:200
Sox2	IHC/IF	Goat	R&D	AF2018	1:250/1:200
tRFP	IHC	Rabbit	Evrogen	AB233	1:1000

15. Mount the coverslip with FluorSave. After the FluorSave has dried, image immediately or store the slide at 4 °C protected from light (Note 35).

3.10 Immunohisto-chemistry

See Fig. 7.5 for examples.

1. Place slides in slide staining dishes, making sure not to accidentally damage the tissue.

2. Deparaffinize and hydrate the tissue sections according to the following steps:

    (a) Incubate sections in xylene for 5 min (three times).

    (b) Incubate sections in 100% ethanol for 10 min (two times).

    (c) Incubate sections in 95% ethanol for 10 min (two times).

    (d) Wash sections in distilled H_2O for 5 min (two times).

3. Unmask antigens by submerging the slides in 1X citrate unmasking solution (diluted in distilled H_2O) and heating in a microwave until boiling. Incubate slides at sub-boiling temperature until the total time heated is 10 min (Notes 36 and 37).

4. After heating for 10 min, let the slides cool in the same 1X citrate unmasking solution at room temperature for 30 min.

5. Wash slides in distilled H_2O for 5 min (three times).

6. Incubate slides in 3% hydrogen peroxide for 5 min.

7. Wash slides in distilled H_2O for 5 min (two times).

8. Wash slides in 1XTBST for 5 min.

9. Remove slides from 1XTBST and dry around the tissue sections using a folded Kimwipe (Note 38).

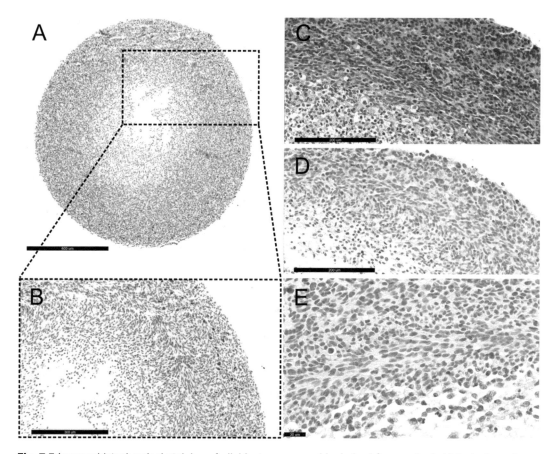

Fig. 7.5 Immunohistochemical staining of glioblastoma organoids derived from a single biological specimen. Representative immunohistochemical images of GBM organoids generated from a single patient specimen. (**a**, **b**) Power slide scans of phospho-histone H3 staining: a marker of active proliferation. The majority of highly proliferative cells are found at the organoid perimeter, with a decrease in the number of positively stained cells in the organoid core. Metaphase plates are visible in multiple cells, demonstrating accurate staining during mitosis. Scale bars are 600 μm and 300 μm, respectively. Power slide scans of (**c**) hematoxylin and eosin, (**d**) cleaved caspase 3, and (**e**) SOX2 staining. Cleaved caspase 3 staining was performed on organoids treated with 3 Gy radiation and 250 μM TMZ and marks apoptotic cells. SOX2 staining was performed on treatment-naïve organoids to label CSCs. Scale bars are 200, 200, and 30 μm, respectively

10. When the slide is sufficiently dry, circle the tissue sections with a PAP pen (Note 39).

11. Block each section with 100–400 μL of preferred blocking solution for 1 h at room temperature (Note 40).

12. Remove blocking solution and add 100–400 μL of primary antibody diluted to desired concentration in SignalStain® Antibody Diluent to each section (Note 41).

13. Incubate sections with primary antibody at 4 °C overnight.

14. After overnight incubation, remove primary antibody solution, and wash each section with 1XTBST for 5 min (three times).

15. During wash steps, make ABC reagent by mixing 100 μL reagent A, 100 μL reagent B, and 10 mL ABC reagent buffer and incubating for 30 min at room temperature.

16. Add 100–400 μL biotinylated secondary antibody (1:1000 dilution) to each section, and incubate at room temperature for 1 h.

17. Wash each section with 1XTBST for 5 min (three times).

18. Add 100–400 μL of ABC reagent to each section and incubate at room temperature for 30 min.

19. Wash sections in 1XTBST for 5 min (three times).

20. Add 1 drop (~30 μL) of SignalStain DAB Chromogen Concentrate to 1 mL of SignalStain DAB Diluent, and mix well by inverting the tube ten times.

21. Add 100–400 μL of the DAB solution to each tissue section and monitor closely. Between 1 and 10 min of staining with DAB should be sufficient to provide acceptable staining intensity (Note 42).

22. Once the desired staining intensity has been achieved, immerse the slide in distilled H_2O (Note 43).

23. Transfer the slides to a slide stain dipper and perform hematoxylin staining according to the following dipping protocol:

    (a) 2 min in Gills 2 Hematoxylin.

    (b) 2 min in distilled H_2O.

    (c) 1 min in clarifier.

    (d) 1 min distilled H_2O.

    (e) 1 min in bluing reagent.

    (f) 2 min in distilled H_2O.

    (g) 1 min in 70% ethanol.

    (h) 1 min in 100% ethanol (three times) (Note 44).

    (i) 2 min in Clear-Rite (three times) (Note 45).

24. Mount the coverslip on the stained tissue section using mounting medium and allow to dry (Notes 46 and 47).

4   Notes and Troubleshooting

1. pH should be 7.2–7.4, and adjust with HCl.

2. Numerous options for orbital shakers exist depending on the specific goals:

    (a) *Short term/exploratory*: A standard orbital shaker (i.e., for western blots) will suffice, although will rust and eventually break down in a high-humidity CO_2 environment over time.

(b) *Long term (>1 year)*: Specific cell culture orbital shaker (i.e., VWR^2 Advanced Dura-Shaker for extreme environments).

(c) *Dedicated shaking incubator*: Benchtop CO_2 incubator with built-in shaker (i.e., New Brunswick™ S41i by Eppendorf).

(d) *Spinning bioreactor*: A spinning bioreactor or spinner flask on a magnetic stir plate can be used for bulk culture of organoids.

(e) *Microwell bioreactors*: Spinner paddles and drive units can be 3D printed for use in 12-well plates and can enable many parallel organoid experiments to be conducted without a shaking apparatus [30].

3. Xylene may be used in place of Clear-Rite.

4. Although parafilm does not need to be sterilized, make sure the side of the parafilm with wax paper, which is the cleaner side, will form the interior of the wells as this surface directly contacts the nascent organoids.

5. If parafilm is not taken off bottom of 96-well PCR plate quickly enough, the parafilm will warm up rapidly and the wells will invert as they are removed.

6. Parafilm molds can be formed in advance as long as sterility is maintained.

7. Finishing the cut with a sliding motion against the plate can help cut fibrous or tough pieces of tissue.

8. Do not vortex suspension to mix.

9. Do not incubate for >15 min as this will cause lysis of tumor cells as well.

10. Although not necessary, this allows for:

(a) Characterization of live/dead ratio.

(b) Control of seeding density.

(c) Isolation of specific cell populations from primary tumor via magnetic or fluorescence activated cell sorting.

11. Addition of this extra manipulation step results in loss of sample that is unable to pass through the filter.

12. Organoids can also be formed from CSCs grown adherently in stem cell condition [12].

13. Although generation of a single-cell suspension is unnecessary to form organoids, it allows for cell counting and live/dead analysis and thus more reproducible formation. If organoids will be formed with intact tumor spheres, proceed to step 8.

14. GBM organoids can be successfully formed with varying numbers of cells; however 10,000 cells per organoid have produced consistent and reproducible results. Counterintuitively, very high numbers of cells per organoid can be detrimental to successful organoid formation.

15. Make sure not to introduce air bubbles into cell-Matrigel mixture when pipetting up and down.

16. Pipette tips should be cold when working with Matrigel. Do not hold Matrigel in the pipette tip for long at room temperature as it will polymerize in the narrow tip region first, clogging the tip.

17. Do not hold body of tube in hand since Matrigel will polymerize as it warms.

18. The 20 µL should form a pearl-like droplet in the hydrophobic parafilm mold. See Fig. 7.1.

19. Use p1000 to gently flush organoids out of mold with Neurobasal complete medium into a 10 cm cell culture dish filled with Neurobasal complete medium.

20. The organoids will be difficult to see as they are still transparent at this point. Be careful when changing the medium. We recommend removing the old medium with a controllable pipette and not a vacuum aspirator.

21. Some recommendations for volumes of medium to use in different cell culture layouts are as follows:

    (a) 12 well: 2 mL medium per well (recommended to only culture one organoid per well).

    (b) 6 well: 4 mL medium per well.

    (c) 10 cm plate: 20 mL (30 mL for specimens that grow very large).

22. Although budding of tumor spheres from the edges of the organoid surface is normal, excessive fraying could indicate poor health and excessive cell death. This indicates the need to increase the frequency of medium changes and/or transfer organoids to individual wells.

23. Significant buildup of cellular material at the edges of the plate indicates a need to transfer the organoids to a new plate.

24. Although passaging of organoids is not needed for most samples if expansion is not desired, organoids can be dissociated as described in Sect. 3.5, then passaged, and reformed.

25. Some specimens can be efficiently passaged by cutting into several pieces using a sterile blade and allowing the pieces to regrow. However, the ability to do this varies on a case-by-case basis and should be tested for each individual sample.

26. If clumps and aggregates increase over time rather than decrease, this is likely due to release of DNA from dying cells. Dilute and remove Accutase immediately.

27. Dissociated cells can also be used to reform new organoids.

28. If possible, we recommend testing both methods to see which works best for each individual laboratory and specimen.

29. Make sure there is minimal fluid present on the organoid. Fluid on the organoid surface can create a boundary between the organoid and the OCT, causing the tissue to fall out during the sectioning process. Gently touching the organoid to a folded Kimwipe may help.

30. Be careful to minimize bubbles as this will cause issues during the sectioning process.

31. Do not touch the tissue sections or allow them to dry completely. Tissue drying will affect the staining quality. The folded Kimwipe will wick away the 1XPBS without the need to get too close to the tissue.

32. Using four straight lines to draw a box around the tissue section as opposed to a circle helps ensure there are no areas for liquid to escape.

33. When adding blocking solution or primary antibody to the tissue, pipette gently and not directly on the tissue. Excessive force will cause the tissue section to detach from the slide and affect staining.

34. For antibody recommendations, see Table 7.1.

35. Imaging the slide within a few days after mounting is ideal, although within 2 weeks will still yield good results.

36. Slides can be maintained at sub-boiling temperature after initial boiling has been reached by heating for the remainder of the 10 min at low power. This process should be optimized for each laboratory's individual microwave.

37. Make sure the citrate solution does not vigorously boil during the 10 min incubation. If too much citrate solution evaporates, the tissue may dry and impact downstream staining.

38. Do not touch the tissue sections or allow them to dry completely. Tissue drying will affect the staining quality. The folded Kimwipe will wick away the 1XTBST without the need to get too close to the tissue.

39. Using four straight lines to draw a box around the tissue section as opposed to a circle helps ensure there are no areas for liquid to escape.

40. Some recommended blocking solutions are:

(a) 1XTBST, 10% serum (corresponding to animal host of the secondary antibody).

(b) Background Sniper (Biocare Medical).

41. For antibody recommendations, see Table 7.1.

42. When using a new primary antibody or performing this step for the first time, we recommend monitoring the staining in real time under a dissection microscope. This will allow you to determine how much time is required to obtain the desired staining intensity.

43. This wash step should be done in the sink using running distilled H_2O. This provides a constant stream of fresh water for optimal washing.

44. Make sure each new wash is in a different container of 100% ethanol, or the hematoxylin staining will be darker than expected.

45. Make sure each new wash is in a different container of 100% Clear-Rite.

46. In order to minimize the amount of mounting medium used, dry the slides by placing them on a paper towel and propping them at a 45° angle against a pipette tip box. This will allow removal of excess mounting medium, which might otherwise interfere with imaging.

47. We have given our basic IHC protocol for individual slide use, and this should be accessible to most labs. However, it is common practice in our lab to modify the above steps/volumes for use in a Shandon Sequenza slide staining apparatus, allowing higher throughput of parallel slides with smaller fluid volumes. All concentrations used in this case are identical to those given above.

5   Future Directions

As the field of cancer organoid research continues to advance, it is imperative that we learn from past studies and analogous methods in order to maximize the benefits gained from brain tumor organoids. Efforts to make brain tumor organoid biobanks should be made to serve as repositories of human tissue that can easily be perturbed to gain biological insight into the impact of inter-patient variation on the disease. Although patient-derived xenografts are the current standard for creating patient avatars to guide clinical treatment, patient-derived organoids serve as an attractive alternative due to their low cost, ease of generation/maintenance, reduced regulatory hurdles, and controllability. Even when formed from patient-derived sphere cultures, GBM organoids show heterogeneity absent in these traditional culture methods that will likely result in differential response to drug treatment [17]. Performing drug and genetic screens in tumor organoids could allow us to parse region-specific vulnerabilities. This avenue of research is especially important due to the role of cellular and microenvironmental heterogeneity in patient response to treatment.

With the advent of single-cell genomic technologies for looking at cellular heterogeneity at the transcriptional (scRNA-seq) [31], genomic (scCNV, scWGS, and scWES), and epigenetic levels (scATAC-seq) [32], the ability to comprehensively characterize GBM organoids is becoming increasingly feasible. Tissue clearing methods in combination with immunohistochemistry and in situ hybridization/sequencing will allow researchers to probe these three-dimensional structures in their native states under various perturbations [33–36].

GBM organoids display markedly different characteristics from GBM CSCs or non-CSCs grown in traditional in vitro systems when implanted intracranially. The resulting xenografts from GBM organoids can display a diffuse and invasive morphology commonly seen in clinical presentations of GBM, whereas the same specimens grown in traditional in vitro systems may result in circumscribed tumors with limited infiltration [17]. Further investigation into which aspects of GBM organoids promote the invasive characteristics of downstream tumors is necessary to gain further insight into the biological factors that lead to the aggressive behavior of GBM tumors that ultimately results in patient mortality.

Cellular heterogeneity is observable throughout GBM organoid structures, even when formed from homogenous culture-selected or FACS-purified populations of GBM CSCs. Microenvironmental gradients present within a GBM organoid likely contribute to the differing phenotypes of its cell populations, but the exact mechanisms through which these divergent populations arise are unknown. Lineage tracing experiments to characterize the phenotypic plasticity of different organoid cell populations would help pinpoint how different microenvironmental factors contribute to various aspects of the disease and how disruption of these might lead to predictable changes in tumor biology.

A significant gap in our current organoid models is the lack or paucity of normal stromal cell populations within the tumorous structures. We believe the next frontier of organoid research relies on the incorporation of additional clinically relevant cell populations, specifically non-cancer cells including immune and vascular populations. The immune content of the tumor microenvironment can drastically impact the aggressiveness of the tumor and even impact response to treatment [37]. Vasculature supplies the tumor with nutrients and oxygen throughout the three-dimensional structure and can provide cancer cells with a route for distal metastasis or paracrine cell-to-cell communication [38]. Recent studies have demonstrated the feasibility of modeling disease with vascularized and immune-competent organoids, with strategies applicable to GBM [39, 40]. Human brain organoids can even be fully vascularized in vivo through transplantation into the brains of mice [41]. For fully human models, matched patient-derived induced pluripotent stem cells may allow the possibility of incorporating neuronal, glial, vascular, or other cell populations into GBM organoids derived from the same patient to comprehensively recapitulate the cerebral microenvironment containing the tumor. The inclusion of these components to GBM organoids could create ex vivo systems that even more accurately represent patient GBM biology than current technologies.

While cancer organoid research opens the door to ask many exciting and previously unanswerable questions, there are also outstanding issues that need to be addressed to move the field forward. For instance: how similar are cancer organoids to the original tumor and to alternative models? Which cell populations from the tumor are maintained ex vivo, which are lost, and what new populations are gained that were never present in the patient? How does the evolution and therapeutic response of the tumor ex vivo mimic and differ from the residual disease in the patient? Answering these questions will improve our ability to translate biological discoveries from organoid research to improved patient care.

6   Conclusion

Despite the demonstration that *inter*tumoral and *intra*tumoral heterogeneities in GBM are important for patient response to therapy, these attributes are frequently overlooked during in vitro study of the disease. GBM organoids are a powerful system to mimic the extensive microenvironmental and cellular heterogeneity of patient tumors ex vivo, permitting study of GBM biology in previously unachievable ways. Besides a dedicated orbital shaker, there is little specialized equipment needed to make GBM organoids from cultured cell models or primary patient samples. This gives the system a low barrier to entry for laboratories wishing to bring the technology in house, as long as they are willing to accept the inherent longer timescales and initial learning curves.

The list of applications for GBM organoids is long and constantly growing. It has already been shown that GBM organoids give rise to intracranial xenografts with different characteristics than those from CSC sphere cultures grown from the same specimen. In addition, the CSCs present within GBM organoids show resistance to radiotherapy, while non-CSCs present within the same region do not. The ease of manipulation of this system will allow for adaption to technologies such as high-content drug screening, gene editing, and single-cell omics.

We see organoid culture becoming a standard method of studying GBM and CSC biology that is complementary to the standard methods of in vitro CSC culture and in vivo intracranial xenografts. Broad application of this technology will reveal deep insights into the importance of heterogeneity for disease progression and contribute to the discovery of novel therapeutic targets based on a more complete understanding of GBM biology.

References

1. Stupp R, Hegi ME, Mason WP et al (2009) Effects of radiotherapy with concomitant and adjuvant temozolomide versus radiotherapy alone on survival in glioblastoma in a randomised phase III study: 5-year analysis of the EORTC-NCIC trial. Lancet Oncol 10:459–466. https://doi.org/10.1016/S1470-2045(09)70025-7

2. Calabrese C, Poppleton H, Kocak M et al (2007) A perivascular niche for brain tumor stem cells. Cancer Cell 11:69–82. https://doi.org/10.1016/j.ccr.2006.11.020

3. Mohyeldin A, Garzón-Muvdi T, Quiñones-Hinojosa A (2010) Oxygen in stem cell biology: a critical component of the stem cell niche. Cell Stem Cell 7:150–161. https://doi.org/10.1016/j.stem.2010.07.007

4. Li Z, Bao S, Wu Q et al (2009) Hypoxia-inducible factors regulate tumorigenic capacity of glioma stem cells. Cancer Cell 15:501–513. https://doi.org/10.1016/j.ccr.2009.03.018

5. Jin X, Kim LJY, Wu Q et al (2017) Targeting glioma stem cells through combined BMI1 and EZH2 inhibition. Nat Med. https://doi.org/10.1038/nm.4415

6. Bao S, Wu Q, Sathornsumetee S et al (2006) Stem cell–like glioma cells promote tumor angiogenesis through vascular endothelial growth factor. Cancer Res 66:7843–7848. https://doi.org/10.1158/0008-5472.CAN-06-1010

7. Alvarado AG, Thiagarajan PS, Mulkearns-Hubert EE et al (2017) Glioblastoma cancer stem cells evade innate immune suppression of self-renewal through reduced TLR4 expression. Cell Stem Cell 20:450–461.e4. https://doi.org/10.1016/j.stem.2016.12.001

8. Wang X, Prager BC, Wu Q et al (2018) Reciprocal signaling between glioblastoma stem cells and differentiated tumor cells promotes malignant progression. Cell Stem Cell 22:514–528.e5. https://doi.org/10.1016/j.stem.2018.03.011

9. Cheng L, Huang Z, Zhou W et al (2013) Glioblastoma stem cells generate vascular pericytes to support vessel function and tumor growth. Cell 153:139–152. https://doi.org/10.1016/j.cell.2013.02.021

10. Lee J, Kotliarova S, Kotliarov Y et al (2006) Tumor stem cells derived from glioblastomas cultured in bFGF and EGF more closely mirror the phenotype and genotype of primary tumors than do serum-cultured cell lines. Cancer Cell 9:391–403. https://doi.org/10.1016/j.ccr.2006.03.030

11. Li A, Walling J, Kotliarov Y et al (2008) Genomic changes and gene expression profiles reveal that established glioma cell lines are poorly representative of primary human gliomas. Mol Cancer Res 6:21–30. https://doi.org/10.1158/1541-7786.MCR-07-0280

12. Pollard SM, Yoshikawa K, Clarke ID et al (2009) Glioma stem cell lines expanded in adherent culture have tumor-specific phenotypes and are suitable for chemical and genetic screens. Cell Stem Cell 4:568–580. https://doi.org/10.1016/j.stem.2009.03.014

13. Kim S-J, Lee HJ, Kim MS et al (2015) Macitentan, a dual endothelin receptor antagonist, in combination with temozolomide leads to glioblastoma regression and long-term survival in mice. Clin Cancer Res 21:4630–4641. https://doi.org/10.1158/1078-0432.CCR-14-3195

14. Quintiliani M (1986) The oxygen effect in radiation inactivation of DNA and enzymes. Int J Radiat Biol Relat Stud Phys Chem Med 50:573–594. https://doi.org/10.1080/09553008614550981

15. Friedmann-Morvinski D (2014) Glioblastoma heterogeneity and cancer cell plasticity. Crit Rev Oncog 19:327–336. https://doi.org/10.1615/CritRevOncog.2014011777

16. van de Wetering M, Francies HE, Francis JM et al (2015) Prospective derivation of a living organoid biobank of colorectal cancer patients. Cell 161:933–945. https://doi.org/10.1016/j.cell.2015.03.053

17. Hubert CG, Rivera M, Spangler LC et al (2016) A three-dimensional organoid culture system derived from human glioblastomas recapitulates the hypoxic gradients and cancer stem cell heterogeneity of tumors found In Vivo. Cancer Res 76:2465–2477. https://doi.org/10.1158/0008-5472.CAN-15-2402

18. Sachs N, de Ligt J, Kopper O et al (2018) A living biobank of breast cancer organoids captures disease heterogeneity. Cell 172:373–386.e10. https://doi.org/10.1016/j.cell.2017.11.010

19. Lee SH, Hu W, Matulay JT et al (2018) Tumor evolution and drug response in patient-derived organoid models of bladder cancer. Cell 173:515–528.e17. https://doi.org/10.1016/j.cell.2018.03.017

20. Ben-David U, Siranosian B, Ha G et al (2018) Genetic and transcriptional evolution alters cancer cell line drug response. Nature 560:325–330. https://doi.org/10.1038/s41586-018-0409-3

21. Belderbos ME, Koster T, Ausema B et al (2017) Clonal selection and asymmetric distribution of human leukemia in murine xenografts revealed by cellular barcoding. Blood 129:3210–3220. https://doi.org/10.1182/blood-2016-12-758250

22. Bian S, Repic M, Guo Z et al (2018) Genetically engineered cerebral organoids model brain tumor formation. Nat Methods 15:631–639. https://doi.org/10.1038/s41592-018-0070-7

23. Ogawa J, Pao GM, Shokhirev MN, Verma IM (2018) Glioblastoma model using human cerebral organoids. Cell Rep 23:1220–1229. https://doi.org/10.1016/j.celrep.2018.03.105

24. Lancaster MA, Renner M, Martin C-A et al (2013) Cerebral organoids model human brain development and microcephaly. Nature 501:373–379. https://doi.org/10.1038/nature12517

25. Bershteyn M, Nowakowski TJ, Pollen AA et al (2017) Human iPSC-derived cerebral organoids model cellular features of lissencephaly and reveal prolonged mitosis of outer radial glia. Cell Stem Cell 20:435–449.e4. https://doi.org/10.1016/j.stem.2016.12.007

26. Madhavan M, Nevin ZS, Shick HE et al (2018) Induction of myelinating oligodendrocytes in human cortical spheroids. Nat Methods 15:700–706. https://doi.org/10.1038/s41592-018-0081-4

27. Lancaster MA, Knoblich JA (2014) Generation of cerebral organoids from human pluripotent stem cells. Nat Protoc 9:2329–2340. https://doi.org/10.1038/nprot.2014.158

28. Hambardzumyan D, Bergers G (2015) Glioblastoma: defining tumor niches. Trends Cancer 1:252–265. https://doi.org/10.1016/j.trecan.2015.10.009

29. Ozawa T, James CD (2010) Establishing intracranial brain tumor xenografts with subsequent analysis of tumor growth and response to therapy using bioluminescence imaging. J Vis Exp:e1986. https://doi.org/10.3791/1986

30. Qian X, Nguyen HN, Song MM et al (2016) Brain-region-specific organoids using mini-bioreactors for modeling ZIKV exposure. Cell 165:1238–1254. https://doi.org/10.1016/j.cell.2016.04.032

31. Macosko EZ, Basu A, Satija R et al (2015) Highly parallel genome-wide expression profiling of individual cells using nanoliter droplets. Cell 161:1202–1214. https://doi.org/10.1016/j.cell.2015.05.002

32. Buenrostro JD, Wu B, Litzenburger UM et al (2015) Single-cell chromatin accessibility reveals principles of regulatory variation. Nature 523:486–490. https://doi.org/10.1038/nature14590

33. Chen F, Wassie AT, Cote AJ et al (2016) Nanoscale imaging of RNA with expansion microscopy. Nat Methods 13:679–684. https://doi.org/10.1038/nmeth.3899

34. Chung K, Wallace J, Kim S-Y et al (2013) Structural and molecular interrogation of intact biological systems. Nature 497:332–337. https://doi.org/10.1038/nature12107

35. Lee JH, Daugharthy ER, Scheiman J et al (2015) Fluorescent in situ sequencing (FISSEQ) of RNA for gene expression profiling in intact cells and tissues. Nat Protoc 10:442–458. https://doi.org/10.1038/nprot.2014.191

36. Wang X, Allen WE, Wright MA et al (2018) Three-dimensional intact-tissue sequencing of single-cell transcriptional states. Science 361:eaat5691. https://doi.org/10.1126/science.aat5691

37. Wang M, Zhao J, Zhang L, et al (2017) Role of tumor microenvironment in tumorigenesis. J Cancer 8:761–773. https://doi.org/10.7150/jca.17648

38. Bielenberg DR, Zetter BR (2015) The contribution of angiogenesis to the process of metastasis. Cancer J 21:267. https://doi.org/10.1097/PPO.0000000000000138

39. Neal JT, Li X, Zhu J et al (2018) Organoid Modeling of the tumor immune microenvironment. Cell 175:1972–1988.e16. https://doi.org/10.1016/j.cell.2018.11.021

40. Wimmer RA, Leopoldi A, Aichinger M et al (2019) Human blood vessel organoids as a model of diabetic vasculopathy. Nature 565:505–510. https://doi.org/10.1038/s41586-018-0858-8

41. Mansour AA, Gonçalves JT, Bloyd CW et al (2018) An in vivo model of functional and vascularized human brain organoids. Nat Biotechnol 36:432–441. https://doi.org/10.1038/nbt.4127

Chapter 8

In Vitro Mechanobiology of Glioma: Mimicking the Brain Blood Vessels and White Matter Tracts Invasion Paths

Pascale Monzo, Michele Crestani, and Nils C. Gauthier

Abstract

Invasive glioma is migrating mostly along the white matter tracts and the abluminal walls of the brain blood vessels, representing hundreds of miles of linear highways that glioma cells can use to invade away from the tumor core, seeding secondary gliomas. Several mechanobiology techniques have been used to mimic these linear structures. In this chapter, we describe in details how to use two specific biophysical tools to recapitulate and study glioma motility along the blood vessels and white matter tracts in controlled environments: micropatterning allowing to print linear patterns and electrospinning to generate nanofibers to study glioma linear motility in 3D environments. Our goal is to provide the reader with guidelines on how to prepare these substrates and how to image them and to highlight advantages and limitations of the different approaches.

Key words Migration, Micropatterning, Electrospinning, Laminin, Total internal reflection fluorescence microscopy (TIRFM), Mechanobiology, Glioma

1 Introduction: Glioma Linear Migration in the Brain

Glioma cells originate in the brain and rarely metastasize outside the brain [1, 2]. These cells rapidly invade and destroy the brain ultimately killing the patient. Targeting specifically glioma motility would prevent the formation of a diffuse tumor and would allow effective treatment with radiation and complete neurosurgery. Several targets have been identified in the invasion of tumor cells [3–6], including metalloproteases and integrin-targeting drugs. Unfortunately, none of them has showed improvement for GBM, probably due to the fact that these cells display specific motility modes that need to be precisely defined in order to identify effective molecular targets.

At the opposite of most metastatic cells, glioma cells do not use the blood circulation to disseminate, probably because they fail at crossing the blood-brain barrier [7]. Instead, they typically travel

Giorgio Seano (ed.), *Brain Tumors*, Neuromethods, vol. 158,
https://doi.org/10.1007/978-1-0716-0856-2_8, © Springer Science+Business Media, LLC, part of Springer Nature 2021

along the wall of the brain blood vessels and along the white matter tracts, both linear structures known as structures of Scherer [8–10].

The brain blood vessels constitute efficient highways for invading glioma cells. In human, they spread over hundreds of miles of linear tracks where matrix proteins, which support migration and adhesion, are found [11–14], and they provide glucose and nutrients, constituting a powerful niche for glioma stem cells [15–17].

In many experimental setups, glioma cells have been shown to rapidly gather, proliferate, and migrate along the brain blood vessels. With time, the massive accumulation of glioma cells around the blood vessels can lead to vessel encasement and damaging [18–25].

Glioma cells, like many other tumor cells, display a significant plasticity in their mode of motility, allowing them to adapt and successfully invade various environments [26–28]. Such plasticity has been evidenced in live imaging experiments, where C6 glioma cells have been observed switching from a slow random migration coupled to a multipolar morphology when located in the parenchyma to an efficient linear motility coupled to an unipolar elongated morphology when moving along the blood vessels [18, 24, 25, 29].

The molecular pathways regulating this plasticity needs to be elucidated in order to define new targets. Several studies point toward the factors that regulate the balance of RhoA/Rac1 activity, cell-matrix adhesions, the balance of Arp2/3 and formin, and actomyosin contractility [25, 28–34].

Our understanding of how the extracellular inputs from the microenvironment are processed and integrated by glioma cells to enable efficient migration depends on our ability to dissect them and reproduce them with fidelity in controllable systems. This is where the emerging field of mechanobiology enters into the picture. Mechanobiology tools can indeed recapitulate in vivo features with biomaterials and engineered biomimetic devices to precisely study how specific parameters of the brain environment will affect glioma motility [35–42]. In the following sections, we will describe (1) what physical and chemical parameters of the brain environment can be tuned in vitro and (2) what type of biomaterials and engineered tools have been used, so far, to address these parameters. We will then detail in the material and method section how to prepare and use two specific tools, micropatterning and nanofibers, that are used to mimic blood vessels and white matter tracts, respectively.

1.1 Invasive Behaviors Can Be Tuned by Three Main Parameters of the Substrate: Topography, Mechanical Properties (i.e., Stiffness), and Matrix Composition

Tumor microenvironment plays a critical role in cancer cell progression and survival [26–28, 43]. Because tumor microenvironment is extremely complex, it is almost impossible to determine how individual feature (such as matrix component, stiffness, topography, neighboring cells, confinement, etc.) influences tumor progression in vivo.

The contribution of each of these parameters can be defined only by reductionist systems, which offer additionally high degree of control and reproducibility compared to in vivo systems.

We need to understand how topography, stiffness, and ECM composition impacts glioma invasion. A number of studies have addressed one or several of these parameters in the recent years [35, 36, 41].

1.1.1 Topography

Glioma cells are known to invade onto key tracks known as the structures of Scherer [8–10], corresponding to the blood vessel walls and the white matter tracts. The architecture of the white matter tracts corresponds to fibers with a nanosized diameter and high aspect ratio [10, 44], while the architecture of the brain blood vessels corresponds to larger tubes (diameter varying between 5 and 50 μm) on which collections of glioma cells can accumulate [19–21, 23]. These long uninterrupted tracks are ideal to promote an efficient cell migration [18, 24, 25, 45, 46].

1.1.2 Mechanical Properties of the Brain Environment (Stiffness)

The mechanoproperties of the healthy brain: The brain parenchyma presents a severe mechanical challenge to migrating glioma. Cell processes are tightly packed with sub-micrometer pore sizes [47], and glioma cells must deform their nuclei to allow migration through the pores [29]. However, the brain is also extremely compliant with a young's modulus ranging from several hundred pascals to several kilopascals, depending on the testing conditions and on the regions of the brain [48–50]. Interestingly, some structures of Scherer (i.e., the perivascular region) have been found to be stiffer than the surrounding parenchyma, suggesting that glioma cell guidance is influenced by their mechanoproperties, a process called durotaxis [51]. In accordance, glioma cells have been shown to migrate faster on stiffer substrates than on soft substrate [52, 53], the optimal stiffness for U251 glioma cell migration ranging around 100 kPa (100-fold higher than in the parenchyma) [54].

The mechanoproperties of the glioma brain: Generally, tumors increase the stiffness of their environment, and reciprocally stiffer environments increase the invasive properties of cancer cells [55]. These observations appear to be also true for brain tumors. Ultrasound imaging during neurosurgery suggests that GBM tumors are stiffer than the rest of the brain. Glioma could stiffen the brain environment by secreting elastic fibril and by increasing the pressure due to their growth and the resistance of the skull [35, 56]. Also, the leaking vasculature of GBM tumors and the compression of the vessels can alter the mechanical environment by inducing additional shear stresses [57, 58].

An accurate measurement of the stiffness of the brain that is invaded by glioma is necessary to clearly appreciate the real stiffness of glioma environment. Moreover, different glioma cells from patient respond differently to substrate stiffness; this could be subclass specific with proneural subclass being independent on rigidity [59].

1.1.3 Matrix Composition   The brain environment in which glioma cells evolve is composed of various matrix proteins and glycosaminoglycans (GAGs), which can be found in specific locations in the brain. Moreover, glioma cells can use the surface of other cells to adhere and migrate [44, 60, 61]. Hyaluronic acid (HA) is present in the brain parenchyma, while laminin and collagen IV are mostly found around the blood vessels. Collagen I is absent in the brain, and fibronectin appears around the blood vessels that have been invaded by glioma [44, 60–62]. The presence of fibronectin on glioma-invaded blood vessels could be due to the induction of repair mechanisms in response to vessel injury caused by the glioma cells and could be produced directly by the endothelial cells [63–65].

Brain parenchyma: Hyaluronic acid and CD44. HA is a high-molecular-weight glycosaminoglycan made of D-glucuronic acid and N-acetyl-D-glucosamine. It is one of the main matrix proteins of the brain [66–68]. Glioma cells can also produce HA, and high levels of HA have been correlated with increase invasion and proliferation [69, 70]. CD44, the receptor of HA, is also overexpressed by glioma cells particularly in cells localized at the margin of the tumor, suggesting that CD44 promotes invasion [71, 72]. It has been shown that CD44 contributes to mechanosensing and invasive motility of glioma cells [73–75]. This contribution could occur via the activation of key protumorigenic signals, such as Rho GTPAses and PI3-kinase, in response to the binding of CD44 to HA [76–78]. As a consequence, CD44 constitutes an important therapeutic target for glioblastoma (GBM) treatment [79].

Blood vessels: Laminin, collagen IV, collagen V, fibronectin, and integrins. On the surface of the blood vessels, various matrix proteins that can promote adhesion and migration can be detected [14, 80–82]. The basolateral membrane of the blood vessels contains matrix proteins, such as laminin, collagen IV, and collagen V. Fibronectin is also found around GBM-associated blood vessels [13, 44, 83]. Glioma cell can stimulate the production of laminin, collagen, and fibronectin [83, 84]. These matrix proteins enhance cell survival, proliferation, and migration in vivo and in vitro [84–88] via their attachment to various integrins [82, 89]. Several studies have looked at the role of these various ECM proteins on the migratory properties of different glioma cell lines on 2D supports [33, 90, 91]. Laminin has been found to be the most permissive substrate for glioma cell migration compared to other matrix proteins [33, 90] and appears as a substrate of choice for glioma motility. Laminin is produced by both endothelial cells and astrocytes. In GBM patient and GBM-bearing animals, laminin has also been found in the parenchyma, at the margin of the tumor, independently of its association with blood vessels [92]. Moreover, laminin receptors α2β1, α3β1, α6β1, and α6β4 have potential roles in glioma invasion as revealed by their expression and localization [82, 83, 93–100].

Other matrix proteins: Collagen I, tenascin, and myelin. Although collagen I is not found in the brain environment, it can stimulate migration of several glioma cell lines in vitro and has been used in a number of assays [101–104]. Tenascin C is a proteoglycan. It is produced by glioma cells and it is found in the parenchyma [105] and in and around the walls of hyperplastic blood vessels, suggesting a role in promoting angiogenesis [106]. It may also promote the motility of glioma cells [90]. Myelin could also be used by glioma cells to migrate along the white matter tracts [44].

Glioma cells: Glioma cells can migrate on top of each other [90] probably using cell-cell contacts as suggested in [107]. These interactions could help them to generate antiparallel flows as observed with C6 glioma migrating on laminin micropatterns [33] and with neuroblast cells migrating in the brain or at high density in vitro [108, 109].

1.2 Biomaterials and Engineered Microenvironment to Study the Mechanoproperties of Glioma

The development of anti-invasive approaches for gliomas has been largely hampered by the difficulty to model glioma cell migration appropriately in vitro. A number of in vitro migration assays are of common use, but significant limitations restrict their potential to predict glioma cell behavior in vivo. Most available models use homogenous matrices and fail to mimic oriented motility along the linear structures. Most of these studies have used matrix proteins relevant to other solid tumors but failed to recapitulate the matrix organization and composition of the brain environment. As a consequence, anti-migratory approaches against gliomas have targeted specific cell adhesion molecules or tumor-associated proteases following anti-metastatic strategies used in other solid tumors. These strategies failed in clinical trials underscoring the need for additional studies to identify the master regulators of locomotion specific to glioma cells [110].

Tissue engineering and biomaterials allow the fabrication of sophisticated models of human cancer in vitro recapitulating tumor microenvironment with greater fidelity while allowing the use of suitable human cells [111–115] (see Table 8.1). Some systems, such as 3D hydrogels and transwell, do not require any microfabrication and allow the study of 3D migration. On top of controlling the matrix composition, microfabrication techniques allow to

Table 8.1
Main biomaterials and engineered microenvironment to study the mechanoproperties of glioma

Biophysical tools	Biophysical properties addressed	References
2D hydrogels	Substrate rigidity	[52–54, 128–130].
3D hydrogels	Substrate rigidity in 3D	[53, 101, 103, 104, 133, 136, 138–149, 208]
Transwell system	Pore size and chemoattraction	[29, 152, 153]
Linear grooves	Linearity of white matter tracts	[33, 156–159]
Microchannels	Linearity and confinement	[161, 162]
Micropatterning	Linearity of brain blood vessels or white matter tracts	[33, 84, 176]
Nanofibers	Linearity, curvature, and 3D aspect of white matter tracts	[102, 177, 188, 192, 193, 196, 202, 206–209]
BBB mimetic devices	Contact with the components of the BBB	[152, 153, 243, 252]

control and recapitulate physical features observed in the brain environment, such as topography, roughness, and elasticity [36, 116–118] (see Table 8.1).

1.2.1 Hydrogels to Study the Effect of Rigidity on Glioma Behavior

The most common hydrogels are made of polyacrylamide (PAA) and were originally introduced by Yu Li Wang and co-workers in studies establishing the role of the ECM stiffness in fibroblast motility [119–121]. They can also be made of other material such as hyaluronic acid [53, 122] (Fig. 8.1a, Table 8.1).

Advantages. Hydrogels can be functionalized with the same matrix proteins found in the brain parenchyma or around the blood vessels, and their elastic modulus can be controlled by tuning the cross-linking of the hydrogel matrix [123–127]. These gels can be used to understand how glioma cells sense and integrate the mechanical signals from their environment [35, 36, 41].

Limitations. These hydrogels do not recapitulate the topology that glioma cells follow in the brain. In these systems, cells evolve in a 2D plane, which is very far from the brain environment conditions.

Hydrogels and glioma. With these hydrogels, glioma cells have been shown to migrate faster on stiffer substrates [52, 53], with an optimal stiffness for U251 glioma cell migration ranging around 100 kPa [54]. The sensitivity to stiffness has been shown to depend on myosin II, alpha actinin, talin, and RhoA [128, 129]. These gels have been also used to mimic the interface of blood vessel/parenchyma by sandwiching glioma cells between fibronectin surface mimicking blood vessel and HA soft surface mimicking the parenchyma. In this assay, glioma cells migrated most efficiently

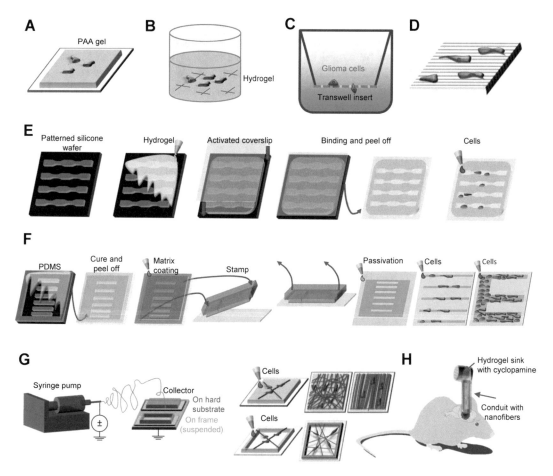

Fig. 8.1 Overview of different mechanobiology techniques used to study glioma motility. (**a**) 2D hydrogels to study the effect of substrate rigidity and measure traction forces. (**b**) 3D hydrogels to study the effects of substrate rigidity and measure traction forces in 3D. (**c**) Transwell system to study chemoattraction and motility through 3D matrix. (**d**) Linear grooves (=gratings) to mimic white matter tracts. (**e**) Microchannels to mimic blood vessels and white matter tracts upon confinement. (**f**) Microcontact printing to mimic blood vessels and white matter tracts. (**g**) Nanofibers to mimic white matter tracts. (**h**) Nanofibers to build up a glioma trap

when adhesion to the dorsal surface was weak, whether that adhesion was mediated by CD44 or integrins [130].

1.2.2   ECM Gels to Study Glioma Motility in 3D

3D matrices hydrogels (Fig. 8.1b, Table 8.1) are made of native biopolymers such as collagen I, HA, or matrigel. Matrigel is a laminin-rich ECM extracted from mouse sarcoma tumors. Synthetic polymers such as agarose can also be used to prepare 3D gels [131–133]. With these synthetic polymers, ECM stiffness and biochemical ligand can be adjusted independently [37, 38, 133].

In 3D ECM gels, cell shape and adhesions are similar to those observed in vivo.

Advantages. Cells in 3D ECM adopt shapes and adhesions that are much closer to those described in vivo than when grown on 2D matrices [134]. They can be used to study matrix remodeling and proteolytic degradation [135]. The porosity of some of these gels can be tuned by temperature (e.g., collagen matrices nucleated at 22 °C instead of 37 °C exhibits a more porous architecture), allowing to study the effect of matrix porosity in glioma invasion [136].

Limitations. While some gels, such as matrigel or HA hydrogels, contain matrix proteins that are found along the blood brain vessels (laminin and collagen IV in the matrigel) or in the brain parenchyma (HA), these systems lack the topographical features, which are important to GBM migration.

Invasion through collagen 1 base matrix has been used in many studies but appears inappropriate since collagen 1 is absent from the brain [61, 137].

3D gels and glioma. 3D hydrogels have been used to model GBM migration in several studies [133, 138–147]. Using matrigel, it has been shown that U87 glioma growing spheroids exerted compressive forces, whereas invading cells exerted traction forces on the ECM [148]. Studies with collagen 1 gels showed that spheroid growth was facilitated by increasing the collagen concentration, due to an increase of fibers available for adhesion and traction, and that the cellular levels of cadherin and matrix metalloprotease were impacting on the invasive pattern [101, 103, 104]. Also, increasing the porosity of the collagen gel facilitated glioma invasion [136]. Using agarose gel mixed with collagen 1, it has been shown that increasing the stiffness in 3D inhibited glioma spheroid invasion, a result opposite to that observed in 2D in which increasing the stiffness of the gel was increasing glioma migration speed [52, 133]. This could be due to the fact that in 3D, glioma cells need to remodel the collagen matrix to move, a process that can be impeded by the increased stiffness due to the agarose [149]. Also increasing the density of the 3D matrix led to a switch of motility from an elongated mesenchymal mode to an amoeboid mode. HA gels functionalized with RGD sequence and cross-linked with dithiothreitol (DTT) have been used to better mimic the brain parenchyma, allowing the formation of high-density gels with submicron porosity similar to the brain matrix and devoid of the fibrillary structures of the collagen [47, 53, 150]. In these gels, U373-MG spheroid motility was prevented by increasing the ECM stiffness and displayed a phenotype similar to that previously observed on brain slices [53].

1.2.3 The Transwell System to Study Nucleus Deformation and Chemoattraction

The transwell system (=filter membrane migration assay, transwell migration assay, chemotaxis assay, or Boyden chamber) is a tool of reference to test the invasion capability of cells to migrate through a 3D meshwork (invasion) (Fig. 8.1c, Table 8.1). It can additionally be used to study the chemotaxis of the cells toward a given attractant. It was initially introduced by Boyden to study leukocyte chemotaxis. It is also the most widely used in vitro tool to reproduce the brain-vessel interface. In a transwell system, a polymeric porous insert (of defined pore size) is placed in a standard well plate. Glioma cells migrate through the porous insert that can be coated with various matrix proteins [151].

Advantages. Transwell systems are easy to use and commercially available (Merck Millipore, Corning® Transwell®, Falcon HTS FluoroBlok Insert); they can be found with various pore size and can be coated with any matrix protein.

Limitations. Cell migration per se cannot be imaged; migration efficiency is evaluated by counting the number of cells that have passed to the other side of the insert by DAPI staining, and this method cannot really distinguish between migration and proliferation. Moreover, even if acquiring data is easy and microfabrication techniques are avoided, the linear architecture of blood vessel or white matter tracts is not replicated.

Transwell and glioma. Many studies have used this assay to study glioma invasion properties. Notably, the transwell system has been used to show that glioma cells were attracted toward endothelial cells or endothelial cell conditioned medium [152, 153]. The transwell system has also been used to test the ability of glioma cells to squeeze their nuclei through the pores of the insert and to show that myosin II was required to squeeze through pores that were smaller than cell nuclei [29].

1.2.4 Linear Grooves

Periodic linear grooves can be prepared by direct laser irradiation on polystyrene film or by soft lithography molding of polydimethylsiloxane (PDMS) [154, 155] (Fig. 8.1d, Table 8.1).

Advantages. This technique allows the control of the size of the grooves (deepness, width); they are easy to use and can be functionalized with any matrix protein.

Limitation. Imaging is complicated by the fact that polystyrene is not transparent. In these systems, all the grooves and ridges are coated with the same matrix; hence, cells are usually migrating on several ridges at the time, and they do not necessarily adopt the same shape as when migrating along the brain blood vessel and white matter [33].

Grooves and glioma. This technique has been used in several studies [33, 156–159]. Glioma cells (C6, U87, primary GBM) have been shown to align and move along the axis of the grooves, a phenomenon called contact guidance [33, 156–158]. This technique has been proposed as clinical prognostic platform [159].

1.2.5 Microchannels

Microchannels can be prepared by molding grooves in polyacrylamide gels or PDMS and sticking them onto a flat surface (Fig. 8.1e, Table 8.1). They allow to study the motility of cells migrating upon 3D confinement.

Advantages. They allow the combination of a linear topology and mechanical confinement. The stiffness of the walls of the channels can be tuned, and the walls and the bottom surface can be functionalized with different matrix protein. The diameter of the channels can vary to study the squeezing of cell nuclei *in live*.

Limitation. While cells migrate along the linear tracks in these systems, the curvature of the blood vessels or the white matter tracts is not recapitulated.

Channels and Gliomas. Glioma migration has been studied using channels with various dimensions and coated with different matrices [160–162]. In microchannels coated with collagen IV, U87 glioma cells spontaneously migrated without any stimulation [160]. A study using microfabricated polyacrylamide channels with independently tunable width and wall stiffness has shown that confinement in narrow channels (10 μm) potentiated glioma motility. This migration was also dependent on myosin II [161].

1.2.6 Micropatterning to Mimic Brain Blood Vessel Geometry

Micropatterning is a daily used technique in our laboratory. Micropatterning techniques allow to "print" matrix protein on any surface (plastic, glass, pillars, PDMS, acrylamide gels) to restrain cell adhesion onto a specific pattern [163–167] (Fig. 8.1f, Table 8.1). Several techniques can be used to micropattern surfaces, such as microcontact printing or deep UV lithography [168, 169]. We will detail one microprinting technique in the Material and Method sections.

Advantages. Micropatterning allows to control the geometry of the adhesive surface. Hence, linear tracks can be easily printed, and motility switches (e.g., from 2D to linear tracks) can be studied in details [170–172]. Lines of different width can be printed, allowing to mimic the motility of glioma cells on large or thin blood vessels [33]. Also, because micropatterning can be prepared on glass and that nothing covers the upper surface of the cells, the imaging techniques are virtually unlimited. It allows the use of sophisticated microscopy methods, such as TIRF microscopy, FRAP experiments, super-resolution microscopy, and electron microscopy, and other mechanobiology tools can be used in combination of micropatterning such as AFM measurements, traction force microscopy, and confinement tools. Moreover, micropatterning techniques can be pushed to more complicated levels. For example, adhesion can be controlled both temporally and spatially, with heat or electrical charges [173–175].

Limitation. Similar to microchannels, micropatterning has few limitations. While cells migrate along the linear tracks in these systems, the curvature of the blood vessels or the white matter tracts is not recapitulated.

Micropatterning and glioma. Micropatterning was first used with C6 glioma cells seeded on 10–20 μm wide stripes coated with poly-L-lysin or fibronectin. In these conditions, cells didn't migrate but displayed an autoreverse nuclear motion, allowing to study the machinery involved in nucleokinesis [176]. Micropatterning has also been used to show that upon confinement on fibronectin island, glioma cells expressed collagen types IV and VI and the collagen cross-linking enzyme lysyl oxidase and upregulated the expression of the angiogenic factor vascular endothelial growth factor [84]. Micropatterned laminin linear tracks were used in our lab to dissect the motility of C6 cells and GSC isolated from patients in great details [33]. Migration of glioma cells on large lines mimicking large blood vessel walls was dependent on their density. As density increased, cells were able to switch from a 2D stochastic motility to an efficient linear migration similar to their migration in vivo [33]. At a specific cell density threshold, antiparallel flows of cells were observed along the axis of the laminin tracks. These flows were similar to the migration of neuroprogenitor cells upon density increase [108]. Within these flows, each individual cell is moved by a two-phase saltatory process that depended on microtubules and contractile actin bundles. Individual cell migration was also studied using thinner lines (4 μm width) on which single cells could migrate, allowing to detail cycles of leading edge protrusion and tail retractions (see kymograph in Fig. 8.5). Also, because micropatterning is applicable on glass, TIRF microscopy could be used to follow the formation and removal of adhesions as the cells were migrating along these tracks. Importantly, we found that glioma linear migration was dependent on formin (specifically FHOD3 was involved) and independent of Arp2/3-mediated actin polymerization [33].

1.2.7 Nanofibers to Mimic White Matter Tracts Geometry

Nanofibers are usually manufactured by electrospinning polymer solutions (Fig. 8.1g, h, Table 8.1). They can also be manufactured by non-electrospinning STEP (spinneret-based tunable engineered parameters) technology, which allows high spinnability of polystyrene in the 400–900 nm fiber diameter range [177–181]. Electrospinning uses electric force to elongate fibers from polymer solutions. Electrospinning allows to prepare fibers of diameters ranging from several hundred nanometers to a few micrometers. They can be aligned as they are collected and functionalized with matrix protein to permit adhesion. They have been exploited as tissue-engineered scaffolds, specifically in the field of neural regeneration. They have been used as substrate to study and promote neural stem cell differentiation and Schwann cell maturation and as guide for neural migration following repair from injury [102, 182–188].

Advantages. Nanofibers offer a unique tool to model the white matter tracts for glioma migration. Their anisotropic elongated structure and nanotopography reflect precisely the mechanical and structural cues present in the brain. Moreover, their manufacturing allows to tune their properties and confers reproducibility. Nanofibers are usually under or around 1 μm in diameter and display nearly identical topographical features to white matter tracts

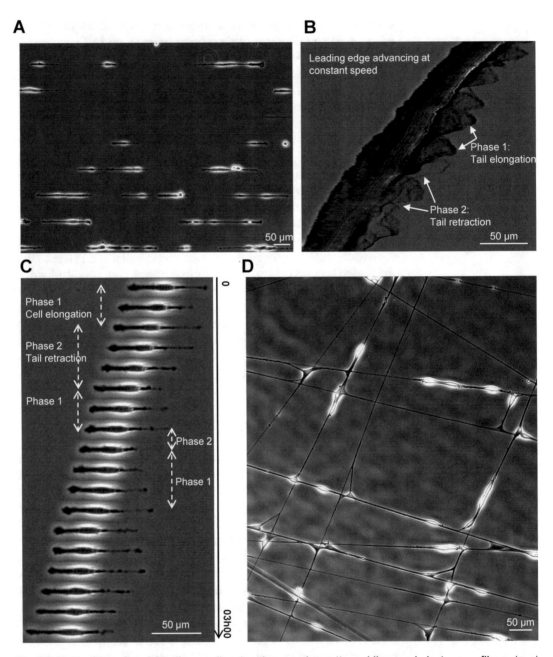

Fig. 8.5 Live cell imaging of C6 glioma cells migrating on micropatterned lines and electrospun fibers. (**a–c**) C6 glioma cells were seeded on laminin micropatterned lines of 3 μm width and imaged every 30 s with a 10× objective. (**a**) Snapshot of the entire field of view. (**b**) Kymograph corresponds to 3 h total time; frames are 30 s apart. (**c**) Montage corresponds to 3 h total time. (**d**) Snapshot of a large field of view of C6 migrating on laminin coated nanofibers (1.3 μm diameter) and imaged with a 10× objective

(diameter 0.5–7 μm) [189–196]. Most of nanofibers are made of a polymer named polycaprolactone (PCL), which degrades slowly and is biocompatible [197]. They have a low tensile modulus (~7 Mpa for PCL fibers) comparable with what is known about the stiffness of white matter tracts [198–200]. This is a great improvement over tissue culture plastic dish or glass coverslips, which have a modulus of >100 MPa [201]. Nanofiber mechanical properties can be further tuned from 2 to 30 MPa by varying their core materials (gelatin, PDMS, PES, chitosan) [177, 196, 202–205]. Fastest migration speeds were observed for aligned fibers of intermediate modulus (11 μm/h for 8 MPa PCL fibers) [202]. Nanofibers can be coated with various matrix proteins that are found in the brain environment, such as hyaluronan, laminin, and collagen IV [44, 202]. Other matrix proteins that are less present in the brain, such as fibronectin and collagen I, have been used as well to study glioma motility. The nanofibers are usually deposited on hard surfaces, aligned or not, at given fiber densities in range similar to white matter tracts in vivo (10,000–30,000 fibers/mm²) [190, 191, 206, 207]. It has been shown in multiple studies that glioma cells recapitulated their "in vivo" morphology when fibers are aligned [188]. These fibers can also be suspended, allowing to observe cells migrating and rotating around single fibers [177, 195]. In our lab, suspended fibers are prepared to precisely observe a true 3D migration of cells [195]. In this setting, glioma and fibroblast are migrating much faster than on aligned fibers (~50 μm/h vs 11 μm/h) [177, 195].

Limitation. Nanofibers are too thin to mimic bigger blood vessels. Some studies have shown that it is possible to increase their diameters up to 10–20 μm, but such fibers have never been used to mimic brain blood vessel topographical features to study glioma motility. Also, imaging on nanofibers is limited compared to micropatterning.

Nanofibers and glioma. Nanofibers have been used to study glioma motility in different settings, allowing single cells to migrate or aggregates of cells to disperse [102, 177, 188, 192, 193, 196, 202, 206–209]. On aligned nanofibers, glioma cells elongate, exhibiting a bipolar spindle-shaped morphology similar to their morphology observed in vivo [44, 45, 102, 193, 196, 202, 206, 207, 209]. This is a common observation for aligned nanotopographies [39]. In most of assays, glioma cells migrated or at least upregulated key migratory genes on aligned fibers [188, 192, 206, 209]. So far, all these assays have been performed on fiber scaffolds not suspended and cells adhered on several fibers at the time; this is very similar to the groove system. Migration speeds depended on the configuration of the fibers (aligned, nonaligned, suspended), their mechanical properties (they migrate better at ~8 MPA), the matrix protein used as coating [102, 196, 202, 206], and the diameter of the fibers [196]. Coculture experiments have shown that astrocytes or astrocyte-conditioned medium promotes glioma migration on nanofibers [207]. Also, glioma cells have been found to be softer when cultured on aligned nanofibers compared to nonaligned fibers or 2D substrates, supporting a role for the topography of the environment in controlling the mechanoproperties of glioma cells [192]. On aligned nanotopography, glioma cells have been found to upregulate key migratory genes, such as FAK, MLC2, TGF-beta, STAT3, and Twist and downregulate key proliferative genes [192, 193, 196, 202]. They displayed actin fibers that aligned along the axis of the fibers, and most focal adhesions were observed in the poles of the cells [177]. Their migration depended on STAT3 signaling [193].

Notably, nanofibers have been used to build up a trap to kill U87 glioma cells in vivo. A conduit containing aligned nanofibers coated with laminin and connected to a cytotoxic collagen gel was inserted inside the brain of glioma-bearing animals. U87 cells were attracted by the fibers and migrated out of the tumor core to die in the extracortical cytotoxic gel [206] (Fig. 8.1H).

1.2.8 Microvessel Bioengineering

A perfect blood vessel mimetic tool for glioma motility studies would involve not only the correct topography, the presence of the correct matrix protein, and the mechanoproperties similar to living blood vessels but also the presence of the cells that compose the perivascular niche of glioma, such as endothelial cells, pericytes, and astrocytes, i.e., the component of the blood-brain barrier (BBB) [60]. Moreover, the first cells to invade away from the tumor must be following naïve blood vessels; hence, a proper "healthy" BBB should be reconstituted to appreciate the motility of these highly invasive cells. However, with time, glioma is modifying the vasculature triggering endothelial cell proliferation and hyperplasia as well as smooth muscle cells and pericyte recruitment, which are necessary for the survival of tumor-associated endothelial cells and the formation of a proper perivascular niche [210–212]. Additionally, glioblastoma stem cells themselves can transdifferentiate to become endothelial cells and connect with the vasculature [213–215]. Also, glioma cells are modifying astrocytes to their own advantage [216–218], and reactive astrocytes are often associated with the glioma residing in the perivascular niche. These astrocytes can promote tumor growth and survival by secreting factors, such as SDF1 and AEG1 [219–223]. All these additional factors should be considered when designing artificial BBB.

The BBB is formed by a layer of highly functionalized endothelial cells (ECs), building the vessel walls that interact mainly with astrocytes and pericytes. The BBB maintains the homeostasis of the brain interstitial fluids necessary for optimal neuronal signaling [224, 225]. Several bioengineering strategies have been developed to reproduce the blood-brain barrier in vitro. They are presented in Fig. 8.2. They include the following:

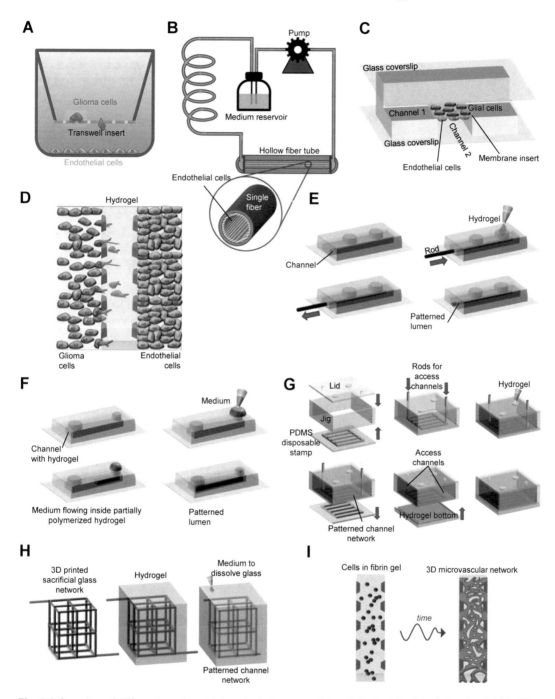

Fig. 8.2 Overview of different mechanobiology techniques used to mimic the blood-brain barrier. (**a**) Modified transwell system; (**b**) dynamic in vitro model utilizing hollow fiber tubes; (**c**) vessel-like microfluidic channels; (**d**) compartmentalized microfluidic models; (**e**) single vessel patterning in hydrogel matrix; (**f**) viscous fingering; (**g**) micromolded vessel networks; (**h**) 3D printed vessel networks; (**i**) organotypic microvascular networks

(a) Adaptations of the transwell system [226, 227]

(b) Dynamic in vitro models utilizing hollow fiber tubes, allowing to study the impact of shear stress on endothelial cells, monocyte extravasation in the brain, or drug screening [228–230]

(c) Vessel-like microfluidic channels [152, 231–235]

(d) Compartmentalized microfluidic models providing a dynamic flow on the luminal side [236–239]

(e) Single vessel patterning in hydrogel matrix allowing the study of invasion and intravasation of cancer cells and brain microvasculature disruption followed by recovery [240–243]

(f) Viscous fingering in which a cylindrical hollow shape serving as a substrate for endothelial cell adhesion is patterned within a hydrogel, allowing live cell studies with a geometry really close to in vivo vessels [244–246]

(g) Micromolded vessel networks in which hollow microchannel networks are patterned between two collagen layers, allowing the study of angiogenic and thrombotic phenomena [247]

(h) 3D printed vessel networks obtained by 3D printing a rigid filament network made of an engineered carbohydrate glass, which is dissolvable in culture medium leaving tangled patterns inside a hydrogel matrix that can be filled with endothelial cells [248]

(i) Organotypic microvascular networks which are engineered in microfluidic platforms where ECs are included in a fibrin gel which spontaneously promote vasculogenesis in vitro [249, 250]

Adopting this approach, the brain-vessel interface can be modeled in a human microfluidic model comprising astrocytes and pericytes, with diameters typical of human arterioles and venules [251]. Notably, patient-derived glioma cells migrated with behavior and morphology resembling in vivo data when cultured in such microvascular networks [252].

Bioengineered microvessels and glioma. Only some of these tools have been used to study glioma invasion properties. The vessel-like microfluidic channels and the transwell assay with endothelial cells [152] have been used to show that endothelial cells or endothelial-conditioned medium attracted glioma cells. However, the migration per se was done through collagen 1 [152]. Single vessel patterning in hydrogel matrix has been used to investigate the influence of glioma cells on angiogenesis [243]. Notably in organotypic microvascular networks, patient-derived glioma cells migrated with behavior and morphology resembling in vivo data when cultured in such microvascular networks [252].

The strategies and the biophysical tools elaborated to reproduce in vitro the brain-vessel interface can offer information more relevant as technology improves, aiming to mimic closer and closer the physiological glioma environment.

In the following chapter, we will detail how to prepare and use two specific tools, micropatterning and nanofibers, that are used to mimic blood vessels and white matter tracts, respectively.

2 Materials

2.1 Microcontact Printing

Chemical: Hexamethyldisilazane (Sigma #440191); Sylgard 184 Silicone Elastomer Kit (Dow Corning); Pluronic® F-127 0.2% solution (Sigma # P2443-250G); PLL(20)-g[3.5]-PEG(2) (SuSoS); laminin isolated from the Engelbreth-Holm-Swarm (EHS) sarcoma (Invitrogen); 3-(trimethoxysilyl)propyl methacrylate (Sigma).

Specific equipment and materials: Compressed inert gas such as nitrogen; vacuum desiccator (Sigma # Z119008-1EA); plasma cleaner (Harrick Plasma), compressed air; silicon wafer with specific designs (note 2), scalpel or razor blade, tweezers with curve tip (EMS high precision and ultrafine tweezers EMS 2AB, SA).

2.2 Electrospun Fiber Preparation

Chemicals: Trifluoroethanol (TFE) (Sigma-Aldrich, cat. No. T63002-100G); polycaprolactone (PCL), (Sigma-Aldrich, cat. No. 440744-5G); laminin isolated from the Engelbreth-Holm-Swarm (EHS) sarcoma (Invitrogen).

Specific equipment and materials: Coverslips, Marienfeld, cat. No 0117640; compressed air; electrospinning machine (see Fig. 8.4) comprising one syringe pump (kd Scientific, cat. No. 78–8100), a power supply (Gamma High Voltage Research Inc., model No. D-ES30PN-5 W/QDPM), two clip electrodes, a vertically adjustable stage); imaging chambers to accommodate suspended fibers (we use Chamlide CMB 35 mm dish type 1-well magnetic chambers for round coverslip: http://www.quorum-technologies.com); precision cover glasses thickness No. 1.5H (tol. ±5 μm) for high-performance microscopes, 24 mm Ø, (Marienfeld); 1 mL plastic syringe; voltage-sensing frames (11 cm × 3.5 cm); 1 small step (smaller than the frame); syringe blunt needle (30Ga).

2.3 Microscopy and Tracking

Long-term imaging: Microscope equipped with temperature, humidity, and CO_2 control (Olympus IX81) or a BioStation (Nikon); Hoechst 33342 (Sigma-Aldrich).

TIRF imaging: iLas2 TIRF system connected to an Olympus IX81 microscope equipped with temperature, humidity, and CO_2 control.

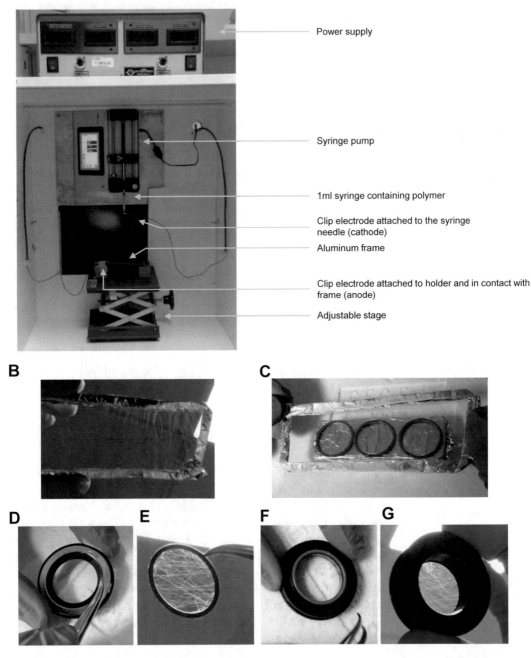

Fig. 8.4 Main steps of electrospinning. (**a**) The electrospinning machine is assembled in a dedicated cabinet. On top of the cabinet stands, the power supply. At the back of the cabinet, two holes allows the passage of the wires connecting the electrodes to the power supply. The syringe pump allows the control of the flow rate. The 1 mL syringe loaded with the PCL solution is connected to a blunt needle (30 gauge) to which the cathode is connected. The adjustable stage allows to position the aluminum frame (11 cm × 3.5 cm) horizontally at 12 cm distance from the tip of the needle. The frame is connected to the anode. (**b**) After electrospinning, fibers are observed aligned across the frame. (**c**) The nanofibers are transferred to three rubber rings that will be as

Tracking: Manual tracking plug-in of Fiji (ImageJ; National Institute of Health, Bethesda, MD).

2.4 Cell Culture and Transfections

Cell lines: C6 rat glioma cells (ATCC), grown in high glucose DMEM supplemented with 10% heat inactivated FBS (HI-FBS) and glutamine (Invitrogen). Human glioma-propagating cells (hGPCs) isolated and cultured as tumor spheres in high glucose DMEM/F12 (Invitrogen) supplemented with sodium pyruvate, nonessential amino acid, penicillin/streptomycin, B27 supplement (Invitrogen), bFGF (20 ng/mL), EGF (20 ng/mL) (gene-ethics-asia.com), and heparin (5 μg/mL) (Sigma).

Transfection: Neon electroporator (Invitrogen); mCherry-paxillin cDNA (Clare Waterman; Bethesda, MD, USA).

3 Methods

3.1 Microcontact Printing
3.1.1 Silanization of Silicon Wafers

Laminin coated lines of 3–400 μm are printed as detailed below and previously described [33, 168, 253] (see Fig. 8.3).

Before the first use, silicon wafers must be silanized. After this step, wafers can be used multiple times without the need to be silanized again. Silane solution is withdrawn from the bottle by pocking a 1-mL-syringe needle through the rubber cap. An equivalent quantity of inert gas must be added first.

Procedure:

1. Place the silicon wafer inside a plastic petri dish with the cap of an Eppendorf tube next to it.

2. Place the petri dish in the desiccator.

3. Add the compressed inert gas in a beaker sealed with parafilm (keep adding the gas while aspirating with the syringe) and aspirate 100–200 μL of gas in the syringe.

4. Punch the rubber cap of the hexamethyldisilazane bottle with the syringe, turn the bottle upside down, inject the gas, and remove 100–200 μL of hexamethyldisilazane.

5. Place the hexamethyldisilazane in the cap of the Eppendorf tube.

Fig. 8.4 (continued) sembled in the imaging chamber. (**d**) A clean 24 mm diameter glass coverslip is placed at the bottom of the chamber. (**e**) Five series of suspended fibers have been transferred on the same rubber ring at different angles to prepare an array of interconnected fibers. (**f**) The ring with the fibers is mounted on the upper part of the magnetic chamber. (**g**) The two parts of the magnetic chambers are assembled. The fibers are on the upper side of the rubber ring; the glass coverslip is on the lower side of the rubber ring, so the fibers do not touch the glass and stay suspended

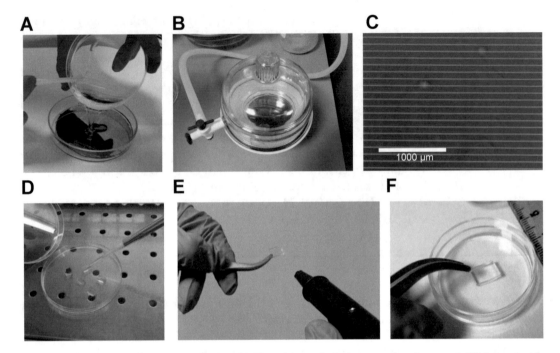

Fig. 8.3 Main steps of microcontact printing. (**a**) The mixture of elastomer and curing agent (10:1, previously degassed) is poured on the silicon wafer (which has been silanized). (**b**) The mixture on the silicon wafer is kept in a vacuum desiccator to remove all bubbles. (**c**) After curing the PDMS, a piece of PDMS line is imaged with a 4X objective. 5 μm lines separated by 75 μm gap are clearly visible. (**d**) Laminin is deposited on stamps after plasma cleaning and incubated for 20–30 min. (**e**) Laminin is air blown of the stamp. (**f**) Stamp is applied against the surface of the imaging dish

6. Cover the desiccator; apply vacuum for 1–2 h.

7. The wafer is then transferred to a glass petri dish ready to use (note 17).

3.1.2 Polydimethyl-siloxane (PDMS) Preparation

Stamps are prepared using the Sylgard 184 Silicone Elastomer Kit (see note 1) and a silanized silicon wafer (see note 2). This kit is composed of two solutions: an elastomer solution and a curing agent that must be mixed at a 10 parts to 1 part ratio (10:1) (w/w).

Procedure:

1. In a weighting board, weight out around 10 g of elastomer solution by pouring the solution directly, and add 1 g of curing agent using a plastic Pasteur pipette.

2. Stir vigorously the mixture using the same plastic Pasteur pipette.

3. Degas the mixture by placing the weighting board into a vacuum desiccator for 30 min.

4. Verify that no more bubbles are present and pour gently all the solution on the wafer.

5. Degas again for 30 min.

6. Verify that no more bubbles are present and place the wafer at 70 °C to cure for 2–4 h (see note 3).

7. Unmold the PDMS and place it face up in a 15 cm plate.

3.1.3 Stamping and Passivation

During this procedure, the stamp is covered with a solution of laminin, excess of laminin solution is then removed, and the stamp is pressed on the surface of the imaging dish to imprint the feature onto the dish (note 5). The rest of the surface is then passivated to prevent the cells from attaching outside of the stamped feature.

1. Using a scalpel, cut out the desired feature (stamps are around 1 cm), place it in a small culture dish (the feature must be facing up), and plasma clean for 3 min.

2. Add a drop of laminin solution (50 µg/mL in dPBS) on top of the stamp, and incubate at room temperature for 20–30 min.

3. Using a tweezer, hold the stamp and air blow dry all the laminin; turn it upside down onto the imaging plastic dish.

4. Press gently the stamp against the surface of the dish and remove it at once (note 4).

5. Add 2 mL 0.2% Pluronic solution and incubate 1 h at room temperature (note 6).

3.1.4 Rinsing Dishes and Cell Seeding

1. Rinse dishes five times with PBS and replace with culture medium.

2. Detach, count glioma cells, and seed between 5000 and 100,000 cells, depending on the size of the lines.

3. Incubate at 37 °C for 1–2 h and rinse again with warmed medium to get rid of the floating cells before imaging.

3.1.5 Micropatterning on Glass Bottom Dishes

To micropattern on glass (dishes or coverslips), the overall technique is the same except that the glass needs to be silanized with 3-(trimethoxysilyl)propyl methacrylate and the passivation is done with PEG-PLL.

Procedure:

1. Incubate the glass bottom dishes in silane solution (1.2 µL of 3-(trimethoxysilyl)propyl methacrylate diluted in 2 mL methanol per dish) for 1 h.

2. Remove silane solution and rinse three times with methanol for 5 min each time on a rocker.

3. Air-dry the dishes under the chemical hood to evaporate the leftover methanol.

4. Proceed with the stamping procedure as indicated in Sect. 3.1.3 (steps 1–4).

5. Instead of passivating with Pluronic, add 300 μL of PLL(20)-g[3.5]-PEG(2) (0.5 mg/mL in dPBS) on the stamped area.

6. Incubate 1 h at room temperature.

7. Proceed as indicated in Sect. 3.1.4.

3.2 Electrospun Fiber Preparation

Nanofibers are prepared as previously described [195]. The different steps are illustrated in Fig. 8.4. Alternatively, aligned nanofibers can be purchased from Nanofiber Solutions TM (https://nanofibersolutions.com/products/).

3.2.1 Polymer Dissolution

To prepare 1300 nm diameter fibers, polycaprolactone (PCL) at 12% (weight/weight) in 2,2,2-trifluoroethanol (TFE) solution is prepared as follow: 1 g of PCL is diluted in 5.26 mL TFE (corresponding to 7.327 g) at room temperature, preferably the day before electrospinning (note 7) in a 15 mL falcon tube.

3.2.2 Electrospinning Machine Preparation

1. Load the PCL solution inside a 1 mL syringe (note 8).

2. Equip the syringe with a blunt needle (30G), making sure to prime the needle volume.

3. Lock the syringe inside the syringe pump.

4. Place the frame (note 9) on a gap where the anode electrode is connected to. The gap can be formed by any material in the lab (e.g., we used two coverslip boxes).

5. Clip the cathode electrode to the syringe needle.

3.2.3 Electrospinning Parameter Adjustments

Three parameters can be adjusted to tune the diameter of the fibers:

- d: distance between the needle and the frame
- Q: flow rate of the syringe pump
- V: voltage of the power supply

To make 1300 nm fibers, the settings are as follows: $d = 12$ cm, $Q = 0.5$ mL/h, $V = 5$ kV (we set 5 kV by imposing 2 kV at the anode and 7 kV at the cathode).

3.2.4 Electrospinning Fibers

Once the syringe has been loaded, the frame and electrodes positioned, and the parameters adjusted (see Fig. 8.4), electrospinning can start following these steps in order (note 10):

1. Clean off any droplet of polymer from the needle tip with a cotton swab.

2. Place a paper sheet between the needle and the frame (note 10).

3. Turn on the anode switch of the power supply.

4. Turn on the cathode switch of the power supply.

5. Run the syringe pump.

6. After 10–15 s, remove the paper, and wait for 20–30 s.

7. Place the paper again.

8. Turn off the syringe pump.

9. Turn off cathode switch.

10. Turn off anode switch.

   See safety issue in note 11.

11. Once all switches have been turned off, the frame can be removed and fibers can be observed under light. Frame with suspended fibers can be kept in boxes or directly mounted on imaging chambers.

3.2.5 Preparing Arrays of Suspended Fibers in Imaging Chambers

We use Chamlide CMB 35 mm dish type 1-well magnetic chambers for round coverslip (see Fig. 8.4). These chambers are formed by four elements: (1) A 24 mm diameter round coverslip is hosted in a (2) thin metallic ring, forming the bottom of the chamber. The presence of the coverslip as bottom surface makes the system compatible with imaging techniques. The metallic ring is coupled with the upper part of the chamber via magnetic forces. In fact, the upper part of the chamber (3) is constituted by another thicker metallic ring embedding a magnet, having on the bottom surface (facing the bottom of the chamber) a round groove. This groove is conceived to host a rubber ring across which fibers will be suspended. The plastic ring can be placed inside the groove by gently applying mechanical pressure.

Procedure:

1. Place a clean cover slip (D 24 mm) on the metallic ring forming the bottom chamber (note 12).

2. Place the chamber's rubber ring on a small step on the bench (see Fig. 8.4).

3. Depose gently the fibers on the rubber ring (note 13).

4. Insert the orange ring in the thick upper metal ring of the chamber by applying pressure and check that it is flat; otherwise, gently flatten it with tweezers.

5. Clip with the bottom ring; the chamber containing fibers is ready and can be kept for days in a tissue culture dish.

3.2.6 Fiber Coating and Cell Seeding

We use laminin to coat our suspended nanofibers. The following steps are carried under a tissue culture hood to maintain sterile conditions.

Procedure to coat one chamber:

1. Prepare 2 mL of laminin at 10 μg/mL in dPBS.

2. Add gently 1 mL of laminin solution inside the chamber. The solution should be added slowly against the side of the dish, not directly on the fibers. Then, add gently another 1 mL.

3. Incubate at 37 °C for 1 h.

4. Rinse fibers with 4 mL of dPBS (extra care must be taken at this stage; see note 14).

5. Rinse fibers with 4 mL of warmed culture medium (extra care must be taken at this stage; see note 14).

6. Detach, count glioma cells, and seed between 30,000 and 100,000 cells in the chamber (note 15).

7. Incubate at 37 °C for 1–2 h before imaging.

3.3 Live Cell Imaging and Cell Tracking

3.3.1 Migration Assays

For migration assays (measure of speed and persistence), cells on micropatterned lines or electrospun fibers are imaged in phase contrast using a 10X objective on a microscope equipped with temperature, humidity, and CO_2 control. Acquisitions are typically obtained over a period of 12–24 h. For analysis of large population of streaming cells, Hoechst 33342 (Sigma-Aldrich) is added to the medium to label the nuclei. To keep the focus on the suspended fibers, see note 16. To analyze the cycle of protrusion retractions of the tail and leading edge, images are taken every 30 s and kymographs and montages are built using ImageJ (see Fig. 8.5b, c).

3.3.2 Imaging Live Focal Adhesions

To image focal adhesion dynamics, cells must be transfected with focal adhesion markers such as vinculin or paxillin coupled to GFP or mCherry. Microcontact printing must be done on glass bottom dishes or coverslips as indicated in method Sect. 3.1.5. TIRF images are obtained through a 60X oil immersion objective with TIRF system (we use an iLas2 TIRF system) connected to a microscope (we use Olympus IX81) equipped with temperature, humidity, and CO_2 control. For a better imaging, medium is replaced with DMEM without red phenol complemented with serum or B27 depending on the cells used.

3.3.3 Tracking

Because glioma cells are moving in antiparallel streams, change shapes, and bump frequently in each other, tracking programs cannot be used to determine speed and persistence. In consequence, cell bodies are manually tracked using the manual tracking plug-in of Fiji (ImageJ; National Institute of Health, Bethesda, MD). Lengths of tracks are used to calculate the mean speeds (distance/time).

4   Notes

1. All MSDS and chemical property information along with further instructions for using the Sylgard 184 Silicone Elastomer Kit can be obtained from the manufacturer, Dow Corning's website: http://www.dowcorning.com/applications/search/

default.aspx?R=131EN. The elastomer solution is extremely viscous. It is easier to pour out the solution directly from the bottle instead of trying to pipet it out. To cover a silicon wafer, 10–15 g of solution is sufficient. The quantity of curing agent is adapted to the quantity of elastomer solution that has been weighted.

2. Photomask to produce wafers with the proper design can be ordered through different companies such as Delta Mask, Enschede, Netherlands; Advance Reproductions, North Andover, MA, USA; and Microtronics, Newton, PA, USA. Silicon wafer can also be directly obtained (Biotry, Lyon, France; Mechanobiology Institute, Singapore). The design of the patterns is usually made with specialized software, such as L-Edit or Clewin, that produce file formats directly compatible with the manufacturer's machine, but it can also be made with any drawing software, when all the necessary information for size of patterns is indicated [168].

3. PDMS can be cured overnight as well.

4. This is the delicate part of the procedure. Extra care must be given to this step. If the stamp falls back onto the dish, then the feature will be printed several times, ruining the stamping procedure.

5. To monitor migrating cells along the laminin linear tracks, we use untreated bacterial plastic dishes (35 mm), which allow a better adsorption of the matrix proteins and a better passivation with the Pluronic solution.

6. Pluronic solution (0.2% in Milli-Q water) solubilizes better when left at 4 °C overnight. The solution is then filtered (0.45 μm pore) and kept at room temperature.

7. PCL is difficult to solubilize in TFE. For a better dissolution, leave the solution at room temperature in a 15 mL tube, overnight. Make sure to seal the container of the solution to avoid TFE evaporation, which would change the concentration of the solution. Before use, vortex the solution for 10 min.

8. The PCL solution is extremely viscous. The solution is aspirated directly in the syringe (without the needle). After electrospinning, the PCL solution left in the syringe can be recycled in the original tube and stored at room temperature.

9. The frame can be simply obtained by wrapping aluminum foil. Its size should be 11 cm × 3.5 cm (Fig. 8.4).

10. We place a piece of paper between the needle and the frame during the first seconds of electrospinning to give the time to the machine to draw fibers evenly. Once the flow is well started, we remove the piece of paper to let the fibers lay on the frame.

After the fibers have been drawn (20–30 s), the paper is placed again between the needle and the frame to nicely cut the fiber tread.

11. The electrospinning is occurring at very high voltage. Extreme care must be taken before touching the frame; all switches must be off.

12. We usually acid wash our coverslips before use by incubating them in a 20% nitric acid solution for 2 h. Coverslips are then rinse three times in distilled water and kept in a solution of ethanol 90%. We use coverslips that are compatible with TIRF microscopy: precision cover glasses thickness No. 1.5H (tol. ±5 μm) for high-performance microscopes, 24 mm Ø, from Marienfeld (Germany).

13. To prepare networks of fibers, we depose fibers 4–8 times on the same rubber ring, gently rotating the rubber ring with tweezers by 45° between each deposition.

14. To rinse fibers, extreme care must be taken since fibers must stay immersed the whole time. From the 2 mL solution of laminin in the imaging chamber, 1 mL is removed and 1 mL of dPBS is gently added. Then, again 1 mL of the solution is removed from the imaging chamber and 1 mL of dPBS is gently added. The same method is applied until the 4 mL of dPBS has been used. Then, the same method is applied to exchange the dPBS with 4 mL of medium.

15. The number of cells to add on the fibers greatly depends on the ability of the cells to attach and migrate. For example, C6 cells attach rapidly, so we add 30,000 cells. For cells that are less prone to attach and migrate, we seed 100,000 cells to increase the chance of the cells to attach on fibers while descending the medium column.

16. Imaging fibers can be problematic, especially when multiple position is required. Since fibers are suspended, they are prone to move around when the position of the microscope stage is changing. To make sure to keep the focus on the cell of interest, we take two images of the same field of view for each time point. The second image is the one of interest.

17. We always test if the silanization was successful by adding a drop of PDMS on a corner of the wafer and curing it for 2–4 h. The PDMS should be easy to peel off. If not, the silanization procedure must be repeated.

References

1. Beauchesne P (2011) Extra-neural metastases of malignant gliomas: myth or reality? Cancers 3:461–477. https://doi.org/10.3390/cancers3010461

2. Hamilton JD et al (2014) Glioblastoma multiforme metastasis outside the CNS: three case reports and possible mechanisms of escape. J Clin Oncol 32:e80–e84. https://doi.org/10.1200/jco.2013.48.7546

3. Wells A, Grahovac J, Wheeler S, Ma B, Lauffenburger D (2013) Targeting tumor cell motility as a strategy against invasion and metastasis. Trends Pharmacol Sci 34:283–289. https://doi.org/10.1016/j.tips.2013.03.001

4. Veiseh O, Kievit FM, Ellenbogen RG, Zhang M (2011) Cancer cell invasion: treatment and monitoring opportunities in nanomedicine. Adv Drug Deliv Rev 63:582–596. https://doi.org/10.1016/j.addr.2011.01.010

5. Lefranc F, Brotchi J, Kiss R (2005) Possible future issues in the treatment of glioblastomas: special emphasis on cell migration and the resistance of migrating glioblastoma cells to apoptosis. J Clin Oncol 23:2411–2422. https://doi.org/10.1200/jco.2005.03.089

6. Bravo-Cordero JJ, Hodgson L, Condeelis J (2011) Directed cell invasion and migration during metastasis. Curr Opin Cell Biol 24:277–283. https://doi.org/10.1016/j.ceb.2011.12.004

7. Bernstein JJ, Woodard CA (1995) Glioblastoma cells do not intravasate into blood vessels. Neurosurgery 36:124–132. discussion 132

8. Scherer HJ (1938) Structural development in gliomas. Am J Cancer 34:333–351. https://doi.org/10.1158/ajc.1938.333

9. Scherer HJ (1940) The forms of growth in gliomas and their practical significance. Brain 63:1–35. https://doi.org/10.1093/Brain/63.1.1

10. Louis DN (2006) Molecular pathology of malignant gliomas. Annu Rev Pathol 1:97–117. https://doi.org/10.1146/annurev.pathol.1.110304.100043

11. Jones TR, Ruoslahti E, Schold SC, Bigner DD (1982) Fibronectin and glial fibrillary acidic protein expression in normal human brain and anaplastic human gliomas. Cancer Res 42:168–177

12. Bellon G, Caulet T, Cam Y, Pluot M, Poulin G, Pytlinska M, Bernard MH (1985) Immunohistochemical localisation of macromolecules of the basement membrane and extracellular matrix of human gliomas and meningiomas. Acta Neuropathol 66:245–252

13. McComb RD, Bigner DD (1985) Immunolocalization of laminin in neoplasms of the central and peripheral nervous systems. J Neuropathol Exp Neurol 44:242–253. https://doi.org/10.1097/00005072-198505000-00003

14. Gladson CL (1999) The extracellular matrix of gliomas: modulation of cell function. J Neuropathol Exp Neurol 58:1029–1040

15. Calabrese C et al (2007) A perivascular niche for brain tumor stem cells. Cancer Cell 11:69–82. https://doi.org/10.1016/j.ccr.2006.11.020

16. Gilbertson RJ, Rich JN (2007) Making a tumour's bed: glioblastoma stem cells and the vascular niche. Nat Rev Cancer 7:733–736. https://doi.org/10.1038/nrc2246

17. Charles N, Ozawa T, Squatrito M, Bleau AM, Brennan CW, Hambardzumyan D, Holland EC (2010) Perivascular nitric oxide activates notch signaling and promotes stem-like character in PDGF-induced glioma cells. Cell Stem Cell 6:141–152. https://doi.org/10.1016/j.stem.2010.01.001

18. Farin A, Suzuki SO, Weiker M, Goldman JE, Bruce JN, Canoll P (2006) Transplanted glioma cells migrate and proliferate on host brain vasculature: a dynamic analysis. Glia 53:799–808. https://doi.org/10.1002/glia.20334

19. Burden-Gulley SM et al (2011) Novel cryo-imaging of the glioma tumor microenvironment reveals migration and dispersal pathways in vivid three-dimensional detail. Cancer Res 71:5932–5940. https://doi.org/10.1158/0008-5472.can-11-1553

20. Zagzag D et al (2000) Vascular apoptosis and involution in gliomas precede neovascularization: a novel concept for glioma growth and angiogenesis. Lab Investig 80:837–849

21. Lugassy C et al (2002) Pericytic-like angiotropism of glioma and melanoma cells. Am J Dermatopathol 24:473–478

22. Holash J et al (1999) Vessel cooption, regression, and growth in tumors mediated by angiopoietins and VEGF. Science 284:1994–1998

23. Nagano N, Sasaki H, Aoyagi M, Hirakawa K (1993) Invasion of experimental rat brain tumor: early morphological changes following microinjection of C6 glioma cells. Acta Neuropathol 86:117–125

24. Winkler F et al (2009) Imaging glioma cell invasion in vivo reveals mechanisms of dissemination and peritumoral angiogenesis. Glia 57:1306–1315. https://doi.org/10.1002/glia.20850

25. Hirata E et al (2012) In vivo fluorescence resonance energy transfer imaging reveals differential activation of rho-family GTPases in glioblastoma cell invasion. J Cell Sci 125:858–868. https://doi.org/10.1242/jcs.089995

26. Friedl P, Alexander S (2011) Cancer invasion and the microenvironment: plasticity and reciprocity. Cell 147:992–1009. https://doi.org/10.1016/j.cell.2011.11.016

27. Friedl P, Wolf K (2010) Plasticity of cell migration: a multiscale tuning model. J Cell Biol 188:11–19. https://doi.org/10.1083/jcb.200909003

28. Petrie RJ, Yamada KM (2016) Multiple mechanisms of 3D migration: the origins of plasticity. Curr Opin Cell Biol 42:7–12. https://doi.org/10.1016/j.ceb.2016.03.025

29. Beadle C, Assanah MC, Monzo P, Vallee R, Rosenfeld SS, Canoll P (2008) The role of myosin II in glioma invasion of the brain. Mol Biol Cell 19:3357–3368. https://doi.org/10.1091/mbc.E08-03-0319

30. Petrie RJ, Gavara N, Chadwick RS, Yamada KM (2012) Nonpolarized signaling reveals two distinct modes of 3D cell migration. J Cell Biol 197:439–455. https://doi.org/10.1083/jcb.201201124

31. Weeks A, Okolowsky N, Golbourn B, Ivanchuk S, Smith C, Rutka JT (2012) ECT2 and RASAL2 mediate mesenchymal-amoeboid transition in human astrocytoma cells American. J Pathol 181:662–674. https://doi.org/10.1016/j.ajpath.2012.04.011

32. Yamazaki D, Kurisu S, Takenawa T (2009) Involvement of Rac and Rho signaling in cancer cell motility in 3D substrates. Oncogene 28:1570–1583. https://doi.org/10.1038/onc.2009.2

33. Monzo P et al (2016) Mechanical confinement triggers glioma linear migration dependent on formin, FHOD3. Mol Biol Cell 27:1246–1261. https://doi.org/10.1091/mbc.E15-08-0565

34. Wolf K et al (2013) Physical limits of cell migration: control by ECM space and nuclear deformation and tuning by proteolysis and traction force. J Cell Biol 201:1069–1084. https://doi.org/10.1083/jcb.201210152

35. Mair DB, Ames HM, Li R (2018) Mechanisms of invasion and motility of high-grade gliomas in the brain. Mol Biol Cell 29:2509–2515. https://doi.org/10.1091/mbc.E18-02-0123

36. Rape A, Ananthanarayanan B, Kumar S (2014) Engineering strategies to mimic the glioblastoma microenvironment. Adv Drug Deliv Rev 79-80:172–183. https://doi.org/10.1016/j.addr.2014.08.012

37. Butcher DT, Alliston T, Weaver VM (2009) A tense situation: forcing tumour progression. Nat Rev Cancer 9:108–122. https://doi.org/10.1038/nrc2544

38. Pedersen JA, Swartz MA (2005) Mechanobiology in the third dimension. Ann Biomed Eng 33:1469–1490. https://doi.org/10.1007/s10439-005-8159-4

39. Bettinger CJ, Langer R, Borenstein JT (2009) Engineering substrate topography at the micro- and nanoscale to control cell function. Angew Chem Int Ed Engl 48:5406–5415. https://doi.org/10.1002/anie.200805179

40. de Gooijer MC, Guillen Navarro M, Bernards R, Wurdinger T, van Tellingen O (2018) An experimenter's guide to glioblastoma invasion pathways. Trends Mol Med 24:763–780. https://doi.org/10.1016/j.molmed.2018.07.003

41. Cha J, Kim P (2017) Biomimetic strategies for the glioblastoma microenvironment. Front Mater 4:1–8. https://doi.org/10.3389/fmats.2017.00045

42. Barnes JM, Przybyla L, Weaver VM (2017) Tissue mechanics regulate brain development, homeostasis and disease. J Cell Sci 130:71–82. https://doi.org/10.1242/jcs.191742

43. Hanahan D, Weinberg RA (2011) Hallmarks of cancer: the next generation. Cell 144:646–674. https://doi.org/10.1016/j.cell.2011.02.013

44. Giese A, Westphal M (1996) Glioma invasion in the central nervous system. Neurosurgery 39:235–252. https://doi.org/10.1097/00006123-199608000-00001

45. Guillamo JS, Lisovoski F, Christov C, Le Guerinel C, Defer GL, Peschanski M, Lefrancois T (2001) Migration pathways of human glioblastoma cells xenografted into the immunosuppressed rat brain. J Neuro-Oncol 52:205–215. https://doi.org/10.1023/a:1010620420241

46. Griveau A et al (2018) A glial signature and Wnt7 Signaling regulate glioma-vascular interactions and tumor microenvironment. Cancer Cell 33:874–889.e877. https://doi.org/10.1016/j.ccell.2018.03.020

47. Thorne RG, Nicholson C (2006) In vivo diffusion analysis with quantum dots and dextrans predicts the width of brain extracellular space. Proc Natl Acad Sci U S A 103:5567–5572. https://doi.org/10.1073/pnas.0509425103

48. Elkin BS, Azeloglu EU, Costa KD, Morrison B III (2007) Mechanical heterogeneity of the rat hippocampus measured by atomic force microscope indentation. J Neurotrauma 24:812–822. https://doi.org/10.1089/neu.2006.0169

49. Hrapko M, van Dommelen JA, Peters GW, Wismans JS (2008) The influence of test conditions on characterization of the mechani-

cal properties of brain tissue. J Biomech Eng 130:031003–031010. https://doi.org/10.1115/1.2907746

50. Franze K (2013) The mechanical control of nervous system development. Development 140:3069–3077. https://doi.org/10.1242/dev.079145

51. Candiello J, Balasubramani M, Schreiber EM, Cole GJ, Mayer U, Halfter W, Lin H (2007) Biomechanical properties of native basement membranes. FEBS J 274:2897–2908. https://doi.org/10.1111/j.1742-4658.2007.05823.x

52. Ulrich TA, de Juan Pardo EM, Kumar S (2009) The mechanical rigidity of the extracellular matrix regulates the structure, motility, and proliferation of glioma cells. Cancer Res 69:4167–4174. https://doi.org/10.1158/0008-5472.CAN-08-4859

53. Ananthanarayanan B, Kim Y, Kumar S (2011) Elucidating the mechanobiology of malignant brain tumors using a brain matrix-mimetic hyaluronic acid hydrogel platform. Biomaterials 32:7913–7923. https://doi.org/10.1016/j.biomaterials.2011.07.005

54. Bangasser BL et al (2017) Shifting the optimal stiffness for cell migration. Nat Commun 8:1–10. https://doi.org/10.1038/ncomms15313

55. Paszek MJ, Zahir N, Johnson KR, Lakins JN, Rozenberg GI, Gefen A, Reinhart-King CA, Margulies SS, Dembmo M, Boettiger D, Hammer DA, Weaver VM (2005) Tensional homeostasis and the malignant phenotype. Cancer Cell 8:241–254. https://doi.org/10.1016/j.ccr.2005.08.010

56. Weinberg SH, Mair DB, Lemmon CA (2017) Mechanotransduction dynamics at the cell-matrix Interface. Biophys J 112:1962–1974. https://doi.org/10.1016/j.bpj.2017.02.027

57. Boucher Y, Salehi H, Witwer B, Harsh GR, Jain RK (1997) Interstitial fluid pressure in intracranial tumours in patients and in rodents. Br J Cancer 75:829–836

58. Jain RK, di Tomaso E, Duda DG, Loeffler JS, Sorensen AG, Batchelor TT (2007) Angiogenesis in brain tumours. Nat Rev Neurosci 8:610–622. https://doi.org/10.1038/nrn2175

59. Grundy TJ et al (2016) Differential response of patient-derived primary glioblastoma cells to environmental stiffness. Sci Rep 6:23353. https://doi.org/10.1038/srep23353

60. Charles NA, Holland EC, Gilbertson R, Glass R, Kettenmann H (2012) The brain tumor microenvironment. Glia 60:502–514. https://doi.org/10.1002/glia.21136

61. Bellail AC, Hunter SB, Brat DJ, Tan C, Van Meir EG (2004) Microregional extracellular matrix heterogeneity in brain modulates glioma cell invasion. Int J Biochem Cell Biol 36:1046–1069. https://doi.org/10.1016/j.biocel.2004.01.013

62. Paulus W, Roggendorf W, Schuppan D (1988) Immunohistochemical investigation of collagen subtypes in human glioblastomas. Virchows Arch A Pathol Anat Histopathol 413:325–332

63. Clark RA, Quinn JH, Winn HJ, Lanigan JM, Dellepella P, Colvin RB (1982) Fibronectin is produced by blood vessels in response to injury. J Exp Med 156:646–651

64. Rhodes JM, Simons M (2007) The extracellular matrix and blood vessel formation: not just a scaffold. J Cell Mol Med 11:176–205. https://doi.org/10.1111/j.1582-4934.2007.00031.x

65. Xu J, Shi G-P (2014) Vascular wall extracellular matrix proteins and vascular diseases. Biochim Biophys Acta 1842:2106–2119. https://doi.org/10.1016/j.bbadis.2014.07.008

66. Laurent TC, Fraser JR (1992) Hyaluronan. FASEB J 6:2397–2404

67. Toole BP (2004) Hyaluronan: from extracellular glue to pericellular cue. Nat Rev Cancer 4:528–539. https://doi.org/10.1038/nrc1391

68. Toole BP (2009) Hyaluronan-CD44 interactions in cancer: paradoxes and possibilities. Clin Cancer Res 15:7462–7468. https://doi.org/10.1158/1078-0432.ccr-09-0479

69. Wiranowska M, Ladd S, Moscinski LC, Hill B, Haller E, Mikecz K, Plaas A (2010) Modulation of hyaluronan production by CD44 positive glioma cells. Int J Cancer 127:532–542. https://doi.org/10.1002/ijc.25085

70. Delpech B et al (1993) Hyaluronan and hyaluronectin in the extracellular matrix of human brain tumour stroma. Eur J Cancer 29a:1012–1017

71. Wiranowska M, Ladd S, Smith SR, Gottschall PE (2006) CD44 adhesion molecule and neuro-glial proteoglycan NG2 as invasive markers of glioma. Brain Cell Biol 35:159–172. https://doi.org/10.1007/s11068-007-9009-0

72. Ariza A et al (1995) Role of CD44 in the invasiveness of glioblastoma-multiforme and the noninvasiveness of meningioma—an immunohistochemistry study. Hum Pathol 26:1144–1147. https://doi.org/10.1016/0046-8177(95)90278-3

73. Kim Y, Kumar S (2014) CD44-mediated adhesion to hyaluronic acid contributes to mechanosensing and invasive motility. Mol Cancer Res 12:1416–1429. https://doi.org/10.1158/1541-7786.mcr-13-0629

74. Kosaki R, Watanabe K, Yamaguchi Y (1999) Overproduction of hyaluronan by expression of the Hyaluronan synthase Has2 enhances anchorage-independent growth and Tumorigenicity. Cancer Res 59:1141–1145

75. Novak U, Slylli SS, Kaye AH, Lepperdinger G (1999) Hyaluronidase-2 overexpression accelerates intracerebral but not subcutaneous tumor formation of murine astrocytoma cells. Cancer Res 59:6246–6250

76. Bourguignon LY (2008) Hyaluronan-mediated CD44 activation of Rho GTPase signaling and cytoskeleton function promotes tumor progression. Semin Cancer Biol 18:251–259. https://doi.org/10.1016/j.semcancer.2008.03.007

77. Bourguignon LY, Zhu H, Shao L, Chen YW (2000) CD44 interaction with tiam1 promotes Rac1 signaling and hyaluronic acid-mediated breast tumor cell migration. J Biol Chem 275:1829–1838

78. Herishanu Y et al (2011) Activation of CD44, a receptor for extracellular matrix components, protects chronic lymphocytic leukemia cells from spontaneous and drug induced apoptosis through MCL-1. Leuk Lymphoma 52:1758–1769. https://doi.org/10.3109/10428194.2011.569962

79. Xu Y, Stamenkovic I, Yu Q (2010) CD44 attenuates activation of the hippo signaling pathway and is a prime therapeutic target for glioblastoma. Cancer Res 70:2455–2464. https://doi.org/10.1158/0008-5472.can-09-2505

80. Goldbrunner RH, Bernstein JJ, Tonn JC (1999) Cell-extracellular matrix interaction in glioma invasion. Acta Neurochir 141:295–305. discussion 304-295

81. Uhm JH, Gladson CL, Rao JS (1999) The role of integrins in the malignant phenotype of gliomas. Front Biosci 4:D188–D199

82. Lathia JD et al (2010) Integrin alpha 6 regulates glioblastoma stem cells. Cell Stem Cell 6:421–432. https://doi.org/10.1016/j.stem.2010.02.018

83. Knott JC et al (1998) Stimulation of extracellular matrix components in the normal brain by invading glioma cells. Int J Cancer 75:864–872. https://doi.org/10.1002/(sici)1097-0215(19980316)75:6<864::aid-ijc8>3.0.co;2-t

84. Mammoto T, Jiang A, Jiang E, Panigrahy D, Kieran MW, Mammoto A (2013) Role of collagen matrix in tumor angiogenesis and glioblastoma multiforme progression. Am J Pathol 183:1293–1305. https://doi.org/10.1016/j.ajpath.2013.06.026

85. Demuth T, Berens ME (2004) Molecular mechanisms of glioma cell migration and invasion. J Neuro-Oncol 70:217–228. https://doi.org/10.1007/s11060-004-2751-6

86. Kawataki T, Yamane T, Naganuma H, Rousselle P, Anduren I, Tryggvason K, Patarroyo M (2007) Laminin isoforms and their integrin receptors in glioma cell migration and invasiveness: evidence for a role of alpha 5-laminin(s) and alpha 3 beta 1 integrin. Exp Cell Res 313:3819–3831. https://doi.org/10.1016/j.yexcr.2007.07.038

87. Lathia JD et al (2012) Laminin alpha 2 enables glioblastoma stem cell growth. Ann Neurol 72:766–778. https://doi.org/10.1002/ana.23674

88. Ohnishi T, Hiraga S, Izumoto S, Matsumura H, Kanemura Y, Arita N, Hayakawa T (1998) Role of fibronectin-stimulated tumor cell migration in glioma invasion in vivo: clinical significance of fibronectin and fibronectin receptor expressed in human glioma tissues. Clin Exp Metastasis 16:729–741. https://doi.org/10.1023/a:1006532812408

89. Hehlgans S, Haase M, Cordes N (2007) Signalling via integrins: implications for cell survival and anticancer strategies. Biochim Biophys Acta 1775:163–180. https://doi.org/10.1016/j.bbcan.2006.09.001

90. Giese A, Loo MA, Rief MD, Tran N, Berens ME (1995) Substrates for astrocytoma invasion. Neurosurgery 37:294–301. https://doi.org/10.1097/00006123-199508000-00015

91. Koochekpour S, Pilkington GJ, Merzak A (1995) Hyaluronic-acid CD44H interaction induces cell detachment and stimulates migration and invasion of human GLIOMA-cells in-vitro. Int J Cancer 63:450–454. https://doi.org/10.1002/ijc.2910630325

92. Tysnes BB et al (1999) Laminin expression by glial fibrillary acidic protein positive cells in human gliomas. Int J Dev Neurosci 17:531–539. https://doi.org/10.1016/s0736-5748(99)00055-6

93. Paulus W, Baur I, Schuppan D, Roggendorf W (1993) Characterization of integrin receptors in normal and neoplastic human brain. Am J Pathol 143:154–163

94. Gingras MC, Roussel E, Bruner JM, Branch CD, Moser RP (1995) Comparison of cell adhesion molecule expression between glioblastoma multiforme and autologous normal brain tissue. J Neuroimmunol 57:143–153. https://doi.org/10.1016/0165-5728(94)00178-Q

95. Chintala SK, Gokaslan ZL, Go Y, Sawaya R, Nicolson GL, Rao JS (1996) Role of extracellular matrix proteins in regulation of human glioma cell invasion in vitro. Clin Exp Metastasis 14:358–366. https://doi.org/10.1007/BF00123395

96. Previtali S, Quattrini A, Nemni R, Truci G, Ducati A, Wrabetz L, Canal N (1996) Alpha6 beta4 and alpha6 beta1 integrins in astrocytomas and other CNS tumors. J Neuropathol Exp Neurol 55:456–465

97. Tysnes BB, Larsen LF, Ness GO, Mahesparan R, Edvardsen K, Garcia-Cabrera I, Bjerkvig R (1996) Stimulation of glioma-cell migration by laminin and inhibition by anti-alpha3 and anti-beta1 integrin antibodies. Int J Cancer 67:777–784. https://doi.org/10.1002/(SICI)1097-0215(19960917)67:6<777::AID-IJC5>3.0.CO;2-O

98. Haugland HK, Tysnes BB, Tysnes OB (1997) Adhesion and migration of human glioma cells are differently dependent on extracellular matrix molecules. Anticancer Res 17:1035–1042

99. Belot N et al (2001) Molecular characterization of cell substratum attachments in human glial tumors relates to prognostic features. Glia 36:375–390. https://doi.org/10.1002/glia.1124

100. Delamarre E et al (2009) Expression of integrin alpha6beta1 enhances tumorigenesis in glioma cells. Am J Pathol 175:844–855. https://doi.org/10.2353/ajpath.2009.080920

101. Kaufman LJ, Brangwynne CP, Kasza KE, Filippidi E, Gordon VD, Deisboeck TS, Weitz DA (2005) Glioma expansion in collagen I matrices: analyzing collagen concentration-dependent growth and motility patterns. Biophys J 89:635–650. https://doi.org/10.1529/biophysj.105.061994

102. Gerardo-Nava J et al (2009) Human neural cell interactions with orientated electrospun nanofibers in vitro. Nanomedicine (Lond) 4:11–30. https://doi.org/10.2217/17435889.4.1.11

103. Hegedus B, Marga F, Jakab K, Sharpe-Timms KL, Forgacs G (2006) The interplay of cell-cell and cell-matrix interactions in the invasive properties of brain tumors. Biophys J 91:2708–2716. https://doi.org/10.1529/biophysj.105.077834

104. Kim HD, Guo TW, Wu AP, Wells A, Gertler FB, Lauffenburger DA (2008) Epidermal growth factor-induced enhancement of glioblastoma cell migration in 3D arises from an intrinsic increase in speed but an extrinsic matrix- and proteolysis-dependent increase in persistence. Mol Biol Cell 19:4249–4259. https://doi.org/10.1091/mbc.E08-05-0501

105. Mahesparan R, Read TA, Lund-Johansen M, Skaftnesmo KO, Bjerkvig R, Engebraaten O (2003) Expression of extracellular matrix components in a highly infiltrative in vivo glioma model. Acta Neuropathol 105:49–57. https://doi.org/10.1007/s00401-002-0610-0

106. Zagzag D et al (1995) Tenascin expression in astrocytomas correlates with angiogenesis. Cancer Res 55:907–914

107. Peglion F, Llense F, Etienne-Manneville S (2014) Adherens junction treadmilling during collective migration. Nat Cell Biol 16:639–651. https://doi.org/10.1038/ncb2985

108. Kawaguchi K, Kageyama R, Sano M (2017) Topological defects control collective dynamics in neural progenitor cell cultures. Nature 545:327–331. https://doi.org/10.1038/nature22321

109. Lois C, Garcia-Verdugo JM, Alvarez-Buylla A (1996) Chain migration of neuronal precursors. Science 271:978–981. https://doi.org/10.1126/science.271.5251.978

110. Stupp R et al (2014) Cilengitide combined with standard treatment for patients with newly diagnosed glioblastoma with methylated MGMT promoter (CENTRIC EORTC 26071-22072 study): a multicentre, randomised, open-label, phase 3 trial. Lancet Oncol 15:1100–1108. https://doi.org/10.1016/s1470-2045(14)70379-1

111. Hutmacher DW (2010) Biomaterials offer cancer research the third dimension. Nat Mater 9:90–93. https://doi.org/10.1038/nmat2619

112. Fischbach C, Chen R, Matsumoto T, Schmelzle T, Brugge JS, Polverini PJ, Mooney DJ (2007) Engineering tumors with 3D scaffolds. Nat Methods 4:855–860. https://doi.org/10.1038/nmeth1085

113. Liu Y, Shu XZ, Prestwich GD (2007) Tumor engineering: orthotopic cancer models in mice using cell-loaded, injectable, cross-linked hyaluronan-derived hydrogels. Tissue Eng 13:1091–1101. https://doi.org/10.1089/ten.2006.0297

114. Griffith LG, Swartz MA (2006) Capturing complex 3D tissue physiology in vitro. Nat Rev Mol Cell Biol 7:211–224. https://doi.org/10.1038/nrm1858

115. Yamada KM, Cukierman E (2007) Modeling tissue morphogenesis and cancer in 3D. Cell 130:601–610. https://doi.org/10.1016/j.cell.2007.08.006

116. Khademhosseini A, Langer R, Borenstein J, Vacanti JP (2006) Microscale technologies for tissue engineering and biology. Proc Natl Acad Sci U S A 103:2480–2487. https://doi.org/10.1073/pnas.0507681102

117. Ross AM, Jiang Z, Bastmeyer M, Lahann J (2012) Physical aspects of cell culture substrates: topography, roughness, and elasticity. Small 8:336–355. https://doi.org/10.1002/smll.201100934

118. Polacheck WJ, Zervantonakis IK, Kamm RD (2013) Tumor cell migration in complex microenvironments. Cell Mol Life Sci 70:1335–1356. https://doi.org/10.1007/s00018-012-1115-1

119. Pelham RJ Jr, Wang Y (1997) Cell locomotion and focal adhesions are regulated by substrate flexibility. Proc Natl Acad Sci U S A 94:13661–13665. https://doi.org/10.1073/pnas.94.25.13661

120. Lo CM, Wang HB, Dembo M, Wang YL (2000) Cell movement is guided by the rigidity of the substrate. Biophys J 79:144–152. https://doi.org/10.1016/S0006-3495(00)76279-5

121. Tse JR, Engler AJ (2010) Preparation of hydrogel substrates with tunable mechanical properties. Curr Protoc Cell Biol. Chapter 10:Unit 10.16. https://doi.org/10.1002/0471143030.cb1016s47

122. Smeds KA, Pfister-Serres A, Miki D, Dastgheib K, Inoue M, Hatchell DL, Grinstaff MW (2001) Photocrosslinkable polysaccharides for in situ hydrogel formation. J Biomed Mater Res 54:115–121. https://doi.org/10.1002/1097-4636(200105)55:2<254::AID-JBM1012>3.0.CO;2-5

123. Cretu A, Castagnino P, Assoian R (2010) Studying the effects of matrix stiffness on cellular function using acrylamide-based hydrogels. J Vis Exp 42:e2089. https://doi.org/10.3791/2089

124. Engler AJ, Sen S, Sweeney HL, Discher DE (2006) Matrix elasticity directs stem cell lineage specification. Cell 126:677–689. https://doi.org/10.1016/j.cell.2006.06.044

125. Engler A, Bacakova L, Newman C, Hategan A, Griffin M, Discher D (2004) Substrate compliance versus ligand density in cell on gel responses. Biophys J 86:617–628. https://doi.org/10.1016/s0006-3495(04)74140-5

126. Johnson KR, Leight JL, Weaver VM (2007) Demystifying the effects of a three-dimensional microenvironment in tissue morphogenesis. Methods Cell Biol 83:547–583. https://doi.org/10.1016/s0091-679x(07)83023-8

127. Yeung T, Georges PC, Flanagan LA, Marg B, Ortiz M, Funakii M, Zahir N, Ming W, Weaver V, Jammey PA (2005) Effects of substrate stiffness on cell morphology, cytoskeletal structure, and adhesion. Cell Motil Cytoskeleton 60:24–34. https://doi.org/10.1002/cm.20041

128. Sen S, Dong M, Kumar S (2009) Isoform-specific contributions of alpha-actinin to glioma cell mechanobiology. PLoS One 4:e8427. https://doi.org/10.1371/journal.pone.0008427

129. Sen S, Ng WP, Kumar S (2012) Contributions of Talin-1 to glioma cell-matrix tensional

130. Rape AD, Kumar S (2014) A composite hydrogel platform for the dissection of tumor cell migration at tissue interfaces. Biomaterials 35:8846–8853. https://doi.org/10.1016/j.biomaterials.2014.07.003

131. Seliktar D (2012) Designing cell-compatible hydrogels for biomedical applications. Science 336:1124–1128. https://doi.org/10.1126/science.1214804

132. Lutolf MP, Hubbell JA (2005) Synthetic biomaterials as instructive extracellular microenvironments for morphogenesis in tissue engineering. Nat Biotechnol 23:47–55. https://doi.org/10.1038/nbt1055

133. Ulrich TA, Jain A, Tanner K, MacKay JL, Kumar S (2010) Probing cellular mechanobiology in three-dimensional culture with collagen-agarose matrices. Biomaterials 31:1875–1884. https://doi.org/10.1016/j.biomaterials.2009.10.047

134. Cukierman E, Pankov R, Stevens DR, Yamada KM (2001) Taking cell-matrix adhesions to the third dimension. Science 294:1708–1712. https://doi.org/10.1126/science.1064829

135. Lu P, Weaver VM, Werb Z (2012) The extracellular matrix: a dynamic niche in cancer progression. J Cell Biol 196:395–406. https://doi.org/10.1083/jcb.201102147

136. Yang YL, Motte S, Kaufman LJ (2010) Pore size variable type I collagen gels and their interaction with glioma cells. Biomaterials 31:5678–5688. https://doi.org/10.1016/j.biomaterials.2010.03.039

137. Decaestecker C, Debeir O, Van Ham P, Kiss R (2007) Can anti-migratory drugs be screened in vitro? A review of 2D and 3D assays for the quantitative analysis of cell migration. Med Res Rev 27:149–176. https://doi.org/10.1002/med.20078

138. Benton G, George J, Kleinman HK, Arnaoutova IP (2009) Advancing science and technology via 3D culture on basement membrane matrix. J Cell Physiol 221:18–25. https://doi.org/10.1002/jcp.21832

139. Wang C, Tong X, Yang F (2014) Bioengineered 3D brain tumor model to elucidate the effects of matrix stiffness on glioblastoma cell behavior using PEG-based hydrogels. Mol Pharm 11:2115–2125. https://doi.org/10.1021/mp5000828

140. Yang YL, Sun C, Wilhelm ME, Fox LJ, Zhu J, Kaufman LJ (2011) Influence of chondroitin sulfate and hyaluronic acid on structure, mechanical properties, and glioma invasion of collagen I gels. Biomaterials 32:7932–7940. https://doi.org/10.1016/j.biomaterials.2011.07.018

homeostasis. J R Soc Interface 9:1311–1317. https://doi.org/10.1098/rsif.2011.0567

141. Coquerel B et al (2009) Elastin-derived peptides: matrikines critical for glioblastoma cell aggressiveness in a 3-D system. Glia 57:1716–1726. https://doi.org/10.1002/glia.20884

142. David L, Dulong V, Coquerel B, Le Cerf D, Cazin L, Lamacz M, Vannier JP (2008) Collagens, stromal cell-derived factor-1alpha and basic fibroblast growth factor increase cancer cell invasiveness in a hyaluronan hydrogel. Cell Prolif 41:348–364. https://doi.org/10.1111/j.1365-2184.2008.00515.x

143. Rao SS, DeJesus J, Short AR, Otero JJ, Sarkar A, Winter JO (2013) Glioblastoma behaviors in three-dimensional collagen-hyaluronan composite hydrogels. ACS Appl Mater Interfaces 5:9276–9284. https://doi.org/10.1021/am402097j

144. Pedron S, Becka E, Harley BA (2015) Spatially gradated hydrogel platform as a 3D engineered tumor microenvironment. Adv Mater 27:1567–1572. https://doi.org/10.1002/adma.201404896

145. Pedron S, Harley BAC (2013) Impact of the biophysical features of a 3D gelatin microenvironment on glioblastoma malignancy. J Biomed Mater Res A 101:3404–3415. https://doi.org/10.1002/jbm.a.34637

146. Jin S-G, Jeong Y-I, Jung S, Ryu H-H, Jin Y-H, Kim I-Y (2009) The effect of hyaluronic acid on the invasiveness of malignant glioma cells: comparison of invasion potential at hyaluronic acid hydrogel and Matrigel. J Korean Neurosurg Soc 46:472–478. https://doi.org/10.3340/jkns.2009.46.5.472

147. Caspani EM, Echevarria D, Rottner K, Small JV (2006) Live imaging of glioblastoma cells in brain tissue shows requirement of actin bundles for migration. Neuron Glia Biol 2:105–114. https://doi.org/10.1017/s1740925x06000111

148. Gordon VD et al (2003) Measuring the mechanical stress induced by an expanding multicellular tumor system: a case study. Exp Cell Res 289:58–66. https://doi.org/10.1016/s0014-4827(03)00256-8

149. Ulrich TA, Lee TG, Shon HK, Moon DW, Kumar S (2011) Microscale mechanisms of agarose-induced disruption of collagen remodeling. Biomaterials 32:5633–5642. https://doi.org/10.1016/j.biomaterials.2011.04.045

150. Marklein RA, Burdick JA (2010) Spatially controlled hydrogel mechanics to modulate stem cell interactions. Soft Matter 6:136–143. https://doi.org/10.1039/B916933D

151. Chen HC (2005) Boyden chamber assay. Methods Mol Biol 294:15–22. https://doi.org/10.1385/1-59259-860-9:015

152. Chonan Y, Taki S, Sampetrean O, Saya H, Sudo R (2017) Endothelium-induced three-dimensional invasion of heterogeneous glioma initiating cells in a microfluidic coculture platform. Integr Biol (Camb) 9:762–773. https://doi.org/10.1039/c7ib00091j

153. Infanger DW et al (2013) Glioblastoma stem cells are regulated by Interleukin-8 signaling in a tumoral perivascular niche. Cancer Res 73:7079–7089. https://doi.org/10.1158/0008-5472.can-13-1355

154. Kim DH, Han K, Gupta K, Kwon KW, Suh KY, Levchenko A (2009) Mechanosensitivity of fibroblast cell shape and movement to anisotropic substratum topography gradients. Biomaterials 30:5433–5444. https://doi.org/10.1016/j.biomaterials.2009.06.042

155. Kim DH, Seo CH, Han K, Kwon KW, Levchenko A, Suh KY (2009) Guided cell migration on microtextured substrates with variable local density and anisotropy. Adv Funct Mater 19:1579–1586. https://doi.org/10.1002/adfm.200990041

156. Zhu BS, Zhang QQ, Lu QH, Xu YH, Yin J, Hu J, Wang Z (2004) Nanotopographical guidance of C6 glioma cell alignment and oriented growth. Biomaterials 25:4215–4223. https://doi.org/10.1016/j.biomaterials.2003.11.020

157. Cha J, Koh I, Choi Y, Lee J, Choi C, Kim P (2014) Tapered microtract array platform for antimigratory drug screening of human glioblastoma multiforme. Adv Healthc Mater 4:405–411. https://doi.org/10.1002/adhm.201400384

158. Gallego-Perez D, Higuita-Castro N, Denning L, DeJesus J, Dahl K, Sarkar A, Hansford DJ (2012) Microfabricated mimics of in vivo structural cues for the study of guided tumor cell migration. Lab Chip 12:4424–4432. https://doi.org/10.1039/c2lc40726d

159. Smith CL et al (2016) Migration phenotype of brain-cancer cells predicts patient outcomes. Cell Rep 15:2616–2624. https://doi.org/10.1016/j.celrep.2016.05.042

160. Irimia D, Toner M (2009) Spontaneous migration of cancer cells under conditions of mechanical confinement. Integr Biol (Camb) 1:506–512. https://doi.org/10.1039/b908595e

161. Pathak A, Kumar S (2012) Independent regulation of tumor cell migration by matrix stiffness and confinement. Proc Natl Acad Sci 109:10334–10339. https://doi.org/10.1073/pnas.1118073109

162. Pathak A, Kumar S (2013) Transforming potential and matrix stiffness co-regulate confinement sensitivity of tumor cell migration. Integr Biol (Camb) 5:1067–1075. https://doi.org/10.1039/c3ib40017d

163. Chen C, Mrksich M, Huang S, Whitesides G, Ingber D (1997) Geometric control of cell life and death. Science 276:1425–1428. https://doi.org/10.1126/science.276.5317.1425

164. Chen CS, Mrksich M, Huang S, Whitesides GM, Ingber DE (1998) Micropatterned surfaces for control of cell shape, position, and function. Biotechnol Prog 14:356–363. https://doi.org/10.1021/bp980031m

165. Craighead HG, James CD, AMP T (2001) Chemical and topographical patterning for directed cell attachment. Curr Opinion Solid State Mater Sci 5:177–184. https://doi.org/10.1016/S1359-0286(01)00005-5

166. Csucs G, Michel R, Lussi JW, Textor M, Danuser G (2003) Microcontact printing of novel co-polymers in combination with proteins for cell-biological applications. Biomaterials 24:1713–1720. https://doi.org/10.1016/S0142-9612(02)00568-9

167. Singhvi R, Kumar A, Lopez GP, Stephanopoulos GN, Wang DI, Whitesides GM, Ingber DE (1994) Engineering cell shape and function. Science 264:696–698. https://doi.org/10.1126/science.8171320

168. Fink J, Thery M, Azioune A, Dupont R, Chatelain F, Bornens M, Piel M (2007) Comparative study and improvement of current cell micro-patterning techniques. Lab Chip 7:672–680. https://doi.org/10.1039/b618545b

169. Azioune A, Carpi N, Tseng Q, Thery M, Piel M (2010) Protein micropatterns: a direct printing protocol using deep UVs. Methods Cell Biol 97:133–146. https://doi.org/10.1016/S0091-679X(10)97008-8

170. Chang SS, Guo WH, Kim Y, Wang YL (2013) Guidance of cell migration by substrate dimension. Biophys J 104:313–321. https://doi.org/10.1016/j.bpj.2012.12.001

171. Doyle AD, Kutys ML, Conti MA, Matsumoto K, Adelstein RS, Yamada KM (2012) Microenvironmental control of cell migration—myosin IIA is required for efficient migration in fibrillar environments through control of cell adhesion dynamics. J Cell Sci 125:2244–2256. https://doi.org/10.1242/jcs.098806

172. Doyle AD, Wang FW, Matsumoto K, Yamada KM (2009) One-dimensional topography underlies three-dimensional fibrillar cell migration. J Cell Biol 184:481–490. https://doi.org/10.1083/jcb.200810041

173. Kaji H, Kanada M, Oyamatsu D, Matsue T, Nishizawa M (2004) Microelectrochemical approach to induce local cell adhesion and growth on substrates. Langmuir 20:16–19. https://doi.org/10.1021/la035537f

174. Yamato M, Kwon OH, Hirose M, Kikuchi A, Okano T (2001) Novel patterned cell coculture utilizing thermally responsive grafted polymer surfaces. J Biomed Mater Res 55:137–140. https://doi.org/10.1002/1097-4636(200104)55:1<137::AID-JBM180>3.0.CO;2-L

175. Bhatia SN, Yarmush ML, Toner M (1997) Controlling cell interactions by micropatterning in co-cultures: hepatocytes and 3T3 fibroblasts. J Biomed Mater Res 34:189–199. https://doi.org/10.1002/(SICI)1097-4636(199702)34:2<189::AID-JBM8>3.0.CO;2-M

176. Szabo B, Kornyei Z, Zach J, Selmeczi D, Csucs G, Czirok A, Vicsek T (2004) Autoreverse nuclear migration in bipolar mammalian cells on micropatterned surfaces. Cell Motil Cytoskeleton 59:38–49. https://doi.org/10.1002/cm.20022

177. Sharma P, Sheets K, Elankumaran S, Nain AS (2013) The mechanistic influence of aligned nanofibers on cell shape, migration and blebbing dynamics of glioma cells. Integr Biol (Camb) 5:1036–1044. https://doi.org/10.1039/c3ib40073e

178. Nain AS, Sitti M, Jacobson A, Kowalewski T, Amon C (2009) Dry spinning based spinneret based tunable engineered parameters (STEP) technique for controlled and aligned deposition of polymeric Nanofibers. Macromol Rapid Commun 30:1406–1412. https://doi.org/10.1002/marc.200900204

179. Nain AS, Amon C, Sitti M (2006) Proximal probes based Nanorobotic drawing of polymer micro/nanofibers. IEEE Trans Nanotechnol 5:499–510. https://doi.org/10.1109/TNANO.2006.880453

180. Nain AS, Wang J (2013) Polymeric nanofibers: isodiametric design space and methodology for depositing aligned nanofiber arrays in single and multiple layers. Polym J 45:695–700. https://doi.org/10.1038/pj.2013.1

181. Bhardwaj N, Kundu SC (2010) Electrospinning: a fascinating fiber fabrication technique. Biotechnol Adv 28:325–347. https://doi.org/10.1016/j.biotechadv.2010.01.004

182. Corey JM, Gertz CC, Wang BS, Birrell LK, Johnson SL, Martin DC, Feldman EL (2008) The design of electrospun PLLA nanofiber scaffolds compatible with serum-free growth of primary motor and sensory neurons. Acta Biomater 4:863–875. https://doi.org/10.1016/j.actbio.2008.02.020

183. Prabhakaran MP, Venugopal JR, Ramakrishna S (2009) Mesenchymal stem cell differentiation to neuronal cells on electrospun nanofibrous substrates for nerve tissue engineering. Biomaterials 30:4996–5003. https://doi.org/10.1016/j.biomaterials.2009.05.057

184. Schnell E, Klinkhammer K, Balzer S, Brook G, Klee D, Dalton P, Mey J (2007) Guidance of glial cell migration and axonal growth on electrospun nanofibers of poly-epsilon-caprolactone and a collagen/poly-epsilon-caprolactone blend. Biomaterials 28:3012–3025. https://doi.org/10.1016/j.biomaterials.2007.03.009

185. Wang HB, Mullins ME, Cregg JM, Hurtado A, Oudega M, Trombley MT, Gilbert RJ (2009) Creation of highly aligned electrospun poly-L-lactic acid fibers for nerve regeneration applications. J Neural Eng 6:016001. https://doi.org/10.1088/1741-2560/6/1/016001

186. Chew SY, Park TG (2009) Nanofibers in regenerative medicine and drug delivery. Adv Drug Deliv Rev 61:987. https://doi.org/10.1016/j.addr.2009.07.004

187. Chew SY, Mi R, Hoke A, Leong KW (2008) The effect of the alignment of electrospun fibrous scaffolds on Schwann cell maturation. Biomaterials 29:653–661. https://doi.org/10.1016/j.biomaterials.2007.10.025

188. Soliman E, Bianchi F, Sleigh JN, George JH, Cader MZ, Cui Z, Ye H (2018) Aligned electrospun fibers for neural patterning. Biotechnol Lett 40:601–607. https://doi.org/10.1007/s10529-017-2494-z

189. Romano A, Scheel M, Hirsch S, Braun J, Sack I (2012) In vivo waveguide elastography of white matter tracts in the human brain. Magn Reson Med 68:1410–1422. https://doi.org/10.1002/mrm.24141

190. Benninger Y et al (2006) Beta1-integrin signaling mediates premyelinating oligodendrocyte survival but is not required for CNS myelination and remyelination. J Neurosci 26:7665–7673. https://doi.org/10.1523/jneurosci.0444-06.2006

191. Makino M, Mimatsu K, Saito H, Konishi N, Hashizume Y (1996) Morphometric study of myelinated fibers in human cervical spinal cord white matter. Spine (Phila Pa 1976) 21:1010–1016

192. Beliveau A, Thomas G, Gong J, Wen Q, Jain A (2016) Aligned Nanotopography promotes a migratory state in Glioblastoma Multiforme tumor cells. Sci Rep 6:26143. https://doi.org/10.1038/srep26143

193. Agudelo-Garcia PA et al (2011) Glioma cell migration on three-dimensional nanofiber scaffolds is regulated by substrate topography and abolished by inhibition of STAT3 Signaling. Neoplasia 13:831–896. https://doi.org/10.1593/neo.11612

194. Jovanov-Milosevic N, Benjak V, Kostovic I (2006) Transient cellular structures in developing corpus callosum of the human brain. Coll Antropol 30:375–381

195. Guetta-Terrier C et al (2015) Protrusive waves guide 3D cell migration along nanofibers. J Cell Biol 211:683–701. https://doi.org/10.1083/jcb.201501106

196. Kievit FM et al (2013) Aligned chitosan-polycaprolactone polyblend nanofibers promote the migration of glioblastoma cells. Adv Healthc Mater 2:1651–1659. https://doi.org/10.1002/adhm.201300092

197. Duling RR, Dupaix RB, Katsube N, Lannutti J (2008) Mechanical characterization of electrospun polycaprolactone (PCL): a potential scaffold for tissue engineering. J Biomech Eng 130:011006. https://doi.org/10.1115/1.2838033

198. McKee CT, Last JA, Russell P, Murphy CJ (2011) Indentation versus tensile measurements of Young's modulus for soft biological tissues. Tissue Eng Part B Rev 17:155–164. https://doi.org/10.1089/ten.TEB.2010.0520

199. Heredia A, Bui CC, Suter U, Young P, Schaffer TE (2007) AFM combines functional and morphological analysis of peripheral myelinated and demyelinated nerve fibers. NeuroImage 37:1218–1226. https://doi.org/10.1016/j.neuroimage.2007.06.007

200. Moore SW, Sheetz MP (2011) Biophysics of substrate interaction: influence on neural motility, differentiation, and repair. Dev Neurobiol 71:1090–1101. https://doi.org/10.1002/dneu.20947

201. Rao SS et al (2012) Inherent interfacial mechanical gradients in 3D hydrogels influence tumor cell behaviors. PLoS One 7:e35852. https://doi.org/10.1371/journal.pone.0035852

202. Rao SS et al (2013) Mimicking white matter tract topography using core-shell electrospun nanofibers to examine migration of malignant brain tumors. Biomaterials 34:5181–5190. https://doi.org/10.1016/j.biomaterials.2013.03.069

203. Zhang Y, Huang Z-M, Xu X, Lim CT, Ramakrishna S (2004) Preparation of core−shell structured PCL-r-gelatin bi-component nanofibers by coaxial electrospinning. Chem Mater 16:3406–3409. https://doi.org/10.1021/cm049580f

204. Nam J, Johnson J, Lannutti JJ, Agarwal S (2011) Modulation of embryonic mesenchymal progenitor cell differentiation via control over pure mechanical modulus in electrospun nanofibers. Acta Biomater 7:1516–1524. https://doi.org/10.1016/j.actbio.2010.11.022

205. Zhang YZ, Venugopal J, Huang ZM, Lim CT, Ramakrishna S (2005) Characterization of the surface biocompatibility of the electro-

spun PCL-collagen nanofibers using fibroblasts. Biomacromolecules 6:2583–2589. https://doi.org/10.1021/bm050314k

206. Jain A et al (2014) Guiding intracortical brain tumour cells to an extracortical cytotoxic hydrogel using aligned polymeric nanofibres. Nat Mater 13:308–316. https://doi.org/10.1038/nmat3878

207. Grodecki J et al (2015) Glioma-astrocyte interactions on white matter tract-mimetic aligned electrospun nanofibers. Biotechnol Prog 31:1406–1415. https://doi.org/10.1002/btpr.2123

208. Cha J, Kang SG, Kim P (2016) Strategies of mesenchymal invasion of patient-derived brain tumors: microenvironmental adaptation. Sci Rep 6:24912. https://doi.org/10.1038/srep24912

209. Johnson J, Nowicki MO, Lee CH, Chiocca EA, Viapiano MS, Lawler SE, Lannutti JJ (2009) Quantitative analysis of complex glioma cell migration on electrospun polycaprolactone using time-lapse microscopy. Tissue Eng Part C Methods 15:531–540. https://doi.org/10.1089/ten.TEC.2008.0486

210. Song S, Ewald AJ, Stallcup W, Werb Z, Bergers G (2005) PDGFRbeta+ perivascular progenitor cells in tumours regulate pericyte differentiation and vascular survival. Nat Cell Biol 7:870–879. https://doi.org/10.1038/ncb1288

211. Chekenya M et al (2002) The glial precursor proteoglycan, NG2, is expressed on tumour neovasculature by vascular pericytes in human malignant brain tumours. Neuropathol Appl Neurobiol 28:367–380

212. Wen PY, Kesari S (2008) Malignant gliomas in adults. N Engl J Med 359:492–507. https://doi.org/10.1056/NEJMra0708126

213. Ricci-Vitiani L et al (2010) Tumour vascularization via endothelial differentiation of glioblastoma stem-like cells. Nature 468:824–828. https://doi.org/10.1038/nature09557

214. Soda Y et al (2011) Transdifferentiation of glioblastoma cells into vascular endothelial cells. Proc Natl Acad Sci U S A 108:4274–4280. https://doi.org/10.1073/pnas.1016030108

215. Wang R et al (2010) Glioblastoma stem-like cells give rise to tumour endothelium. Nature 468:829–833. https://doi.org/10.1038/nature09624

216. Couldwell WT, Yong VW, Dore-Duffy P, Freedman MS, Antel JP (1992) Production of soluble autocrine inhibitory factors by human glioma cell lines. J Neurol Sci 110:178–185

217. Lal PG, Ghirnikar RS, Eng LF (1996) Astrocyte-astrocytoma cell line interactions in culture. J Neurosci Res 44:216–222. https://doi.org/10.1002/(SICI)1097-4547(19960501)44:3<216::AID-JNR2>3.0.CO;2-J

218. Le DM et al (2003) Exploitation of astrocytes by glioma cells to facilitate invasiveness: a mechanism involving matrix metalloproteinase-2 and the urokinase-type plasminogen activator-plasmin cascade. J Neurosci 23:4034–4043

219. Barbero S et al (2002) Expression of the chemokine receptor CXCR4 and its ligand stromal cell-derived factor 1 in human brain tumors and their involvement in glial proliferation in vitro. Ann N Y Acad Sci 973:60–69

220. Hoelzinger DB, Demuth T, Berens ME (2007) Autocrine factors that sustain glioma invasion and paracrine biology in the brain microenvironment. J Natl Cancer Inst 99:1583–1593. https://doi.org/10.1093/jnci/djm187

221. Rempel SA, Dudas S, Ge SG, Gutierrez JA (2000) Identification and localization of the cytokine SDF1 and its receptor, CXC chemokine receptor 4, to regions of necrosis and angiogenesis in human glioblastoma. Clin Cancer Res 6:102–111

222. Emdad L et al (2007) Astrocyte elevated gene-1: recent insights into a novel gene involved in tumor progression, metastasis and neurodegeneration. Pharmacol Ther 114:155–170. https://doi.org/10.1016/j.pharmthera.2007.01.010

223. Zhou Y, Larsen PH, Hao C, Yong VW (2002) CXCR4 is a major chemokine receptor on glioma cells and mediates their survival. J Biol Chem 277:49481–49487. https://doi.org/10.1074/jbc.M206222200

224. Abbott NJ (2013) Blood-brain barrier structure and function and the challenges for CNS drug delivery. J Inherit Metab Dis 36:437–449. https://doi.org/10.1007/s10545-013-9608-0

225. Abbott NJ, Patabendige AA, Dolman DE, Yusof SR, Begley DJ (2010) Structure and function of the blood-brain barrier. Neurobiol Dis 37:13–25. https://doi.org/10.1016/j.nbd.2009.07.030

226. Czupalla CJ, Liebner S, Devraj K (2014) In vitro models of the blood-brain barrier. Methods Mol Biol 1135:415–437. https://doi.org/10.1007/978-1-4939-0320-7_34

227. Patabendige A, Skinner RA, Morgan L, Abbott NJ (2013) A detailed method for preparation of a functional and flexible blood-brain barrier model using porcine brain endothelial cells. Brain Res 1521:16–30. https://doi.org/10.1016/j.brainres.2013.04.006

228. Cucullo L, Hossain M, Puvenna V, Marchi N, Janigro D (2011) The role of shear stress in blood-brain barrier endothelial physiology. BMC Neurosci 12:40. https://doi.org/10.1186/1471-2202-12-40

229. Cucullo L, Hossain M, Rapp E, Manders T, Marchi N, Janigro D (2007) Development of a humanized in vitro blood-brain barrier model to screen for brain penetration of antiepileptic drugs. Epilepsia 48:505–516. https://doi.org/10.1111/j.1528-1167.2006.00960.x

230. Cucullo L, Marchi N, Hossain M, Janigro D (2011) A dynamic in vitro BBB model for the study of immune cell trafficking into the central nervous system. J Cereb Blood Flow Metab 31:767–777. https://doi.org/10.1038/jcbfm.2010.162

231. Brown JA et al (2015) Recreating blood-brain barrier physiology and structure on chip: a novel neurovascular microfluidic bioreactor. Biomicrofluidics 9:054124. https://doi.org/10.1063/1.4934713

232. Sellgren KL, Hawkins BT, Grego S (2015) An optically transparent membrane supports shear stress studies in a three-dimensional microfluidic neurovascular unit model. Biomicrofluidics 9:061102. https://doi.org/10.1063/1.4935594

233. Adriani G, Ma D, Pavesi A, Kamm RD, Goh EL (2017) A 3D neurovascular microfluidic model consisting of neurons, astrocytes and cerebral endothelial cells as a blood-brain barrier. Lab Chip 17:448–459. https://doi.org/10.1039/c6lc00638h

234. Prabhakarpandian B, Shen MC, Nichols JB, Mills IR, Sidoryk-Wegrzynowicz M, Aschner M, Pant K (2013) SyM-BBB: a microfluidic blood brain barrier model. Lab Chip 13:1093–1101. https://doi.org/10.1039/c2lc41208j

235. Xu H et al (2016) A dynamic in vivo-like organotypic blood-brain barrier model to probe metastatic brain tumors. Sci Rep 6:36670. https://doi.org/10.1038/srep36670

236. Wang YI, Abaci HE, Shuler ML (2017) Microfluidic blood-brain barrier model provides in vivo-like barrier properties for drug permeability screening. Biotechnol Bioeng 114:184–194. https://doi.org/10.1002/bit.26045

237. Griep LM et al (2013) BBB on chip: microfluidic platform to mechanically and biochemically modulate blood-brain barrier function. Biomed Microdevices 15:145–150. https://doi.org/10.1007/s10544-012-9699-7

238. Booth R, Kim H (2012) Characterization of a microfluidic in vitro model of the blood-brain barrier (muBBB). Lab Chip 12:1784–1792. https://doi.org/10.1039/c2lc40094d

239. Booth R, Kim H (2014) Permeability analysis of neuroactive drugs through a dynamic microfluidic in vitro blood-brain barrier model. Ann Biomed Eng 42:2379–2391. https://doi.org/10.1007/s10439-014-1086-5

240. Chrobak KM, Potter DR, Tien J (2006) Formation of perfused, functional microvascular tubes in vitro. Microvasc Res 71:185

241. Kim JA, Kim HN, Im SK, Chung S, Kang JY, Choi N (2015) Collagen-based brain microvasculature model in vitro using three-dimensional printed template. Biomicrofluidics 9:024115. https://doi.org/10.1063/1.4917508

242. Wong AD, Searson PC (2014) Live-cell imaging of invasion and intravasation in an artificial microvessel platform. Cancer Res 74:4937–4945. https://doi.org/10.1158/0008-5472.can-14-1042

243. Cui X et al (2018) Hacking macrophage-associated immunosuppression for regulating glioblastoma angiogenesis. Biomaterials 161:164–178. https://doi.org/10.1016/j.biomaterials.2018.01.053

244. Bischel LL, Lee SH, Beebe DJ (2012) A practical method for patterning lumens through ECM hydrogels via viscous finger patterning. J Lab Autom 17:96–103. https://doi.org/10.1177/2211068211426694

245. Bischel LL, Young EW, Mader BR, Beebe DJ (2013) Tubeless microfluidic angiogenesis assay with three-dimensional endothelial-lined microvessels. Biomaterials 34:1471–1477. https://doi.org/10.1016/j.biomaterials.2012.11.005

246. Herland A, van der Meer AD, FitzGerald EA, Park TE, Sleeboom JJ, Ingber DE (2016) Distinct contributions of astrocytes and pericytes to Neuroinflammation identified in a 3D human blood-brain barrier on a Chip. PLoS One 11:e0150360. https://doi.org/10.1371/journal.pone.0150360

247. Zheng Y et al (2012) In vitro microvessels for the study of angiogenesis and thrombosis. Proc Natl Acad Sci U S A 109:9342–9347. https://doi.org/10.1073/pnas.1201240109

248. Miller JS et al (2012) Rapid casting of patterned vascular networks for perfusable engineered three-dimensional tissues. Nat Mater 11:768. https://doi.org/10.1038/nmat3357

249. Jeon JS, Bersini S, Gilardi M, Dubini G, Charest JL, Moretti M, Kamm RD (2015) Human 3D vascularized organotypic microfluidic assays to study breast cancer cell extravasation. Proc Natl Acad Sci U S A

112:214–219. https://doi.org/10.1073/pnas.1417115112

250. Kim S, Lee H, Chung M, Jeon NL (2013) Engineering of functional, perfusable 3D microvascular networks on a chip. Lab Chip 13:1489–1500. https://doi.org/10.1039/c3lc41320a

251. Campisi M, Shin Y, Osaki T, Hajal C, Chiono V, Kamm RD (2018) 3D self-organized microvascular model of the human blood-brain barrier with endothelial cells, pericytes and astrocytes. Biomaterials 180:117–129. https://doi.org/10.1016/j.biomaterials.2018.07.014

252. Truong D, Fiorelli R, Barrientos ES, Melendez EL, Sanai N, Mehta S, Nikkhah M (2018) A three-dimensional (3D) organotypic microfluidic model for glioma stem cells—vascular interactions. Biomaterials. https://doi.org/10.1016/j.biomaterials.2018.07.048

253. Vedula SR, Leong MC, Lai TL, Hersen P, Kabla AJ, Lim CT, Ladoux B (2012) Emerging modes of collective cell migration induced by geometrical constraints. Proc Natl Acad Sci U S A 109:12974–12979. https://doi.org/10.1073/pnas.1119313109

Part III

Treatments in Mice

Chapter 9

Assessing Neurological Function in Brain Tumor Mouse Model

Xing Gao, Limeng Wu, Raquel D. Thalheimer, Jie Chen, Yao Sun, Grace Y. Lee, Scott R. Plotkin, and Lei Xu

Abstract

Of all cancer types, brain cancers often constitute the most devastating and difficult-to-treat tumors. In the United States, brain cancer accounts for 1 in every 100 cancer diagnoses. Although significant advances have been made in understanding the biology of brain cancers, the mortality rate for brain cancer has remained consistent for more than 30 years. There is an urgent need to better understand the biology of brain tumor progression and the mechanisms of tumor-induced neurological symptoms. A better understanding of these mechanisms is essential to developing novel therapeutic targets to control tumor growth and improve patient's quality of life. Orthotopic animal models have mainly focused on tumor growth and testing of novel therapeutics, while only few investigate the biology and mechanisms of tumor-induced neurological symptoms. In this chapter, we aim to summarize the methods which can fully utilize these orthotopic animal models for the characterization of brain tumor biology and neurological function. By studying brain tumors from every angle, we aim to unravel the basic tumor pathobiological underpinnings of these tumors and develop novel therapeutic approaches.

Keywords Hearing, Ataxia, Motor function, Brain tumor, Mouse model

1 Introduction

Every year, in the United States, 23,880 men and women are diagnosed with cancer of the brain and nervous system, resulting in 16,830 disease-related deaths annually [1]. Although significant advances have been made in understanding the biology of brain cancers—as well as in tumor diagnosis, treatments, and quality of life of patients with the disease—the mortality rate for brain cancer has remained stable for more than 30 years. There is an urgent need to better understand the biology of brain tumor progression and the mechanisms of tumor-induced neurological symptoms. An

---

Xing Gao and Limeng Wu contributed equally to this work.

Giorgio Seano (ed.), *Brain Tumors*, Neuromethods, vol. 158,
https://doi.org/10.1007/978-1-0716-0856-2_9, © Springer Science+Business Media, LLC, part of Springer Nature 2021

enhanced understanding is essential in order to identify novel therapeutic targets to control tumor growth and improve quality of life.

Primary brain tumors originate either in the brain parenchyma (e.g., gliomas, which include astrocytomas, oligodendrogliomas, ependymomas, or medulloblastomas) or in the extraneural structures (e.g., meningiomas, acoustic neuromas, and schwannomas). Secondary brain tumors develop when cancer cells metastasize to the brain (e.g., lung and breast cancer brain metastasis). Patients with brain tumors face serious challenges to maintaining desirable quality of life. These patients experience general symptoms resulting from increased intracranial pressure, such as headache, anorexia, nausea, vomiting, seizures, blurred or double vision, and insomnia [2, 3]. Patients also suffer from symptoms secondary to focal neurologic deterioration including tumors in the cerebrum portion of the brain, which controls movement and can cause weakness or numbness; tumors in the cerebellum that can cause ataxia; and tumors in or near cranial nerves which can lead to hearing loss and tinnitus (cranial nerve VIII, vestibulocochlear), facial weakness, and pain (cranial nerve V, trigeminal and VII, facial) [4–6].

There are few well-tested interventions to improve the neurological function and quality of life among brain cancer patients. The lack of understanding of the mechanisms underlying tumor-induced neurological deficit has significantly impaired the development of novel therapies to improve neurological function. To faithfully recapitulate human disease, several orthotopic animal models have been developed for glioblastoma [7], medulloblastoma [8], and NF2-related vestibular schwannomas [9, 10]. However, most of these preclinical animal studies have focused on tumor growth and testing of novel therapeutics, and only a few studies investigate the biology and mechanisms of tumor-induced neurological symptoms. Below, we summarize the methods to fully utilize these orthotopic animal models to characterize brain tumor biology and neurological function. By studying the disease from every angle, we aim to unravel the basic tumor pathobiological underpinnings of these tumors and develop novel therapeutic approaches.

1.1 Experimental Design

Protocols to establish the orthotopic xenograft model involve three major components.

First, a surgical procedure to implant brain tumor cells into the orthotopic brain area. The procedure is straightforward, and it involves drilling a hole into the skull and stereotactically injecting tumor cells into the corresponding location in the mice brain (Fig. 9.1).

Second, an evaluation of tumor growth and efficacy for potential therapies. During model development and protocol optimiza-

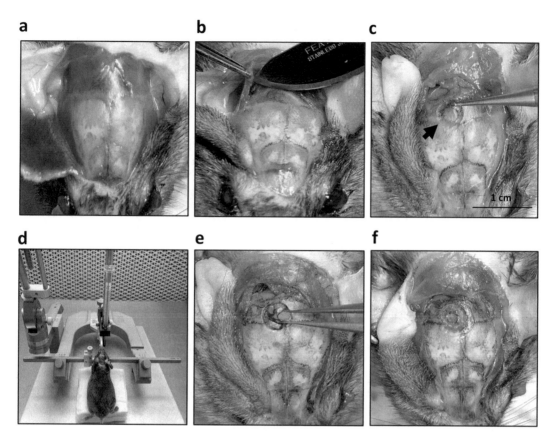

Fig. 9.1 Surgical procedure for stereotactic injection in CPA model. (**a**) Cut and flip the skin flap; expose and clean the periosteum to the temporal crests. (**b**) Detach the cervical trapezius muscle from the skull using a scalpel, and reflect it posteriorly to fully expose the skull above the cerebellum. (**c**) Draw a 2–3 mm circle around the implantation site that is 2.2 mm lateral to the confluence of the sagittal and transverse sinuses and 0.5 mm dorsal pass the transverse sinus. Drill a groove (pointed by the arrow) around the margins of the circle; deepen the groove by repetitive drilling until the bone flap becomes loose. Flip the bone flap carefully without detaching it. (**d**) Fix the mouse in the stereotactic device and inject tumor cells. (**e**) After injection, flip the bone flap back in its anatomic position. (**f**) Apply 1 drop of glue to fix the bone flap. Institutional regulatory board permission was obtained from Massachusetts General Hospital

tion, MR imaging can be used to confirm tumor formation in the targeted area of the brain as early as 7 days post-implantation (Fig. 9.2a) [11]. However, MRI is costly and time-consuming, making serial tumor size measurements infeasible. Plasma *Gaussia* luciferase (GLuc) assay is an inexpensive, rapid, and sensitive alternative method to measure tumor growth [12–14]. In this method, tumor cells are transduced with the secreted bioluminescent GLuc reporter gene, allowing for plasma GLuc activity to be easily tested by collecting a few microliters of blood [11, 14, 15]. To ensure that the changes in Gluc values accurately reflect tumor size changes, concomitant measurements of tumor volume using ultrasound and plasma GLuc assay were also performed. In Fig. 9.2, we dem-

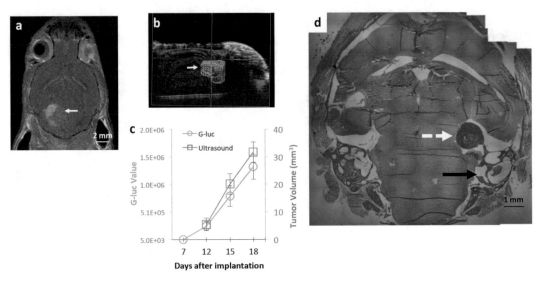

Fig. 9.2 Confirmation of CPA tumor location and measurement of tumor volume. Mouse NF2$^{-/-}$ schwannoma cells are stereotactically injected into the CPA area of mice. (**a**) MR imaging confirms tumor formation in the CPA region 2 weeks after tumor implantation. T1-weighted images were obtained following intravenous administration of Magnevist (0.3 mmol/kg) via the tail vein, using the rapid acquisition relaxation enhanced (RARE) sequence: TR = 873 ms, TE = 17.8 ms, flip angle = 90°, 12 repetitions were acquired and averaged, acquisition time 5 min and 36 s, matrix size 192 x 192, field of view 2.5 × 2.5 cm, slice thickness 0.5 cm, 16 sections acquired used [15]. White arrows point to the CPA tumors. (**b**) High-resolution ultrasound of the brain 12 days after tumor implantation in the CPA. White arrow points to wire mesh of tumors rendering from 3D data reconstruction using Vevo770. (**c**) Plasma GLuc activity of NF2$^{-/-}$ schwannoma grown in the CPA correlates well with the tumor volume measured by ultrasound through a cerebellar window (*N* = 12). (**d**) Representative H&E images of the brain section of a mouse bearing NF2$^{-/-}$ schwannomas (white arrow) located in proximity to the cochlea (black arrow). Institutional regulatory board permission was obtained from MGH IACUC. Adapted with permission from ref. [11]

onstrated a strong correlation between ultrasound-measured tumor volume and plasma GLuc activity in a CPA model of NF2-related vestibular schwannoma (Fig. 9.2b, c), suggesting that the GLuc activity can accurately reflect tumor growth.

Third, an evaluation of the severity of tumor-induced neurological symptoms and the effects of potential treatment on improving neurological function. In order to evaluate these parameters, we summarized three neurological tests: (a) hearing function was measured using distortion-product otoacoustic emissions (DPOAE) and auditory brainstem response (ABR) (Figs. 9.3a and 9.4a) [11], (b) the severity of cerebellar ataxia was assessed using behavior tests including hindlimb clasping, ledge, and gait tests (Figs. 9.3b and 9.4b) [20], and (c) balance, motor strength, and coordination were assessed using rotarod assay [14, 16].

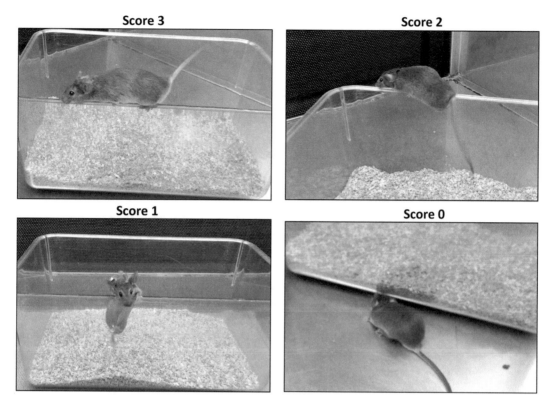

Fig. 9.3 Scoring for ledge test. Lift the mouse from the cage and place it on the cage's ledge. Observe the mouse as it walks along the cage ledge and lowers itself into its cage. If the mouse walks along the ledge without losing balance and lowers itself back into the cage gracefully, it receives a score of 3. If the mouse loses footing while walking along the ledge, but appears coordinated, it receives a score of 2. If the mouse cannot effectively use its hind legs and lands on its head rather than paws when descending into the cage, it receives a score of 1. If the mouse falls off the ledge in less than 1 s, it receives a score of 0

2 Materials

2.1 Equipment

2.1.1 Surgical Supplies

To perform open skull surgery, a bone microdrill for brain surgery (Harvard Apparatus, Holliston, MA) and standard mouse microsurgery tools were required (including straight iris scissors, straight and Dumont forceps, etc.). To perform the stereotactic injection, a stereotactic frame (David Kopf's small-animal stereotactic apparatus, Tujunga, CA), bright-field stereomicroscope (Olympus, Waltham, MA), and Hamilton syringe (10-μL) were used.

2.1.2 Tumor Growth and Imaging Equipment

To confirm tumor formation in the correct location, MRI of the brain was performed on a 4.7 T animal MRI scanner (Bruker, Billerica, MA) with a dedicated mouse head coil. *Gaussia* luciferase (GLuc) assay was used to evaluate tumor growth. GLuc activity was measured using Glomax Microplate Luminometer (Promega, Madison, WI). Ultrasound (Vevo 2100 Imaging System,

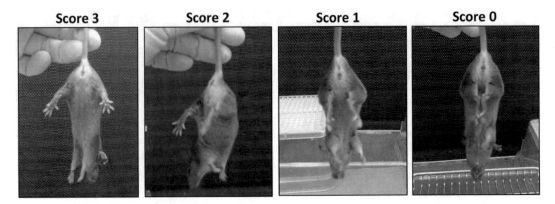

Fig. 9.4 Scoring for hindlimb clasping test. Grasp the tail and lift the mouse clear of all surrounding objects. Observe the hindlimb position for 10 s. If the hindlimbs are consistently splayed outward, away from the abdomen, the mouse receives a score of 3. If one hindlimb is retracted toward the abdomen for more than 50% of the time suspended, the mouse receives a score of 2. If both hindlimbs are partially retracted toward the abdomen for more than 50% of the time suspended, the mouse receives a score of 1. If both hindlimbs are entirely retracted and touching the abdomen for more than 50%, the mouse receives a score of 0

FUJIFILM VisualSonics, Inc., Toronto, ON, Canada) of the brain tumor was used to confirm that GLuc values accurately reflected tumor growth.

2.1.3 Neurological Function Evaluation Equipment

To measure hearing function, DPOAE/ABR tests were performed using Intelligent Hearing Systems (Tucker-Davis Technologies, Alachua, FL). An automated rotarod machine (Rotamex 4/8 4-Lane Treadmill Shock Grid; Columbus Instruments, Columbus, OH) was used to conduct the rotarod test.

2.2 Animal Preparations

2.2.1 Mice

When using human tumor cell lines, immunodeficient mice (e.g., nude mice (nu/nu) or NOD SCID mice) were used. To mimic brain tumors in adult patients, such as GBM and vestibular schwannoma, the mice used were 10–12 weeks old at the time of implantation. To mimic brain tumors in pediatric patients, such as MB and ependymoma, the mice used were 4–6 weeks old at the time of implantation. All animal studies must be reviewed and approved by the relevant animal care committees and must conform to all relevant institutional and national ethics regulations. All animal procedures were performed according to the guidelines of the Public Health Service Policy on Humane Care and Use of Laboratory Animals and are in accordance with protocols approved by the Institutional Animal Care and Use Committee.

2.2.2 Anesthesia

Ketamine (90 mg/kg body weight, Ketaset, Patterson Veterinary, Saint Pau, MN) and xylazine (9 mg/kg body weight, Ketaset, Patterson Veterinary, Saint Pau, MN) were both utilized in this study. Both medications were added to sterile saline to obtain the described concentrations. Caution should be taken since the two agents may cause prolonged respiratory depression. They are controlled substances and should be handled accordingly.

2.3 Surgical Procedures to Implant the Tumor

Anesthetize the animal and ensure that it is sufficiently anesthetized by testing its pedal reflex and monitoring its respiration. Apply an ophthalmic lubricant to the eyes to avoid corneal desiccation. Place the mouse in the stereotactic frame and stabilize the head under the bright-field microscope. Ensure that the position of the mouse is secured. Make a longitudinal incision in the skin between the occiput and the forehead (Fig. 9.1a). Expose and clean the periosteum to the temporal crests and detach the cervical trapezius muscle from the skull using a scalpel and reflect it posteriorly to fully expose the skull above the cerebellum (Fig. 9.1b). Draw a 2 ~ 3 mm circle around the implantation site that is 2.2 mm lateral to the confluence of the sagittal and transverse sinuses and then 0.5 mm dorsal pass the transverse sinus. Use a high-speed bone microdrill with a 1.4 mm diameter burr tip to make a groove around the margins of the circle (Fig. 9.1c). Drill slowly and apply cold saline regularly to avoid thermally induced injury to the brain. Since the bone overlying the cerebellum is thin, drill slowly and carefully to avoid any damage. Gradually deepen the groove by gentle repetitive drilling until the bone flap becomes loose and/or the cerebellar surface vessels become visible. Use a Malis dissector to separate the bone flap from the dura mater underneath. Flip the bone flap carefully without detaching it to expose the brain. Place a piece of gelfoam on the dura mater to stop bleeding and keep the cerebellum moist.

Place a Hamilton syringe with the 28-gauge needle loaded with tumor cells in 1 μL volume on the stereotactic injection device, and adjust it at an angle of 10–15° offset from the vertical plane. Make sure the head of the mouse is fixed in the stereotactic device such that the surface of the cerebellum is parallel to the horizontal plane (Fig. 9.1d). Use the three-dimensional coordination setup of the stereotactic device to place the tip of the Hamilton syringe needle close to the desired injection site. Carefully lower the syringe until it touches the surface of the cerebellum without puncturing it; continue lowering the syringe to the desired depth [8, 10]. Retract the syringe of 1 mm. Inject cell suspension (1 μL) slowly over 30 s and continue to hold the syringe for 60 s. Rinse the surface of the cerebellum with sterile saline to remove any tumor cells leaked at the injection site. Retract the syringe slowly over 30 s until it exits the brain while rinsing continuously with sterile saline. The slow retraction coupled with debridement could mitigate leakage of tumor cells to the brain surface. Flip the bone flap back in its anatomic position (Fig. 9.1e). Use cotton swab to dry the bone flap and surrounding areas, and apply 1 drop of glue to fix the bone flap (Fig. 9.1f). Close the skin using a running 5–0 Ethibond suture. After surgical procedures, mice should be kept in a pathogen-free condition with sufficient supply of food and water. The quality of each surgery and the health of the mouse are key determinants of imaging quality and the accuracy of neurological function evaluation, as hemorrhage or tissue damage can obscure the key features of the brain, tumor vasculature, and brain function.

Notes and Troubleshooting: The following problems may occur during animal surgery; potential reasons and solution for each are provided below:

1. The mouse is not fully anesthetized. This may occur when the anesthetic dose is insufficient or different strains of mice may have different sensitivities to anesthesia.

   Solution: Perform a literature search on the reported anesthetic dose for the corresponding mouse strain, and increase the dose accordingly. If no anesthetic dose has been published, carefully titrate the anesthesia and monitor the respiratory gating for the duration of animal surgery.

2. The mouse dies of anesthesia. This may occur when tumor-bearing mice become moribund. At the end stage, these mice may be more sensitive to anesthesia.

   Solution: Carefully titrate the anesthesia as described above.

3. Hypothermia occurs during general anesthesia. During general anesthesia, mice cannot maintain stable body temperature.

   Solution: Apply a heating pad during surgery.

4. Bleeding from the skull. This may occur when blood vessels were injured during the surgery.

   Solution: Provide hemostasis with gelfoam, and then rinse with sterile saline.

5. Tumor cells leak during the injection. This may occur when injection and retraction of the needle is too fast.

Solution: It is very important to inject the cells slowly over 30 s, hold the syringe for 60 seconds, and retract the syringe over the course of 30 s as described above.

2.4 Confirmation of Tumor Location and Assessment/ Evaluation of Tumor Growth

2.4.1 Confirmation of Tumor Location by MRI

Anesthetize the mice and ensure that the animal is sufficiently anesthetized by testing its pedal reflex and monitoring its respiration. Perform localizer sequences to identify the relevant anatomy. T1-weighted images were obtained before and after the intravenous administration of 0.3 mmol/kg of Magnevist (Bayer, Whippany, NJ) via the tail vein, using the rapid acquisition relaxation enhanced (RARE) sequence: TR = 873 ms, TE = 17.8 ms, flip angle = 90°, 12 repetitions were acquired and averaged, acquisition time 5 min and 36 s, matrix size 192 × 192, field of view 2.5 × 2.5 cm, slice thickness 0.5 cm, 16 sections acquired used [15] (Fig. 9.2a).

Notes and Troubleshooting: With standard MR imaging procedure, tumor mass should be detected readily. If no tumor mass can be detected by MRI, potential reasons include:

1. The timing of MR imaging is not optimal. Consider perform MRI at different time points.

2. The tumor did not form as expected. To ensure tumor formation, consider repeating the in vitro tumor cell culture and counting, and test for potential mycoplasma contamination, which will prevent tumor formation in mice.

2.4.2 Evaluation of Tumor Growth by GLuc Assay

Tumor cells were transduced to express the *Gaussia* luciferase (GLuc) reporter gene, which is a secreted reporter that can be detected in the blood of tumor-bearing animals. To test blood GLuc level, collect 10 μL of blood from the tail vein, and immediately mix the sample with 10 μL of 10 mM EDTA (as an anticoagulant). Measure the GLuc activity using a plate luminometer, which is set to inject 100 μL of the substrate of GLuc—coelenterazine (50 μM)—and to acquire photon counts for 10 s. Analyze the data by plotting the relative light units (RLU; *y*-axis) with respect to time (*x*-axis). Caution should be taken as the coelenterazine is prone to auto-oxidation which results in a decrease in signal over time. To stabilize this substrate, the working concentration of coelenterazine mixture should be prepared and incubated 30 min before use at room temperature in the dark.

Notes and Troubleshooting: GLuc reading may be low and undetectable due to the following reasons:

1. GLuc reporter gene level is low in tumor cells. Consider repeating the in vitro GLuc reporter gene infection/transfection of the tumor cells, and select a high GLuc-transduced tumor cell population using fluorescence-activated cell sorting (FACS).

2. GLuc assay is not optimal. Consider ordering new kit with fresh coelenterazine, and check the luminometer setting.

In the NF2-related vestibular schwannoma CPA model, 7 days following tumor implantation, we detected tumor mass in the CPA region using MR imaging (Fig. 9.2a). Once the tumor formation in the CPA area is confirmed, assessment of tumor growth is continued by testing plasma GLuc level every 3 days (Fig. 9.2b, c). Schwannomas grow and mice become moribund in 3–5 weeks. After the mice were sacrificed, H&E staining of the brain section of a mouse bearing NF2$^{-/-}$ schwannomas confirmed the schwannomas (white arrow) were located in proximity to the cochlea (Fig. 9.2d).

3 Neurological Tests

3.1 Ataxia and Cerebellar Function Test

Brain tumor can cause ataxia—a syndrome characterized by a disturbance of coordination and balance; the effects may result in abnormal stance, gait, and limb incoordination [16].

A group of tests can be combined to evaluate ataxia, including (a) ledge test to measure coordination, which is the symptom most directly related to human signs of cerebellar ataxia (Fig. 9.3); (b) hindlimb clasping test, which has been used to evaluate disease progression in a number of mouse models of neurodegeneration (Fig. 9.4); (c) gait test, which focuses on the measure of coordination and muscle function (Fig. 9.5); and (d) kyphosis test, which assesses for a characteristic dorsal curvature of the spine, a common manifestation caused by a loss of muscle tone in the spinal muscles secondary to neurodegeneration (Fig. 9.6). Each test is scored from 3 to 0, with 3 being normal and 0 being the most severe manifestation of symptoms. The scores from all four tests were combined to create a composite phenotype score (Figs. 9.3–9.6).

Notes and Troubleshooting: In our NF2 CPA model, we found that the ledge test is the most sensitive test and changes in mouse neurologic function can easily be observed in early tumor stage and changes of gait and kyphosis show differences with late- to end-stage tumors.

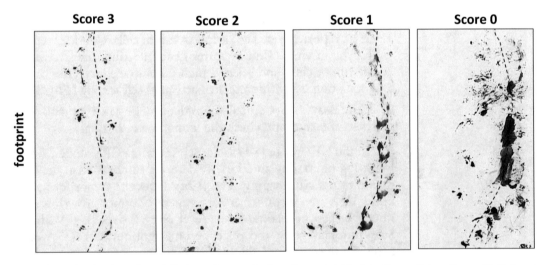

Fig. 9.5 Scoring for the gait test. Remove the mouse from its cage and place it on a flat surface with its head facing away from the investigator. Observe the mouse from behind as it walks. If the mouse walks normally, without the abdomen touching the ground, and both hindlimbs participating evenly, it receives a score of 3. If the mouse tremors or appears to limp while walking, it receives a score of 2. If the mouse exhibits severe tremor and limp, lowered pelvis, or the feet point away from the body during locomotion ("duck feet"), it receives a score of 2. If the mouse fails to move forward and drags the abdomen along the ground, it receives a score of 0

Score 3

Score 2

Score 1

Score 0

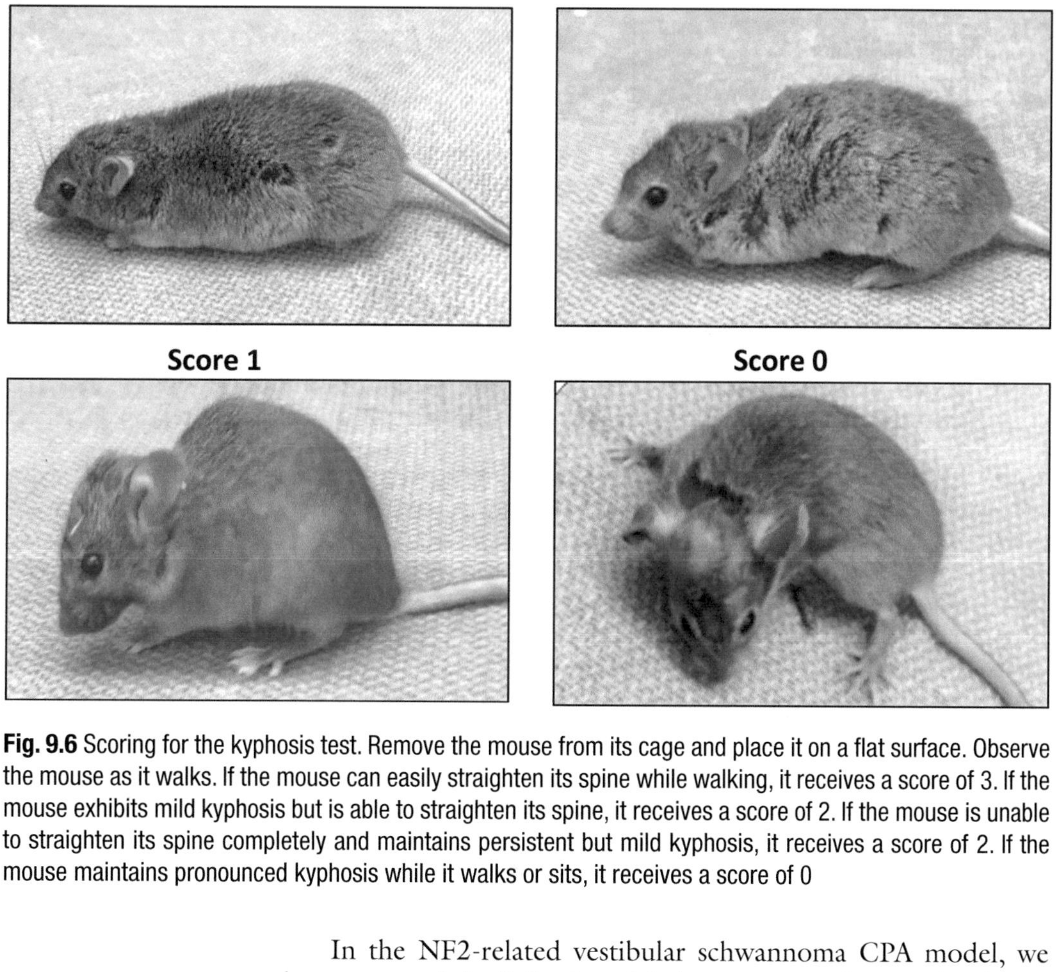

Fig. 9.6 Scoring for the kyphosis test. Remove the mouse from its cage and place it on a flat surface. Observe the mouse as it walks. If the mouse can easily straighten its spine while walking, it receives a score of 3. If the mouse exhibits mild kyphosis but is able to straighten its spine, it receives a score of 2. If the mouse is unable to straighten its spine completely and maintains persistent but mild kyphosis, it receives a score of 2. If the mouse maintains pronounced kyphosis while it walks or sits, it receives a score of 0

In the NF2-related vestibular schwannoma CPA model, we have reported the following: (a) the surgery and tumor implantation procedures do not result in ataxia symptoms (Fig. 9.7a); (b) mice develop symptoms of ataxia 2 weeks after tumor implantation (Fig. 9.7b); and (c) the severity of ataxia deteriorates as the tumors progress, and the ataxia score inversely correlates with tumor size (Fig. 9.7c) [10].

3.2 Hearing Function Test

Otoacoustic emissions (OAEs) and auditory brainstem response (ABR) are tests widely used in hearing screening. ABR audiometry was first described by Jewett and Williston and is the standard method to assess hearing loss in humans and small animals [17]. OAE test was first reported by Kemp; it tests the function of the cochlea and the brain pathways for hearing [18].

To evaluate the hearing function, DPOAE and ABR thresholds are tested in tumor-bearing mice. Before the hearing test, make sure the mouse is not completely deaf by using a finger friction test close to its head and noting the appropriate acoustic startle reflex (i.e., Preyer's reflex). If testing a new mouse strain and

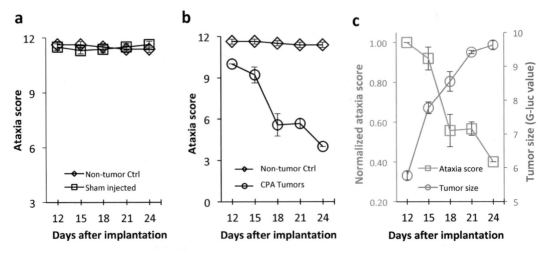

Fig. 9.7 Ataxia test in CPA model of NF2-related vestibular schwannomas. (**a**) Surgery and tumor implantation techniques do not cause ataxia. Ataxia score test and rotarod performance test show no differences between uninjected non-tumor-bearing control and sham groups ($n = 12$). (**b**) CPA tumors cause ataxia in mice. Compared to non-tumor-bearing mice ($n = 12$), mice bearing NF2$^{-/-}$ tumors in the CPA model ($n = 8$) start to show decreased ataxia score 2 weeks after tumor implantation, which deteriorates over time. (**c**) In mice bearing NF2$^{-/-}$ schwannomas in the CPA model, ataxia score inversely correlates with tumor size as measured by plasma GLuc assay. Institutional regulatory board permission was obtained for all procedures performed within this protocol from MGH and MEEI IACUC

establishing reference values for untreated control animals, make sure that rodents are sex- and age-matched with the tumor-bearing mice, because these variables can influence hearing.

After the mouse is sufficiently anesthetized, create a cut (approximately 3 mm long) parallel to the edges of tragus and antitragus to inferiorly extend the external auditory canal in order to create more space for the placement of electrodes and the microphone-speaker probe. Place the mouse in the sound-attenuating chamber. Position sub-dermal electrodes at the lower ipsilateral pinna (positive), the nose bridge (negative), and the junction of the tail and torso (ground) (Fig. 9.8a). While the vertex of the skull in general is often used to place the negative electrode, this area has been affected by the surgery and therefore should be avoided in these animals.

Calibrate the setup. Lower the microphone-speaker probe to a standardized position in the external auditory canal, making sure that the distance between the tympanic membrane and probe is identical between measurements of different ears/animals. Carry out an additional in-ear calibration that matches the other measurements in this experimental series. Close the door of the sound-attenuating chamber, and make sure that the light and surrounding noise are consistent during measurements.

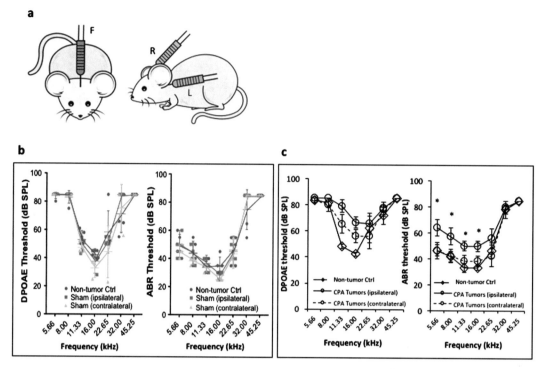

Fig. 9.8 Hearing function test in CPA model of NF2-related vestibular schwannomas. (**a**) Schematic of hearing function test in mouse. Three electrodes were placed in the mouse head, within the soundproof chamber. F indicates forehead, R indicates right ear, and L indicates left ear. (**b**) Surgery and tumor implantation techniques do not affect hearing function. In not-injected non-tumor-bearing control mice ($n = 5$) and in mice that underwent unilateral sham surgery (saline injection) ($n = 3$), DPOAE and ABR tests show no differences in thresholds both ipsilaterally and contralaterally. Data presented are mean ± SEM. Adapted with permission from ref. [11]. (**c**) CPA tumors induce hearing loss in mice. Mice with NF2$^{-/-}$ schwannomas in the CPA model ($n = 6$) have elevated ABR thresholds ipsilateral to the tumor compared to thresholds contralateral to the tumor or thresholds in non-tumor-bearing control mice ($n = 5$). * Significance of results is depicted by asterisks. $P = 0.03$ for 5.6, 8.0, and 22.6 kHz, respectively, and $P = 0.01$ for 11.3 and 16.0 kHz. Although DPOAE thresholds demonstrate a trend for elevation at some frequencies, this trend does not reach statistical significance. Adapted with permission from ref. [11]. Institutional regulatory board permission was obtained for all procedures performed within this protocol from MGH and MEEI IACUC

Measure DPOAEs and ABRs, potentially on both sides with repositioning of the pinna electrode when switching ears. DPOAEs are assessed via simultaneous presentations of two primary tones (f_1 and f_2) with the frequency ratio $f_2/f_1 = 1.2$ at half-octave steps from $f_2 = 5.66$–45.25 kHz. Intensity is increased in 5 dB steps from 15 to 80 dB sound pressure level (SPL). Monitoring of the $2f_1 - f_2$ DPOAE amplitude and surrounding noise floor allows the definition of a threshold as the f2 intensity creating a distortion tone >0 dB SPL (Fig. 9.8).

ABR waveforms are recorded via the sub-dermal electrodes described above. The same frequencies and sound levels as for DPOAE measurements are tested. Responses are amplified

(10,000X), filtered (0.3–3.0 kHz), and averaged (512 repetitions), using a custom LabVIEW data-acquisition software application installed on a PXI chassis (National Instruments Corp., Austin, Texas). Auditory threshold for each frequency is set to the lowest intensity at which recognizable peaks can be detected. In case of no response, 85 dB is chosen as the threshold (i.e., 5 dB above the highest level tested) (Fig. 9.8).

Notes and Troubleshooting:

1. Strong background activity during hearing test. This may occur when the electrodes are positioned incorrectly or anesthetic depth is inadequate. Consider repositioning the electrodes or provide anesthetic boost as needed.

2. Large variation in each DPOAE and ABR measurement. This may occur when the positioning of the electrodes is inconsistent between mice and the recording system is not correctly calibrated. To solve the problem, consistently position electrodes at the same anatomical location and calibrate the system prior to testing.

In the NF2$^{-/-}$ CPA model, the surgical procedure involving skull drilling for a craniotomy and needle injections for tumor cell implantation did not cause any hearing loss (Fig. 9.8b). We found increased DPOAE and ABR thresholds in tumor-bearing mice, indicating tumor-induced hearing loss (Fig. 9.8c).

3.3 Assess Motor Function Using Rotarod Test

The rotarod test was first developed by Dunham and Miya in 1957 to evaluate the level of neurotoxicities from experimental drugs [19]. In the original test, the duration that the animal could maintain its equilibrium on the revolving roller, which was set at constant 5 revolutions per minute (rpm), was recorded and compared between control and treatment arms. In 1968, Jones and Roberts modified the test by setting an initial low rotating speed and then gradually increasing the speed linearly [20]. This modification had made this procedure much more sensitive and accurate. Since then, rotarod test has been applied in a wide range of animal studies to investigate the effects of disease and treatment on balance and coordination (Table 9.1).

Rotarod test is performed using an automated rotarod machine. The device contains a control panel for speed, direction of rotation (forward or reverse), and brake. A typical rotarod test apparatus has four or six isolate chambers separated by clear plexiglass, and each one of the chamber tests one animal in a single test. The rods are set at least 15 inches above the table surface, in order to discourage animals from jumping off the roller [21]. A capture cage below the rod contains an infrared sensor within to intrigue the calculagraph which measures the duration the animals stayed walking on the rod [22, 23].

Table 9.1 Usage of hearing, ataxia, and rotarod tests in animal models of CNS tumors

Hearing test				
Disease	Model	Findings	Methods	References
NF2 vestibular schwannoma	Genetically engineered *Postn-Cre; Nf2^flox/flox* mice	The development of cranial nerve VIII tumors was correlated with impairments in hearing and balance	ABR test	[25]
NF2 vestibular schwannoma	• Athymic immunodeficient nude mice • Mouse Nf2^−/− schwannoma cell line • Tumor cells implanted into the auditory-vestibular nerve complex model	• Models tumor-induced hearing loss in mouse model • Surgery and sham injection caused hearing loss and animals took up to 14 days to recover hearing	ABR test	[26]
NF2 vestibular schwannoma	• Athymic immunodeficient nude mice • Mouse Nf2^−/− schwannoma cell line • Tumor cells implanted into the cerebellopontine angle model	• Models tumor-induced hearing loss in mouse • Tumor size does not correlate with the severity of hearing loss; faithfully models hearing loss in NF2 patients • cMET inhibitor controls tumor growth but has no effect on hearing • Potential mechanisms: cMET inhibitor does not effect the expression of HGF, which has a therapeutic potential for hearing	DPOAE and ABR test	[11]

(continued)

Table 9.1 (continued)

NF2 vestibular schwannoma	• Immunodeficient Rowett nude rats • Mouse merlin$^{-/-}$ MTC-10 Schwann cells • Tumor cells implanted in the facial and cochleovestibular nerves	Models tumor-induced hearing loss in rat	ABR test	[9]
NF2 vestibular schwannoma	• Athymic immunodeficient nude mice • Immune-competent FVB mice • Tumor cells implanted into the cerebellopontine angle model	Surgery and sham implantation does not effect hearing function Models tumor-induced hearing loss in mouse	DPOAE and ABR test	[10]

Ataxia test

Disease	Model	Findings	Methods	References
Medulloblastoma	• Conditional Patched1 knockout mice • Ptch1 haploinsufficiency Pax7-expressing medulloblastoma cell lines	• Qualitative correlation between increasing cerebellar tumor volumes decreasing stride lengths • Stride length analysis is a sensitive method for assessing new therapeutic compounds in mouse model of brain tumors	• Rotarod test: Motor performance • Forced air challenge test: Cranial nerve VII function • 180° horizontal screen test: Coordination • Horizontal wire test: Balance and limb coordination • Painted hind paws prints: Measure stride lengths	[27]

(continued)

Table 9.1 (continued)

Glioma	• BALB/c nude mice • Human U87MG cells • Tumor cells implanted into the cortex	Biopsy of the mouse brain is associated with minimal neurological morbidity	Ten-point neurological severity score for: • Presence of mono or hemiparesis • Inability to walk on beams • Inability to balance on beams • Circle exit test • Gait test: Coordination and muscle function • Startle reflex • Seeking behavior	[28]
Glioma	• Wistar rats • Rat C6 glioma cells • Tumor cells implanted into the right frontal cortex	• Tumor growth changes motor behavior • The results of gait analysis were more sensitive than general motor analysis for the glioma model	• Spontaneous locomotor activity: Actimeter • Gait test	[29]
NF2 vestibular schwannoma	• Athymic immunodeficient nude mice • Immune-competent FVB mice • Tumor cells implanted into the cerebellopontine angle model	• Surgery and sham implantation does not effect neurological function • Two weeks after tumor implantation, neurological function starts to decrease in mice • Ataxia score is inversely correlated with tumor growth	• Ledge test: Coordination • Hindlimb clasping test: Disease progression • Gait test: Coordination and muscle function • Kyphosis: Posture • Rotarod test: Motor function	[10]

(continued)

Table 9.1 (continued)

Rotarod assay				
Disease	**Model**	**Findings**	**Mechanisms**	**References**
NF2 vestibular schwannoma	• Nude mouse • Mouse NF2$^{-/-}$ Schwann cells • Tumor cells implanted in the sciatic nerve	Anti-angiogenic therapy improves rotarod performance	Anti-VEGF therapy improves rotarod performance via reducing tissue edema, decreasing muscle atrophy, and enhancing nerve regeneration	[14]
Glioma	• Sprague-Dawley rats • Rat C6 glioma cell line • Tumor cells implanted in the intracerebral model	Tumor growth reduced rotarod performance	N/A	[30]
Medulloblastoma	• Conditional Patched1 knockout mice • Ptch1 haploinsufficiency Pax7-expressing medulloblastoma cell lines	Tumor growth reduced rotarod performance	N/A	[27]
NF2 vestibular schwannoma	• Athymic immunodeficient nude mice • Immune-competent FVB mice • Tumor cells implanted into the cerebellopontine angle model	Tumor growth reduced rotarod performance	N/A	[10]
Malignant peripheral nerve sheath tumors	• C57BL/6j:RJ mixed strain • Ahr knockout mice • MPNST-derived cell lines STS26T and 90–8	Ahr$^{-/-}$ mice have impaired locomotion and motor coordination compared with wild-type mice	Ahr ablation in mice cause locomotor defects and worsen the rotarod performance	[31]

GBM glioblastoma, *CNF1* Cytotoxic Necrotizing Factor 1, *MB* medulloblastoma-prone, *MPNSTs* malignant peripheral nerve sheath tumors, *Ahr* aryl hydrocarbon receptor

Mice were trained on the rotarod daily for 3 days. Each training session consisted of (a) acclimation period, 30 s at 4 rpm, and (b) acceleration period, increase of 4 rpm every 60 s to a maximum of 40 rpm. The rotarod endurance was determined as the amount of time that elapsed before the mouse fell off [24]. To avoid heterogeneity between animals, the average time to fall from the rotating cylinder was normalized to the value from each mouse on the first day of the training session and presented as relative rotarod endurance. To control the tumor burden, the rotarod test was performed in animals bearing size-matched tumors.

Notes and Troubleshooting: Rotarod test is a relatively straightforward procedure; potential problems include:

1. The mouse holds on to the rod but does not walk (cartwheeling). This may happen when the rod has ridges deep enough to facilitate grip. Control the surface and diameter of the rod so the mouse is not able to grip the rod.

2. The mouse jumps onto the floor or escapes from the rod. Closely observe the mouse during the test; when this happens, data should be excluded from analysis; and rotarod test should be repeated.

3. Significant variation in rotarod duration. The rotarod test performance depends on age, sex, strain, and body weight. It is important to keep these parameters comparable between groups in a single experiment. In addition, it is recommended to perform the rotarod test at the same time of day because physiological and behavior parameters change throughout the day.

In the $NF2^{-/-}$ sciatic nerve model, we have reported that anti-angiogenic therapy using an antibody (B20.4.4, Genentech) that neutralizes both human (tumor) and mouse (host) VEGF significantly improved rotarod performance (Fig. 9.9a, b). Starting as early as 6 h after the first B20 treatment, rotarod duration increased and significantly improved after 24 h compared with mice in control group. To examine the effect of different treatments on neurological function, mice bearing same-sized tumors were treated with control IgG, radiation (5 Gy), B20, or combined B20 and radiation. Before treatment, no significant differences were noted from the rotarod test in mice among different groups. After B20 treatments, mice showed significantly improved rotarod performance compared with mice in control and RT groups (Fig. 9.9b). Furthermore, we found that the degree of tumor edema was significantly inversely correlated with rotarod duration (Fig. 9.9c). Anti-VEGF therapy has been shown to reduce tissue edema by decreasing vessel hyperpermeability, which in turn improves the neurological function [14].

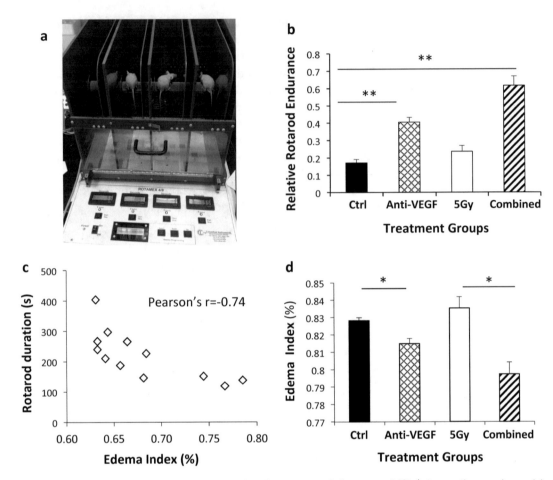

Fig. 9.9 Anti-VEGF treatment improves rotarod performance and decreases $Nf2^{-/-}$ tumor tissue edema. (**a**) Image of mouse on the rotarod machine. The rotarod assay is a classic assay of motor performance and coordination. Mice balanced upon an accelerating, rotating rod. The latency to falling off the beam is recorded as rotarod endurance. (**b**) Rotarod test was carried out in mice bearing size-matched tumors, followed by tumor collection to evaluate edema. The average time to fall from the rotating cylinder was normalized to the value from each mouse on the first day and presented as relative rotarod endurance. Anti-VEGF treatment, as monotherapy as well as in combination with radiation therapy (5Gy), significantly enhanced rotarod performance. (**c**) Edema index significantly inversely correlates with rotarod duration. Tumor edema index was calculated as follows: Tumor edema index = (wet weight − dry weight)/wet weight [14]. Pearson product moment correlation coefficient $r = -0.7416$, $P = 0.006$. (**d**) Anti-VEGF treatment, as monotherapy as well as in combination with radiation therapy (5Gy), significantly decreased tumor edema. Representative of at least three independent experiments ($n = 8$), data presented are mean ± SEM. *$P \leq 0.01$, **$P \leq 0.005$. Adapted with permission from ref. [14]. Institutional regulatory board permission was obtained for all procedures performed within this protocol from MGH and MEEI IACUC

Acknowledgments

I thank Drs. Lukas Landegger and Konstantina Stankovic for their help with Fig. 9.8.

Funding Support
This study was supported by the Department of Defense New Investigator Award (W81XWH-16-1-0219, L.X.), the American Cancer Society Research Scholar Award (RSG-12-199-01-TBG, L.X.), the Children's Tumor Foundation Drug Discovery Initiative (L.X.), and Ira Spiro Award (L.X.).

References

1. Siegel RL, Miller KD, Jemal A (2018) Cancer statistics, 2018. CA Cancer J Clin 68(1):7–30. https://doi.org/10.3322/caac.21442

2. Osoba D, Brada M, Prados MD, Yung WK (2000) Effect of disease burden on health-related quality of life in patients with malignant gliomas. Neuro-Oncology 2(4):221–228. https://doi.org/10.1093/neuonc/2.4.221

3. Heimans JJ, Taphoorn MJ (2002) Impact of brain tumour treatment on quality of life. J Neurol 249(8):955–960. https://doi.org/10.1007/s00415-002-0839-5

4. Kirazli T, Oner K, Bilgen C, Ovul I, Midilli R (2004) Facial nerve neuroma: clinical, diagnostic, and surgical features. Skull Base 14(2):115–120. https://doi.org/10.1055/s-2004-828707

5. Lu-Emerson C, Plotkin SR (2009) The neurofibromatoses. Part 2: NF2 and schwannomatosis. Rev Neurol Dis 6(3):E81–E86

6. Manto M (2018) Cerebellar motor syndrome from children to the elderly. Handb Clin Neurol 154:151–166. https://doi.org/10.1016/B978-0-444-63956-1.00009-6

7. Fukumura D, Duda DG, Munn LL, Jain RK (2010) Tumor microvasculature and microenvironment: novel insights through intravital imaging in pre-clinical models. Microcirculation 17(3):206–225. https://doi.org/10.1111/j.1549-8719.2010.00029.x

8. Askoxylakis V, Badeaux M, Roberge S, Batista A, Kirkpatrick N, Snuderl M, Amoozgar Z, Seano G, Ferraro GB, Chatterjee S, Xu L, Fukumura D, Duda DG, Jain RK (2017) A cerebellar window for intravital imaging of normal and disease states in mice. Nat Protoc 12(11):2251–2262. https://doi.org/10.1038/nprot.2017.101

9. Dinh CT, Bracho O, Mei C, Bas E, Fernandez-Valle C, Telischi F, Liu XZ (2018) A xenograft model of vestibular Schwannoma and hearing loss. Otol Neurotol 39(5):e362–e369. https://doi.org/10.1097/MAO.0000000000001766

10. Chen J, Landegger LD, Sun Y, Ren J, Maimon N, Wu L, Ng MR, Chen J, Zhang N, Zhao Y, Gao X, Fujita T, Roberge S, Huang P, Jain RK, Plotkin S, Stankovic KM, Xu L (2019) A cerebellopontine angle mouse model for the investigation of tumour biology, hearing and neurological function in NF2-related vestibular schwannoma. Nat Protoc 14(2):541–555. https://doi.org/10.1038/s41596-018

11. Zhao Y, Liu P, Zhang N, Chen J, Landegger LD, Zhao F, Zhang J, Fujita T, Stemmer-Rachamimov AO, Zhang Y, Ferraro G, Liu H, Muzikansky A, Plotkin S, Stankovic KM, Jain RK, Xu L (2018) Targeting the cMET pathway augments radiation response without adverse effect on hearing in NF2 schwannoma models. Proc Natl Acad Sci U S A 115(9):E2077–E2084

12. Tannous BA (2009) Gaussia luciferase reporter assay for monitoring biological processes in culture and in vivo. Nat Protoc 4(4):582–591. https://doi.org/10.1038/nprot.2009.28

13. Bovenberg MS, Degeling MH, Tannous BA (2012) Enhanced Gaussia luciferase blood assay for monitoring of in vivo biological processes. Anal Chem 84(2):1189–1192. https://doi.org/10.1021/ac202833r

14. Gao X, Zhao Y, Stemmer-Rachamimov AO, Liu H, Huang P, Chin S, Selig MK, Plotkin SR, Jain RK, Xu L (2015) Anti-VEGF treatment improves neurological function and augments radiation response in NF2 schwannoma model. Proc Natl Acad Sci U S A 112(47):14676–14681. https://doi.org/10.1073/pnas.1512570112

15. McCann CM, Waterman P, Figueiredo JL, Aikawa E, Weissleder R, Chen JW (2009)

Combined magnetic resonance and fluorescence imaging of the living mouse brain reveals glioma response to chemotherapy. NeuroImage 45(2):360–369. https://doi.org/10.1016/j.neuroimage.2008.12.022

16. Guyenet SJ, Furrer SA, Damian VM, Baughan TD, La Spada AR, Garden GA (2010) A simple composite phenotype scoring system for evaluating mouse models of cerebellar ataxia. J Vis Exp 39. https://doi.org/10.3791/1787

17. Henry KR (1979) Auditory brainstem volume-conducted responses: origins in the laboratory mouse. J Am Aud Soc 4(5):173–178

18. Kemp DT (1978) Stimulated acoustic emissions from within the human auditory system. J Acoust Soc Am 64(5):1386–1391

19. Dunham NW, Miya TS (1957) A note on a simple apparatus for detecting neurological deficit in rats and mice. J Am Pharm Assoc Am Pharm Assoc 46(3):208–209

20. Jones BJ, Roberts DJ (1968) A rotarod suitable for quantitative measurements of motor incoordination in naive mice. Naunyn Schmiedebergs Arch Exp Pathol Pharmakol 259(2):211

21. Lee BH, Kim J, Lee RM, Choi SH, Kim HJ, Hwang SH, Lee MK, Bae CS, Kim HC, Rhim H, Lim K, Nah SY (2016) Gintonin enhances performance of mice in rotarod test: involvement of lysophosphatidic acid receptors and catecholamine release. Neurosci Lett 612:256–260. https://doi.org/10.1016/j.neulet.2015.12.026

22. Deacon RM (2013) Measuring motor coordination in mice. J Vis Exp 75:e2609. https://doi.org/10.3791/2609

23. Stroobants S, Gantois I, Pooters T, D'Hooge R (2013) Increased gait variability in mice with small cerebellar cortex lesions and normal rotarod performance. Behav Brain Res 241:32–37. https://doi.org/10.1016/j.bbr.2012.11.034

24. Sharma N, Baxter MG, Petravicz J, Bragg DC, Schienda A, Standaert DG, Breakefield XO (2005) Impaired motor learning in mice expressing torsinA with the DYT1 dystonia mutation. J Neurosci Off J Soc Neurosci 25(22):5351–5355. https://doi.org/10.1523/JNEUROSCI.0855-05.2005

25. Gehlhausen JR, Park SJ, Hickox AE, Shew M, Staser K, Rhodes SD, Menon K, Lajiness JD, Mwanthi M, Yang X, Yuan J, Territo P, Hutchins G, Nalepa G, Yang FC, Conway SJ, Heinz MG, Stemmer-Rachamimov A, Yates CW, Wade Clapp D (2015) A murine model of neurofibromatosis type 2 that accurately phenocopies human schwannoma formation. Hum Mol Genet 24(1):1–8. https://doi.org/10.1093/hmg/ddu414

26. Bonne NX, Vitte J, Chareyre F, Karapetyan G, Khankaldyyan V, Tanaka K, Moats RA, Giovannini M (2016) An allograft mouse model for the study of hearing loss secondary to vestibular schwannoma growth. J Neuro-Oncol 129(1):47–56. https://doi.org/10.1007/s11060-016-2150-9

27. Samano AK, Ohshima-Hosoyama S, Whitney TG, Prajapati SI, Kilcoyne A, Taniguchi E, Morgan WW, Nelon LD, Lin AL, Togao O, Jung I, Rubin BP, Nowak BM, Duong TQ, Keller C (2010) Functional evaluation of therapeutic response for a mouse model of medulloblastoma. Transgenic Res 19(5):829–840. https://doi.org/10.1007/s11248-010-9361-1

28. Rogers S, Hii H, Huang J, Ancliffe M, Gottardo NG, Dallas P, Lee S, Endersby R (2017) A novel technique of serial biopsy in mouse brain tumour models. PLoS One 12(4):e0175169. https://doi.org/10.1371/journal.pone.0175169

29. Souza TKF, Nucci MP, Mamani JB, da Silva HR, Fantacini DMC, de Souza LEB, Picanco-Castro V, Covas DT, Vidoto EL, Tannus A, Gamarra LF (2018) Image and motor behavior for monitoring tumor growth in C6 glioma model. PLoS One 13(7):e0201453. https://doi.org/10.1371/journal.pone.0201453

30. Song T-W, Lee J-K, Lee S-Y, Lian S, Joo S-P, Kim H-S (2016) Establishment of a malignant glioma model in rats. The Nerve 2(2):17–21. https://doi.org/10.21129/nerve.2016.2.2.17

31. Shackleford G, Sampathkumar NK, Hichor M, Weill L, Meffre D, Juricek L, Laurendeau I, Chevallier A, Ortonne N, Larousserie F, Herbin M, Bieche I, Coumoul X, Beraneck M, Baulieu EE, Charbonnier F, Pasmant E, Massaad C (2018) Involvement of Aryl hydrocarbon receptor in myelination and in human nerve sheath tumorigenesis. Proc Natl Acad Sci U S A 115(6):E1319–E1328. https://doi.org/10.1073/pnas.1715999115

Chapter 10

Dynamic Immunotherapy Study in Brain Tumor-Bearing Mice

Luiz Henrique Medeiros Geraldo, Yunling Xu, and Thomas Mathivet

Abstract

We describe here a method to follow mouse glioma progression and its immune profile over time by intravital multiphoton microscopy. After craniotomy of the parietal bone, orthotopic glioma cells are engraft in the brain cortex and sealed with a glass window cemented on the skull, allowing longitudinal imaging of the tumor progression. The expression of various fluorescent reporters in tumor cells and in designated cell types, thanks to numerous available transgenic mouse strains, allows obtaining multicolor 4D images of the glioma. This technique is suitable both to evaluate the effect of immunotherapeutic treatments and to unravel the basic mechanisms of tumor-host interactions.

Keywords Glioma, Intravital imaging, Multiphoton, Immunotherapy

1 Introduction

Every in vitro finding needs to be validated in an in vivo setting, taking into the account the more complex environment of a living organism. Biological imaging is a relevant part of such a task and can nowadays be performed in vivo using a number of imaging techniques (MRI, PET, CT), but many among them lack the resolution required to detect single cells and their interplay during biological processes. Hence, an alternative imaging technique, multiphoton laser scanning microscopy, has emerged as a suitable approach for imaging living animals due to its ability to penetrate tissues in depth and its low phototoxicity [1, 2], and it is therefore not surprising that in the last two decades, this technology has been widely used. During the same time, considerable technological developments by microscope manufacturers have led to the production of ready-to-use systems with more performant lasers, optics, and other hardware specifically designed for this application. Further, a wide range of fluorescent reporter transgenic mouse strains are available [3], enabling the detection of distinct

Giorgio Seano (ed.), *Brain Tumors*, Neuromethods, vol. 158,
https://doi.org/10.1007/978-1-0716-0856-2_10, © Springer Science+Business Media, LLC, part of Springer Nature 2021

features in vivo without the need of antibody labeling; specific probes allow to detect physiological statuses like tissue hypoxia and track their local variation in the tumor over time [4].

The method described here makes use of these available resources in order to follow the progression of brain tumors in mice and associated therapeutic treatments. It consists of implanting glioma cells in the brain cortex and making intracranial tumor growth observable by multiphoton microscopy through a permanent cranial window. The experimental procedure can be divided into three stages: (1) preparation of glioma cells for the implantation, (2) surgery creating the orthotopic glioma and the cranial window, and (3) imaging, which can be performed repeatedly for weeks during tumor progression and treatments. This last task is greatly facilitated by the installation of the permanent cranial window, as each imaging session does not require any further intervention apart from/other than the temporary anesthetization of the animals.

Compared with ex vivo histological analysis, this method is more laborious and requires expensive instrumentation. However, it possesses the great advantage of directly observing changes occurring over time in the same locations within the same tumor, rather than to deduce them by comparing samples from different animals at different tumor stages. A smaller number of animals are thus required in order to obtain statistically significant results. More importantly, the ability to observe the same region of the same tumor repeatedly over time provides unique information of the dynamic behavior of tumor cells and their changing environment. With the purpose of acquiring high-quality images, the imaging depth from the surface is limited to a few hundred micrometers (800 μm) and thus only to the brain cortex region; however, the images are obtained from intact living tissue, without any fixation or sectioning artifact that could alter its morphology. In the case of imaging blood vessels, this also means observing their morphology and cell behavior while being perfused with blood, a condition that critically affects endothelial cell behavior and that is very challenging to mimic in in vitro settings.

In our experiments we use orthotopic glioma cell lines derived from C57BL/6 mice, in order to use a range of fluorescently labeled transgenic mice strains as recipients with the same genetic background and maintain native immune compartment of the host intact. However, the procedure is also suitable for implanting glioma cells of different origins (such as human biopsies) using immunocompromised mice strains (e.g., nude mice) but will then not be compatible with an in-depth immune system analysis. The choice of the cells to implant must be decided according to the aims of the study, as in our experience some cell types better reproduce specific features of human gliomas. For example, on the one hand, the cells mentioned in this protocol are optimal for imaging tumor blood vessels and immune cell behavior, because they form tumors with the typically dense and dysmorphic vasculature observed in high-grade gliomas [5]; on the other hand, these tumors grow rather compact and lack the typical single-cell infiltration of the brain parenchyma often observed in patients. Other cell types reproducing this behavior better, such as those obtained from biopsies, may be more suitable for studying invasion.

The modality of imaging depends on several factors: the features of the microscope used, the fluorophores expressed by the glioma cells and the mouse cells, eventual contrasting agents injected intravenously, and others. Knowledge of how to set up and use a multiphoton microscope is required, as providing such a wide set of information would go beyond the scope of this protocol. We have however included sections dedicated to imaging, where we provide suggestions on how to apply multiphoton laser scanning microscopy to this particular model. This method was originally developed at the Vesalius Research Center of Leuven (VRC) and the Developmental Biology Institute of Marseille (IBDML) [6, 7]; we offer now an update, integrated with some further tips and more recent technical improvements.

2 Materials

2.1 Cell Culture

1. Glioma261 [8], CT-2A [9], or other glioma cells, ideally stably expressing a fluorescent protein, which allows discriminating them from the host cells.

   Considering that we normally use transgenic mice expressing green and/or red fluorescent proteins [10], we routinely mark our glioma cells with the blue fluorescent protein TagBFP [11]; if the microscopy system capabilities allow it (multiphoton laser tuning and emission filter ranges, objective optimized light transmission), the far-red fluorescent protein iRFP is also a suitable candidate for marking the glioma cells [12] (note 1).

2. Common cell culture reagents: medium (DMEM supplemented with 10% fetal bovine serum, 2.0 mM L-glutamine), trypsin/EDTA solution, and phosphate-buffered saline (PBS).

3. A cell culture incubator.

4. Cell culture-coated and non-coated petri dishes and other common cell culture tools.

2.2 Surgery

1. Adult mice genetically compatible with the glioma cells to be implanted. The age of the mice is not strictly important, but better reproducible data are obtained using mice of 8 to 12 weeks (note 2). In our experiments we often use the mouse strain ROSA^{mTmG} [13], which—when crossed with a properly

chosen Cre strain—allows the expression of membrane-targeted EGFP in a specific cell type and membrane-targeted tdTomato in all other cells. An example of this strategy is shown in Fig. 10.4d, where EGFP labeling of myeloid cells was achieved by crossing these mice with the CSF1R-Mer-Cre-Mer strain [14].

2. A head holder for mice (Fig. 10.2a), commercially available from several companies.

3. Basic surgical tools: scalpels, scissors, skin forceps, mini forceps, ultrafine micro knife, syringes, and needles.

4. A micro-drill with carbon steel burrs of diameter 0.5 mm.

5. One upright stereomicroscope mounted on a vertical stand (Fig. 10.1a).

6. A micromanipulator holding a glass capillary connected with a gaz-tight 50 μL Hamilton syringe (Fig. 10.1b). The capillary tip must be severed and grinded (45° angle) close to the shaft so that the opening will be large enough to let pass the glioma spheroids through it (200–250 μm). The size of the capillary opening and other small items can be measured using a calibration slide under the stereomicroscope.

7. Sepharose microspheres of diameter ~ 200 μm, to be used to cover the implantation site (Fig. 10.1c) (note 3).

8. Spheroids of glioma cells genetically compatible with the mouse strain used, prepared as described in Subheading 3.1 (Fig. 10.1d). Tumor spheroids should ideally be kept in a cell incubator close to the surgery to facilitate the implantation procedure.

9. Round coverslips of diameter 5–6 mm, to be used for the window (Fig. 10.3d).

10. Histocompatible glue (cyanoacrylate liquid glue).

11. Dental resin for temporary restorations. For our experiments we use methyl-methacrylate resin, which is easy to apply and of proven durability; other types of dental resins could work equally well.

12. Heated pad fitting the mouse holder during surgery (Fig. 10.2a) and heated pad or blanket or infrared lamp to be used during the recovery phase following the surgical procedure.

13. A solution suitable for skin disinfection (ethanol 70% or iodine solution).

14. Ketamine and xylazine for anesthesia. A working solution of saline containing ketamine 10 mg/mL and xylazine 1 mg/mL can be prepared in advance. Intraperitoneal injection of 10 μL of it per gram of animal weight.

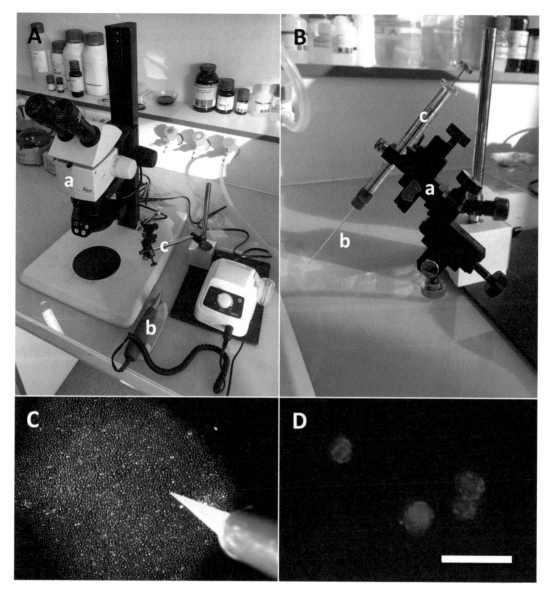

Fig. 10.1 Surgical setting for glioma implantation. (**a**) Stereomicroscope for the craniotomy, tumor spheroid implantation, and mounting of the cranial window (**a**). (**b**) Micro-drill used for craniotomy. (**c**) Micromanipulator for glioma grafting. (**b**) X/Y/Z micromanipulator (**a**) holding a capillary (**b**) mounted on 50 μl Hamilton syringe (**c**) for tumor spheroid implantation. (**c**) Transparent 200 μm diameter Sepharose beads for trans-dura mater point of implantation closure. (**d**) Spheroids of CT2A cells stably expressing TagBFP, seen under a fluorescence stereomicroscope. Scale bar 500 μm

15. Eye ointment, to protect the eyes of the animal from exposure keratitis during anesthesia.

16. Analgesics, to be administered to the animals as recommended by your local veterinarian and ethical committee.

17. Ice-cold phosphate-buffered saline (PBS) complemented with 100 U/mL penicillin and 100 µg/mL streptomycin.

18. Stripes of adsorbent paper, sterilized by autoclaving after preparation.

19. Optional: P20 and P200 micropipettes and capillary micropipette tips, for easier handling of glue.

2.3   Imaging

1. A confocal microscope equipped and optimized for multiphoton imaging. As of today several companies offers such instruments, which can be tailored to specific needs in terms of color channel number, range/power of the excitation laser, and other parameters. For our experiments we are currently using a confocal microscope equipped with a 25x IR-optimized objective, a two-ray laser (one fixed at 1040 nm and one accordable from 680 to 1300 nm), and three non-descanned hybrid detectors. One point to consider when setting up the system is that the sample stage should be suitable for intravital imaging, thus leaving enough space for placing the animal under the objective (Fig. 10.4a).

2. Microscope stage equipped with a plate carrying a head holder and integrated with gas anesthesia/scavenging and thermal control equipment (heating plate/blanket, Fig. 10.4b). Ideally, for easier handling of the animal, it should be possible to extract this plate from under the microscope before and after imaging (Fig. 10.4b). Currently, a ready-to-use device does not exist; thus, single parts should be purchased separately and then assembled together on a plate fitting the microscope stage. Anesthesia scavenging and thermal control equipment are sold by many retailers of veterinary equipment. For our system, we ordered a customized head holder which allows us to easily tilt the mouse head on the X and Y axes of the stage and align the window plane with them (Fig. 10.4b).

3. For gas anesthesia: isoflurane, vaporizer, scavenging apparatus, induction chamber, sources of compressed medical air, and oxygen (Fig. 10.4c), which can be purchased from any company selling veterinary equipment.

4. Eye ointment.

5. Optional: if one needs to mark blood perfusion, fluorescently labeled dextrans of molecular size chosen according to the experimental aims should be used (see Subheading 3.3).

2.4   Immunother apeutic Treatment

Depending on the immune cell population to target according to your experimental aims, the appropriate therapeutic strategy should be determined by the experimenter using monoclonal antibody, peptides, nanoparticles, or alternative compounds.

Our laboratory focuses on myeloid cell-targeted immunotherapy. In this context, we previously used a monoclonal antibody treatment strategy targeting the soluble growth factor CSF1 (clone 5A1), which, by its inhibitory effect on macrophage proliferation, differentiation, and survival, limits the infiltration of tumor-associated macrophages (TAM) in the glioma microenvironment [5]. We also tested with success fluorescently labeled siRNA-loaded nanoparticles [15].

3 Methods

3.1 Cell Culture

In order to be accessible for imaging, the tumor must be positioned a few hundred micrometers under the window surface. It is therefore necessary to implant the glioma cells in the brain cortex and not more in depth as it is done in other brain tumor graft models. In our experience, injecting a cell suspension so superficially is very difficult because single cells are easily flushed out by intracranial pressure; we overcome this problem by implanting a compact mass of cells, obtainable by growing the cells as spheroids in non-adherent conditions.

1. Grow cells as a monolayer in tissue culture-coated T75 cell culture dish and split them so that 2 days prior surgery they will be ~70% confluent.

2. Two days before the surgery, trypsinize each dish of 70% confluent cells, resuspend them in 10–12 mL of complete medium (here DMEM complemented with 10% SVF and 100 U/mL penicillin and 100 μg/mL streptomycin), and plate in a non-coated 10 cm petri dish. Being unable to adhere, the cells will cluster in spheroids that will start to be visible after 24 h.

3. Forty-eight hours after the plating, the spheroids should be ready for implantation (Fig. 10.1d). The time required for the formation of spheroids varies among different glioma cell lines. When using other cell lines than those in this protocol, it is advisable to test the optimal time and conditions required by the cell line to be used (note 4).

3.2 Surgery

The first step of the surgery is to prepare a cranial window; this is performed according to a previously described protocol [10], which our method integrates with the implantation of one or more spheroids of glioma cells. The whole task is far from being extremely difficult and can be performed in less than an hour, but good results require some training. Considering that the quality of the cranial window is essential for obtaining good images, we suggest taking time for practicing this procedure before running any full-scale experiment.

1. Anesthetize the animal by intraperitoneal injection of ketamine to 100 mg/kg and xylazine to 10 mg/kg. Make sure that anesthesia and analgesia are effective by checking for vibrissae movements, toe pinching, etc. Keep monitoring periodically for signs of awakening during the whole time of the surgery; if during the procedure it becomes necessary, re-induce anesthesia by a new injection of ketamine only.

2. Place the animal on the holder and stably fix the head using the mouthpiece and ear bars (Fig. 10.2a). Apply eye ointment, and disinfect the skin. Surgically remove the scalp (Fig. 10.2b), and then use a scalpel to scrape off the connective tissue and periosteum on the area until it appears dry (Fig. 10.2c). Apply histocompatible glue over the whole margin of the cut skin and fix it to the bone (Fig. 10.2d).

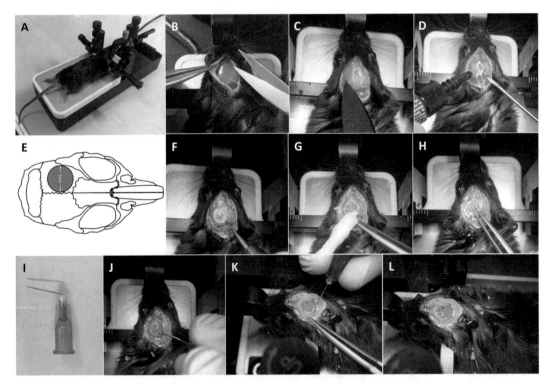

Fig. 10.2 Craniotomy procedure. (**a**) An anesthetized mouse is placed on the head holder equipped with a warming plate, shaved, and disinfected, and (**b**) its scalp is removed. (**c**) The periosteum and other connective tissue above the skull are scraped off using a scalpel blade. (**d**) The margins of the scalp are glued to the bone. (**e**) Schematic of cranial bones and sutures, showing the location of the 5 mm diameter craniotomy on the left parietal bone (gray circle). (**f**) Beginning of craniotomy by drilling and cold PBS washing and drying of the tissue (**g**). (**h**) Once the bone has been drilled to reach the most internal layer, the drilled bone disk should move upon gentle pressure exercise with forceps. (**i**) Bent 26G needle is used to raise the bone, (**j**) allowing the removal of the skullcap with a forceps (**k**). (**l**) Operated mouse exposing its external dura mater following craniotomy, ready for glioma spheroid implantation (keep moisturized with a drop of ice-cold PBS complemented with P/S)

3. Perform the craniotomy by carefully drilling a circle on the parietal bone (Fig. 10.2e, f). Time by time, wash the drilled bone area to avoid overheating and to discard glue and bone dust, and dry it with sterile paper (Fig. 10.2g). Also wash the drill tip with ice-cold PBS. The drilling must continue uniformly until it goes through the external layer of compact bone and the trabecular (spongy) bone under it and stop once reached the most internal, elastic and semitransparent bone layer (tabula vitrea). Periodically verify the status of this task by gently pressing the bone disk inside the drilled circle with small forceps: if it moves downward with respect to the surrounding bone, it is ready for being removed (Fig. 10.2h). Make sure that this happens at every side of the drilled circle: if any side of the bone disk does not elastically move when pushed, it means that further drilling is required on this part of the circle because there is still some trabecular bone to remove. During the drilling, try also to create a flat surface for the window by thinning small areas of the parietal bone, flanking the circle in rostral and caudal positions and proximal to the sagittal suture.

4. Bend a 27-gauge needle (Fig. 10.2i), and, paying extreme attention to avoid penetrating the meninges, insert it sideway in the bone disk (Fig. 10.2j) to hook it and lift it. Remove the bone disk using small forceps, to expose the meninges on top of the brain (Fig. 10.2k). At this stage, minor bleedings may occur, due to the interruption of small blood vessels in the trabecular bone and because of the sudden oxygenation of meninges' superficial vessels, but if the craniotomy has been performed correctly, this will stop after 1 min. Use sterile paper to drain the blood and wash with PBS; check the craniotomy every few minutes, and keep it hydrated with a drop of PBS (Fig. 10.2l). Do not ever let it dry.

5. Place under the capillary a petri dish containing the spheroids of glioma cells (Fig. 10.3a). Set the micromanipulator in order to place the opening of the capillary inside the medium and right in front of a spheroid (Fig. 10.3a), and select those of diameter fitting the opening in the capillary (200–250 μm, Fig. 10.1d). Selecting and implanting a constant number of spheroids of identical size allows to standardize the amount of cells implanted, replacing the counting of cells used in other protocols that inject solutions of single cells. Using the Hamilton syringe, suck the spheroid into the capillary. Let the spheroid(s) falls by gravitation into the capillary shaft. Meanwhile, remove the plate and place the mouse under the capillary.

6. Place the capillary tip close to the right side of the craniotomy (in proximity of the sagittal suture; Fig. 10.3b); using the

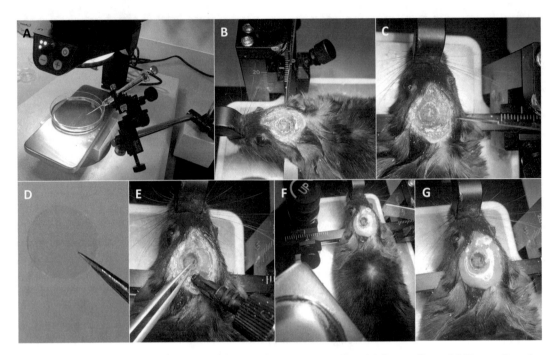

Fig. 10.3 Glioma spheroid implantation and cranial window mounting. (**a**) A smoothened 200 μm diameter capillary is positioned in a non-coated petri dish, pointing at a matching diameter spheroid in order to aspirate it inside. (**b**, **c**) The capillary containing the spheroid is inserted in the brain cortex a few microns under the dura mater, and the spheroid is implanted while the outpouring liquid is drained with sterile paper. After wound closure with a 200 μm diameter Sephadex bead, (**d**) a 6 mm diameter coverslip is (**e**) set on the craniotomy area and glued (it is here important to maintain pressure using forceps not for the glue to enter the wound). (**f**) The edges of the glued coverslip and the surrounding bone are covered with dental resin. (**g**) Operated mice with mounted windows will now be ready for imaging following a week recovery

micromanipulator, deepen the capillary approximately 200–400 μm into the brain (Fig. 10.3b, c). Slowly press on the Hamilton syringe to inject the content of the capillary, until the spheroid(s) enters the brain cortex; at the same time, drain the liquid flowing out from the implantation site using a piece of sterile paper: if this liquid is not completely drained during and after the injection, spheroids may float up into it, leaving the implantation site.

7. Transfer the mouse under the upright stereomicroscope. At this stage, it might be possible to see the spheroid(s) inside the small wound created by the capillary tip. Place one Sepharose microsphere (Fig. 10.1c) on top of the spheroid(s): it will make sure that the tumor cells are at a proper distance from the brain surface without floating up out of the meninges and grow ectopically under the glass coverslip.

8. Gently place the round coverslip (Fig. 10.3d) atop the craniotomy; glue it to the surrounding bone by spreading histocompatible glue all around its border (Fig. 10.3e). Wait a few

minutes until the glue solidifies, then prepare the dental resin, and spread it around the cranial window, covering the whole area of exposed bone (Fig. 10.3f); wait a few minutes until the resin dries (Fig. 10.3g). At this stage, the tumor implantation could be checked (note 5).

9. Release the mouse from the holder, and move it to post-surgery care until it awakens. To this end, we recommend the use of a heating pad, a warming blanket, or an infrared lamp to keep the animal warm. Following the surgery, animals should be inspected daily in order to assess their well-being. In our experience this surgery is quite well tolerated, but precautionary analgesic treatment must be applied the day after the surgery and furthermore, in case of signs of discomfort, as recommended by your local ethical committee.

10. Likewise after any other surgical procedure, an inflammatory reaction would take place in the tissue under the cranial window. Given that, a potential alteration of blood vessel morphology (vascular dilation) and transient local recruitment of innate-immune cells just underneath the coverslip could be observed and would contaminate the area to be imaged. That is why we advise to wait at least 7–10 days before starting imaging for its resolution. This will also give to the tumor the necessary time to start growing and spreading out from the implantation site.

3.3 Imaging

Seven to 10 days following the surgery, a small tumor should be visible around the implantation site; starting from this time, it should be possible to repeatedly image the cranial window. This section focuses on specific aspects of imaging using this particular model, assuming that the experimenter has already a general knowledge of multiphoton microscopy.

1. Turn on the microscope, and prepare the equipment for holding the mouse; remember to preheat the pad used for thermal control (Fig. 10.4a).

2. Anesthetize the mouse in the induction chamber prefilled with an adequate gas mixture; we set our apparatus to isoflurane 2–5% in pure oxygen flow (0.5–1 mL/min) (Fig. 10.4c).

3. Once the mouse is anesthetized, remove it from the chamber, place it on the platform, and firmly fix the head with the holder (Fig. 10.4b). Apply the eye ointment, and check the animal to verify optimal anesthetization and breathing. Set the isoflurane anesthesia for maintenance (isoflurane 1–2%, oxygen and air both at 0.2–0.5 mL/min in a 1:1 ratio). Move the platform under the objective. Look into the microscope binocular, and focus on the brain surface. If movements of the head are visible, check for signs of insufficient or too high anesthesia. If the

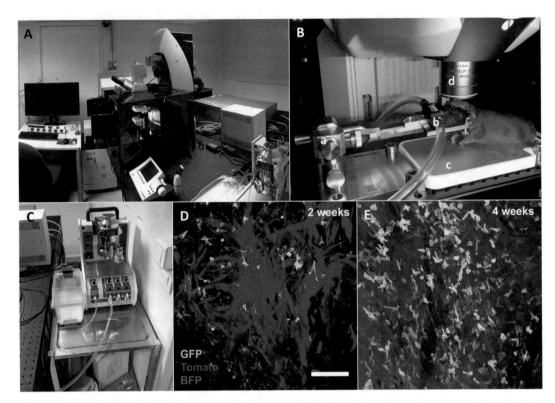

Fig. 10.4 Multiphoton microscopy devices and imaging. (**a**) Multiphoton microscope equipped with a 2-ray laser and non-descanned hybrid detectors. (**b**) Features of the platform holding the mice during imaging, whose single parts have been assembled on a steel plate. (**a**) The head holder is mounted on a micromanipulator which allows aligning the window with the XY plane of the microscope. (**b**) Anesthesia is supplied through a gas mask. (**c**) Animal thermal control is maintained during image acquisition using a warming plate. (**d**) Water immersion 25x objective is installed on top of the cranial window in a waterdrop retained by the hydrophobic properties of the dental cement. (**c**) An isoflurane vaporizer equipped with an air compressor provides anesthesia (and scavenging of anesthesia wastes) to the mouse platform on the microscope and to an induction chamber. (**d**) Longitudinal live imaging of an inducible CSF1R-Mer-Cre-Mer::ROSA^{mTmG} mouse at early (2 weeks, **d**) and late (4 weeks, **e**) tumor growth. This setting allows the visualization of CT2A-BFP tumor expansion, together with the recruitment of GFP-positive bone marrow-derived myeloid cells. Tumor vessels and other host cells penetrating the glioma can be followed in the Tomato red channel. Scale bar 100 μm

breathing is normal but movements are still present, it will be necessary to extract the platform and work on the head holder until a more stable fixation is achieved (note 6).

4. Perform the imaging according to experimental aims (note 7). Experimental timetables including long imaging session (more than 2 h) could require control of additional parameters than temperature (e.g., hydration, blood glucose) [16] (note 8). It is also possible to detect some non-labeled structures by second-harmonic generation (SGH) imaging [17]; the model described here allows detecting SGH from collagen fibers in the meninges. Considering that micro-movements occur in the tissue of living animals, the scanning speed should be set fast enough to

minimize image artifacts arising from this. For the same reason, separating different color channels by using sequential scanning is not advisable in this imaging application: even if the mouse head is firmly fixed, small pulsations of the tissue due to heartbeat and blood circulation will cause micro-movements. If the scanning speed is too low, single images will show waves of positional shifts over the *xy* plane. If this happens, increase the scanning speed until this problem is minimized; with our system we normally use 600 Hz (lines per second) bidirectional scanning for images 1024 × 1024 bits. When imaging multiple-color channels, it is better to optimize their separation by properly setting the emission filters, in order to be able to scan all channels at once.

5. Optional: if willing to visualize blood perfusion and evidence for alteration of the blood/brain barrier (BBB), a solution of fluorescent dextran can be injected intravenously (tail vein or retro-orbital injection) before or during the imaging. Once injected, these markers can reach the tumor vasculature within seconds, to be detected by multiphoton excitation. Fluorescently labeled dextrans are available in different colors and sizes: in order to visualize blood vessel perfusion, favor high molecular weights such as 2.000 kDa as they will hardly leak out of brain vessels. Smaller molecular weight dextrans from 2 to 70 kDa will highlight from subtle BBB alteration to major hotspot of leakage of the vasculature. It is here to note that dextrans will be cleared from the circulation over time proportionally to their molecular weight but will also be scavenged by myeloid cells which then will be labeled by the dextran-linked fluorophore.

6. After imaging, the mice should awake minutes after removal from the stage and discontinuation of anesthesia. They can be put back in their cages, to be eventually imaged again during tumor progression for a few more weeks. The time span available for chronic imaging of this mouse model varies in function of the implanted glioma cell proliferation and expansion into the host brain. In our experience with the cell lines mentioned here, imaging can be performed over 4 weeks without the mice becoming symptomatic. Only very few among them show signs of distress or significant weight loss earlier, thus needing to be withdrawn from the experimentation according to our ethical committee guidelines.

3.4 Immunother apeutic Treatments

The immune system is roughly composed of two main subunits: the innate immune system, where macrophages belong, and the adaptive immune system, which comprises of T and B lymphocytes.

In the following paragraph, we will describe existing strategies to visualize such immune populations in vivo and therapeutic treatments allowing to unravel their role in the tumor microenvironment (by depleting them).

1. Bone marrow myeloid cells: macrophage is the most abundant immune population recruited in the glioblastoma microenvironment, accounting for 20–35% of the tumor mass. Depending on their polarity (Th1–Th2) over glioma growth, myeloid cells will have various biological incidences from tumor cell phagocytosis to tumor support and vascular alterations. Multiple specific Cre-driven recombination mouse strains exist to allow myeloid cells tracking. We favored the usage of the inducible CSF1R-Mer-Cre-Mer strain, which allows the operator to induce a highly efficient recombination (about 95% of the recruited macrophages) at the time desired [5]. Crossing this strain with the ROSA$^{mT/mG}$ strain will induce the expression of GFP in these cells, the other cells of the host staying Tomato positive (Fig. 10.4d).

   To target such population, we advise to use the monoclonal antibody anti-CSF1 (clone 5A1), which will inhibit their proliferation, differentiation from monocytes, and survival and will result in a 50–75% disappearance of the total number of cells [5]. Such strategy will allow you to evaluate the implication of bone marrow-derived macrophages in your biological process.

2. T lymphocytes: like for macrophages, it is now possible to track T cells in vivo using longitudinal multiphoton imaging during glioma progression. Also, a large number of Cre-specific strains exist to fluorescently highlight each T-cell subpopulation (by crossing them with ROSA^{mTmG} mouse strain) or deplete in these cells a gene of interest. To target in a general fashion all mature T cells, the Lck-iCre mouse strain seems to be the most potent one [18].

   For treatment-based approach to target general T-cell population, an anti-CD3ε (clone 145-2C11) could be used or any specific T-cell subpopulation epitope of interest such as anti-CD4 (clone GK1.5) or anti-CD8 (clone 53–6.7) and many others.

3. B lymphocytes: despite a debated and unclear role of the B cells in the glioma context, tools exist to follow them in in vivo live imaging to ultimately unravel their importance in the disease progression. To target in a general fashion all B cells, the CD19^{creERT2} mouse strain seems to be the most interesting one [19].

For specific B-cell depletion in order to evaluate their implication in specific aspects of the glioma progression, a treatment strategy using anti-CD19 monoclonal antibody treatment (clone 1D3) could be considered.

4   Notes and Troubleshooting

1. We present here the engraftment of BFP-positive glioma cells in GFP- and Tomato-positive transgenic mice. It is to note that this fluorophore combination can be adapted to your mice strains and cells with fluorophores from the blue to the far-red end of the spectrum depending on your multiphoton laser tuning and the emission filter range.

2. The mouse age can be adapted to the study. We usually perform this surgery on 8–12-week mice for an optimal reproducibility. It is preferable to use mice of more than 20 g which present a thicker skull for the window installation.

3. During the glioma implantation procedure, the Sepharose bead is meant to be a cork to prevent the tumor spheroid backflow. If the Sepharose bead flows out itself, use a higher Sepharose bead diameter, or carefully glue the site after Sepharose bead placement to prevent the backflow.

4. The spheroid formation presented here works for GL261 and CT-2A cells. Other glioma cell lines could need more time to form proper spheroids. In case of adherence to the plastic dish, pre-coating with agarose can be performed. Also, hanging drop culture can be a solution. These conditions should be optimized in accordance to the cell line to be used.

5. Control your mice implantation. Depending on the brightness of the fluorescent reporter you used, it might be possible to check your glioma spheroid implantation right after the surgery under a performant fluorescent stereo microscope. If your fluorophore is a bit week, check under the multiphoton microscope after 3 weeks. If no tumor growth is visible, it is likely that the implantation failed (or that the growth of your tumor cell type is slower than the ones presented here).

6. Optimize the isoflurane gas anesthesia to your device. For maintenance anesthesia, a too low dose will be evidenced by a high respiratory rhythm and a too high dose by the mouse gasping.

7. For a long-term follow-up of the tumor growth at a frequency of two tumor visualizations a week, we advise not to excide a 2 h anesthesia and imaging for the mice to recover in between sessions and not to alter tumor growth. Some mechanism visualization will need in-depth imaging and so longer sessions. Adapt the imaging repetition frequency to your study.

8. For long session imaging, most of the long-range objectives used in multiphoton microscopy will be water immersion ones. Remember to replace evaporated water over your acquisition. Also, subcutaneous injection of glucose-complemented physiological serum should be considered.

Acknowledgments

The development of the technique described here was supported by the French Association for Cancer (ARC) and the Belgian Cancer Foundation (Stichting Tegen Kanker). T.M. was financed by EMBO long-term fellowship and Lefoulon Delalande foundation. L.H.G.M. was financed by Coordenaçao de Aperfeiçoamento de Pessoal de Ensino Superior (CAPS). We thank Dr. Holger Gerhardt for the initial training in his laboratory. We thank Dr. Till Acker (Institute of Neuropathology, University of Giessen, Germany) for the gift of the GL261 cells. We thank Dr. Thomas N. Seyfried (Biology Department, Boston College, USA) for the gift of the CT2A glioma cells.

References

1. Zipfel WR, Williams RM, Webb WW (2003) Nonlinear magic: multiphoton microscopy in the biosciences. Nat Biotechnol 21:1369–1377. https://doi.org/10.1038/nbt899

2. Helmchen F, Denk W (2005) Deep tissue two-photon microscopy. Nat Methods 2:932–940. https://doi.org/10.1038/nmeth818

3. Abe T, Fujimori T (2013) Reporter mouse lines for fluorescence imaging. Develop Growth Differ 55:390–405. https://doi.org/10.1111/dgd.12062

4. Erapaneedi R, Belousov VV, Schafers M, Kiefer F (2016) A novel family of fluorescent hypoxia sensors reveal strong heterogeneity in tumor hypoxia at the cellular level. EMBO J 35:102–113. https://doi.org/10.15252/embj.201592775

5. Mathivet T, Bouleti C, Van Woensel M et al (2017) Dynamic stroma reorganization drives blood vessel dysmorphia during glioma growth. EMBO Mol Med 9:1629. https://doi.org/10.15252/emmm.201607445

6. Ricard C, Stanchi F, Rodriguez T et al (2013) Dynamic quantitative intravital imaging of glioblastoma progression reveals a lack of correlation between tumor growth and blood vessel density. PLoS One 8:e72655. https://doi.org/10.1371/journal.pone.0072655

7. Ricard C, Stanchi F, Rougon G, Debarbieux F (2014) An orthotopic glioblastoma mouse model maintaining brain parenchymal physical constraints and suitable for intravital two-photon microscopy. J Vis Exp. https://doi.org/10.3791/51108

8. Ausman JI, Shapiro WR, Rall DP (1970) Studies on the chemotherapy of experimental brain tumors: development of an experimental model. Cancer Res 30:2394–2400

9. Seyfried TN, el-Abbadi M, Roy ML (1992) Ganglioside distribution in murine neural tumors. Mol Chem Neuropathol 17:147–167

10. Mostany R, Portera-Cailliau C (2008) A method for 2-photon imaging of blood flow in the neocortex through a cranial window. J Vis Exp. https://doi.org/10.3791/678

11. Subach OM, Gundorov IS, Yoshimura M et al (2008) Conversion of red fluorescent protein into a bright blue probe. Chem Biol 15:1116–1124. https://doi.org/10.1016/j.chembiol.2008.08.006

12. Filonov GS, Piatkevich KD, Ting L-M et al (2011) Bright and stable near-infrared fluorescent protein for in vivo imaging. Nat Biotechnol 29:757–761. https://doi.org/10.1038/nbt.1918

13. Muzumdar MD, Tasic B, Miyamichi K et al (2007) A global double-fluorescent Cre reporter mouse. Genesis 45:593–605. https://doi.org/10.1002/dvg.20335

14. Qian BZ, Li J, Zhang H et al (2011) CCL2 recruits inflammatory monocytes to facilitate breast-tumour metastasis. Nature 475:222–225

15. Van Woensel M, Mathivet T, Wauthoz N et al (2017) Sensitization of glioblastoma tumor micro-environment to chemo- and immuno-therapy by Galectin-1 intranasal knock-down strategy. Sci Rep 7(1):1217. https://doi.org/10.1038/s41598-017-01279-1

16. Tremoleda JL, Kerton A, Gsell W (2012) Anaesthesia and physiological monitoring during in vivo imaging of laboratory rodents: considerations on experimental outcomes and animal welfare. EJNMMI Res 2:44. https://doi.org/10.1186/2191-219X-2-44

17. Zoumi A, Yeh A, Tromberg BJ (2002) Imaging cells and extracellular matrix in vivo by using

second-harmonic generation and two-photon excited fluorescence. Proc Natl Acad Sci U S A 99:11014–11019. https://doi.org/10.1073/pnas.172368799

18. Wang Q, Strong J, Killeen N (2001) Homeostatic competition among T cells revealed by

conditional inactivation of the mouse Cd4 gene. J Exp Med 194(12):1721–1730

19. Yasuda T, Wirtz T, Zhang B et al (2013) Studying Epstein-Barr virus pathologies and immune surveillance by reconstructing EBV infection in mice. Cold Spring Harb Symp Quant Biol 78:259–263

Chapter 11

Experimental and Preclinical Tools to Explore the Main Neurological Impacts of Brain Irradiation: Current Insights and Perspectives

Laura Mouton, Monica Ribeiro, Marc-André Mouthon, Fawzi Boumezbeur, Denis Le Bihan, Damien Ricard, François D. Boussin, and Pierre Verrelle

Abstract

Radiation therapy is a powerful tool in the treatment of primary and metastatic cancers of the brain. However, brain tissue tolerance is limited, and radiation doses must be tailored to minimize deleterious effects on the nervous system. Due to improved treatments, including radiotherapy techniques, many patients with brain tumors survive longer, but they experience late effects of radiotherapy, especially cognitive decline, for which no efficient treatment is currently available. Improving the prevention and treatment of radiation-induced neurological defects first needs to better characterize radiation injuries in brain cells and tissues. Rodent models have been widely used for this.

Here, observations from patients will be reviewed briefly as an introduction, mainly regarding clinical cognitive defects and anatomical alterations using magnetic resonance imaging (MRI). This limited descriptive clinical knowledge addresses many questions that arise in preclinical models regarding understanding the mechanism of radiation-induced brain dysfunction. From this perspective, we next present methods to characterize radiation-induced neurogenesis alterations in adult mice and then detail how MRI could be used as a powerful tool to explore these alterations.

Key words Mouse brain irradiation, Neural progenitors, Fluorescence-activated cell sorting, Immunohistochemistry, Transcriptomics, Magnetic resonance imaging and spectroscopy

---

Laura Mouton and Monica Ribeiro are co-first authors.

---

1 Introduction

Memory is a complex cognitive process implicating different brain areas. Hermann Ebbinghaus (1850–1909) first introduced the concepts of encoding, storing/consolidation, and retrieval. These three subprocesses are the key phases of long-term memory formation and are essential for retrieving information when required.

Giorgio Seano (ed.), *Brain Tumors*, Neuromethods, vol. 158,
https://doi.org/10.1007/978-1-0716-0856-2_11, © Springer Science+Business Media, LLC, part of Springer Nature 2021

Encoding is the first phase of long-term memorization and is impaired when the hippocampus (left or right) is lesioned. The hippocampi are small brain structures located in the ventral temporal lobe. Storage refers to the consolidation of encoded information in the memory system to prevent long-term forgetting and to necessitate the integration of the hippocampus with other cortical zones and their interconnections. Retrieval is the final active phase in which the cognitive system transfers information from long-term memory to working memory and activates many brain areas with many white matter tracks; thus, the function of retrieval is sensitive to white matter alterations, such as those observed in patients with brain tumors [1]. Indeed, following radiotherapy, some long-term survivor patients are at risk of developing neurological deterioration in the absence of tumor recurrence [2]. Distinct magnetic resonance imaging (MRI) cerebral alterations due to brain area irradiation have been observed. Mahajan et al. reported a statistically significant relationship between the maximum irradiation dose in the left hippocampus and cognitive decline (learning, delayed recall), even at 14 months after brain irradiation, in patients treated for high-grade gliomas [3]. Another study in 52 patients treated for primary brain tumors [4] showed that hippocampal volume had decreased 1 year after radiotherapy initiation according to the mean dose in the hippocampus and patient age.

While many articles have dealt with hippocampal exposure to ionizing radiation, cognitive effects could be explained by dosimetric parameters in other healthy cerebral tissues. A retrospective study assessed the radiosensitivity of cerebral cortex regions [5]. The risk of atrophy in the associative cortex, which is mainly involved in cognitive functions, increased significantly 1 year after the start of radiotherapy when these regions were exposed to a mean radiotherapy dose higher than 40 Gy. Moreover, white matter (WM) changes induced by ionizing radiation could also explain cognitive impairment following brain radiotherapy, suggesting a potential clinical effect of the doses delivered to WM [6]. Using diffusion tensor imaging (DTI), Connor et al. observed that mean diffusivity (MD) and fractional anisotropy (FA), parameters reflecting the mobility of water molecules, were significantly modified after treatment depending on radiotherapy dose and time [7]. In addition to the cerebral cortex, sensitivity to radiation exposure seems to differ according to the anatomical location [8]. Decreases in FA, reflecting WM disruption, were reported in the corpus callosum, the cingulum bundle, and the fornix 9–12 months after radiotherapy. The MD also increased significantly in these structures, suggesting a specific dose-dependent radiosensitivity. The ventricular-subventricular zones (V-SVZ) play an essential role in neurogenesis and, consequently, in cognitive neurophysiology, even in adults [9]. However, these areas risk involving high-grade glioma [10, 11]. Regarding radiotherapy, several authors suggested a favorable effect of high total doses delivered to the SVZ in terms of oncological prognosis [12, 13]. The impact of such strategies on cognition remains to be specifically evaluated. Nevertheless, the most frequent delayed treatment-related toxic effect on the brain in long-term survivors has a clinical and MRI presentation of diffuse leukoencephalopathy [2].

This leukoencephalopathy, also observed in other brain tumor patients treated by brain irradiation [2], may describe a wide spectrum of neurological conditions ranging from asymptomatic WM changes in neuroimaging studies to severe dementia, previously reported as decreased memory function, gait disturbance, and urinary incontinence [6, 14, 15]. Particularly in long-term glioma survivors, late-onset treatment-related neurotoxicity represents a substantial reduction in quality of life and may even lead to the full-time need for nursing care. Although the low survival rates for the World Health Organization grade III and grade IV glioma patients are increasing with the use of intensive treatments combining radio- and chemotherapies [16], the question of the late iatrogenic effects is of growing interest. Some individual susceptibility factors have been suggested, such as age, vascular risk factors, and genetic predispositions [17], but these have never convincingly demonstrated.

A few preclinical studies have analyzed the timing of radio-induced brain injury and molecular biomarkers [18–20], but little is known about the cellular and pathophysiological mechanisms [21]. Nevertheless, some theories have emerged, such as vascular damage [22] and damage to oligodendrocytes and neural precursor cells [20, 22–24]. Despite some limitations (e.g., frequent use of a single dose schedule and small doses of irradiation), experimental data consistently show a time-dependent lesion apparition pattern [2, 25]. Histopathological signs of alterations in neural stem cells in the hippocampus of brain-irradiated patients have been convincingly reported [20, 26, 27]. However, hippocampal alterations do not account for cognitive impairments observed in patients, as these individuals do not only display memory disruptions related to diffuse WM alterations in the hippocampus but also in subcortical, predominantly periventricular, regions. Therefore, many questions must be addressed in preclinical models to understand the diffuse radiation-induced lesions and individual sensitivity factors in patients. From this perspective, we present methods to observe radiation-induced brain alterations in several areas of neurogenesis beyond the hippocampus in mouse models. We detail how MRI could be used as a powerful tool to explore the nature and dynamics of radiation-induced brain alterations in animal models.

2   Mouse Models of Adult Radiation-Induced Brain Injuries

2.1   Main Adult Brain Neurogenic Niches

In the adult mammalian brain, neural stem cells (NSCs) persist in specialized niches (Fig. 11.1a): the subgranular zone (SGZ) and the V-SVZ, in which young neurons are generated for the hippocampus and olfactory bulb (OB), respectively [28].

In young adult mice, NSCs in the V-SVZ produce ~10,000 neuroblasts – the precursors of interneurons – every day through the generation of transient-amplifying intermediate progenitors (TAPs) [29]. In the human brain, the production of young neurons rapidly declines after birth, and the persistence of neurogenesis in the adult brain in both the V-SVZ and the SGZ is still under debate [30, 31]. However, cells with NSC features persist in a resting state, forming a ribbon along the ventricles [32].

In the rodent V-SVZ, most adult NSCs are quiescent, in contrast to their progeny, and the tight regulation of the balance between their quiescent and proliferative states appears essential for their long-term maintenance in neurogenic niches [33, 34]. Indeed, genetic manipulation dysregulation and/or a loss of quiescence often results in the premature proliferation of NSCs, ultimately leading to the depletion of neural stem and progenitor cells [35–38].

In addition to the V-SVZ and SGZ neurogenic niches, oligodendrocyte precursor cells (OPCs) are widespread throughout the adult rodent brain and produce remyelinating oligodendrocytes [39]. In an inflammatory demyelinating disease rodent model, NSCs from the V-SVZ are mobilized to undergo oligodendrogenesis [40]. However, this mobilization results in a persistent reduction in OB neurogenesis, leading to impaired long-term olfactory memory [41].

2.2   Assessment of NSCs and Their Progeny by Immunochemistry and Flow Cytometry in the Rodent V-SVZ

At first, young neurons were examined on serial brain slices by immunochemistry after their proliferative precursors had incorporated thymidine analogs, such as bromodeoxyuridine (BrdU). BrdU or Ki67 expression has been advantageously coupled with several specific antibodies to detect different NSC populations and progeny (Fig. 11.1b) (e.g., Note 1). NSCs contact the lateral ventricle and express astroglial markers, such as glial fibrillary acid protein (GFAP) [29], while TAPs express proneural genes such as Mash1 [42].

Currently, several flow cytometry methods have been used with freshly dissociated cells from adult V-SVZ, allowing their quick quantification, as well as their sorting, for subsequent manipulations [43–45]. By using a combination of three membrane markers (CD24, EGF-binding, LeX/SSEA1) (Fig. 11.1c), we set up a method to isolate quiescent NSCs, activated/cycling NSCs, and their progeny, i.e., TAPs and neuroblasts, from the V-SVZ of

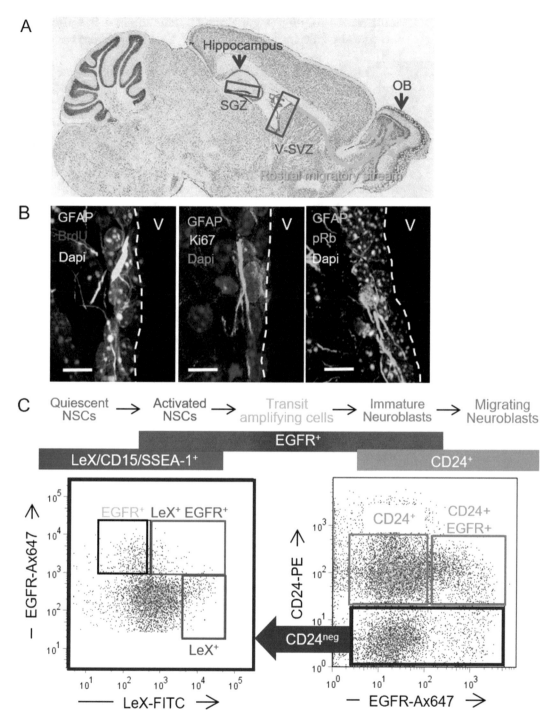

Fig. 11.1 Analysis of NSCs in V-SVZ neurogenic niches. (**a**) Two major neurogenic niches are maintained during adulthood in rodent brains: the subgranular zone (SGZ) in the hippocampus and the ventricular-subventricular zone (V-SVZ). Stained sagittal sections were obtained from www.alleninstitute.org. (**b**) Immunohistochemistry of NSCs in the V-SVZ. NSCs express GFAP and the lateral ventricle (V) just beneath the ependymal cell wall. Proliferating NSCs are labeled for BrdU incorporation, pRb, and Ki67. Scale bars: 5 μm. (**c**) A combination of three membrane markers (CD24, EGF, and LeX/SSEA1) in freshly dissociated V-SVZ cells allows the identification of NSCs and their progeny by FACS

adult mice [46, 47] (e.g., Note 2). This method has also been shown to be efficient in early postnatal V-SVZ (PN8 days or older) [48].

This fluorescence-activated cell sorting (FACS) strategy has been successful in examining the cell cycle alterations in NSCs with aging. An early decline in adult neurogenesis with a dramatic loss of progenitor cells was observed in young, 4-month-old adult mice [49, 50]. Whereas activated and quiescent NSC pools remained stable for up to 12 months, the proliferative status of activated NSCs was altered by 6 months, with an overall extension of the cell cycle resulting from a specific lengthening of G1 [50].

Comparative transcriptomic analyses of quiescent and activated NSCs showed that quiescent NSCs display a distinct molecular signature from activated NSCs [48]. These data highlight the central role of the stem cell microenvironment in the regulation of quiescence in adult neurogenic niches and the expression of numerous membrane receptors on quiescent NSCs.

2.3 High-Dose Radiation Blocks Neurogenesis and Alters Brain Functions

The irradiation of the hippocampus with high doses (10–15 Gy) of X-rays has been shown to disrupt neurogenesis, i.e., neuroblast production, in addition to diminishing memory capacity [51] and decreasing the behavioral response to antidepressants [52].

The irradiation of the hippocampus targets neural stem and progenitor cells, as well as young neuroblasts, that would differentiate and integrate into the neural network. In contrast, the localized irradiation of the V-SVZ allows the OB, where neuroblasts migrate and integrate [53], to be protected from radiation.

Specific radiation injury to the V-SVZ provokes OB-related behavioral effects correlated with a profound reduction in the generation of young neurons in the OB [52, 53]. Indeed, the localized irradiation of the V-SVZ with three doses of 5 Gy with a collimated ^{60}Co source has been shown to alter olfactory memory [53] and social behavior [54]. A similar strategy using X-rays has been reported to disrupt mouse adult neurogenesis and alter OB structure and olfactory fear conditioning [55]. Together, these data clearly show the importance of both SGZ and V-SVZ neurogenesis for the maintenance of adult brain functions.

2.4 Quiescent NSCs Resist Radiation and Restore Neurogenesis After Intermediate-Dose Exposure

2.4.1 Recovery of Neurogenesis in V-SVZ After Intermediate-Dose Irradiation

Single doses of intermediate X-rays (1–3 Gy) have been reported to reversibly affect the levels of NSCs and TAPs in the V-SVZ of juvenile rats that recovered within a week, as shown by nestin immunostaining, which is specific to these cells [56]. After an initial response to radiation, injury is similar in both brain stem cell niches (SGZ and V-SVZ) of juvenile rats [57]. The irradiation (8 Gy) of the juvenile rat brain provokes the long-term deterioration of NSCs and neurogenesis, which is more severe in the SGZ than the V-SVZ, which appears to recover with time [57].

Intermediate-dose exposure also reversibly affects the levels of NSCs and TAPs in the V-SVZ in the adult mouse brain. The FACS quantification of proliferative populations, i.e., cycling NSCs, TAPs, and young neuroblasts, showed that these populations recovered to normal levels in the V-SVZ within 1 week after 4 Gy irradiation, which was confirmed by immunostaining in brain slices for several markers [46]. Supportive results were obtained by analyzing the in vitro proliferative capacity of NSCs and TAPs after irradiation [46].

2.4.2 Irradiation Activates Quiescent NSCs

Radiation exposure induces the rapid death of proliferating V-SVZ cells through a TP53-dependent mechanism [46, 58] but has no such effect on quiescent NSCs. Decreasing the proliferation rate through genetic deficiency in an apoptosis-inducing factor has been shown to protect V-SVZ cells in the juvenile rodent brain against ionizing radiation [59].

After the complete depletion of proliferating cells, V-SVZ neurogenesis appeared to restart 48 h after 4 Gy irradiation, as shown by the detection of proliferating NSCs (by using BrdU incorporation experiments or by immunostaining for phospho-Rb and Ki67), which is consistent with the activation of quiescent NSCs, whose viability was not affected by radiation [46].

Several mechanisms have been involved in the maintenance of NSC quiescence, among which GABA$_A$R signaling exerts negative control over NSC proliferation. The radiation-induced depletion of neuroblasts, the major GABA source within the V-SVZ, appears to release negative retro-control on quiescent NSCs and, therefore, allows their proliferation in the irradiated brain [46].

A comparative analysis of whole transcriptomes of quiescent NSCs sorted by FACS before and 48 h after in vivo 4 Gy radiation exposure revealed the upregulation of genes involved in protein synthesis and translation [48], which was consistent with a low level of protein synthesis in quiescent NSCs. Interestingly, protein synthesis undergoes highly dynamic changes when quiescent NSCs activate and differentiate into neurons in vivo [60].

2.4.3 High-Dose Irradiation of the Brain Perturbs Neurogenic Niches

A single dose of 10 Gy irradiation, specifically delivered to the V-SVZ using computed tomography (CT), eliminated proliferating TAPs and migrating neuroblasts and persisted for a month [61]. Nonetheless, NSCs resisted this radiation regimen and partially repopulated the neurogenic niche 1 year after exposure; these NSCs were also able to respond to a demyelinating lesion [62]. However, higher doses of radiation (≥20 Gy) were associated with irreversible long-term damage to the NSC compartment of the rodent V-SVZ and the loss of OPCs, which was worsened by the fractionation of the dose [63, 64]. Delayed onset demyelination precedes focal necrosis and is likely due to the loss of oligodendrocyte precursors and the inability of the stem cell compartment to compensate for this loss [63].

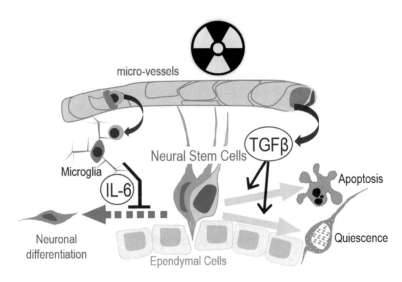

Fig. 11.2 Effects of irradiation on neurogenic niches. Ependymal and endothelial cells nurture NSCs with which they are in contact. After exposure to high-dose radiation, the neurogenic niche is perturbed through the release of the cytokines interleukin-6 (IL-6) [66] and transforming growth factor-β (TGFβ) [65], which block neuronal differentiation and induce apoptosis and the quiescence of NSCs. The activation and/or recruitment of microglial cells and the alteration in vascular cells are processes involved in this dysregulation

After high-dose localized irradiation, the V-SVZ lacked the ability to generate migrating neuroblasts [53, 61], whereas NSCs survived radiation exposure in the V-SVZ [61, 65]. These results indicate that radiation injury also perturbs the microenvironment of V-SVZ neurogenic niches (Fig. 11.2). The detrimental effects of radiation on the neurogenic niche in relation to an inflammatory response have been observed after the irradiation of the hippocampus [26, 66].

The transplantation of NSCs and their progeny, sorted from unirradiated mice using the FACS strategy, into neurogenic niches revealed that the irradiated brain environment is unable to sustain neurogenesis [65]. In addition, we have shown that alterations in transforming growth factor β (TGFβ) signaling were involved in this process and that blocking TGFβ signaling restored the proliferation of NSCs in irradiated V-SVZ niches [65].

2.5 Noninvasive Assessment of Radiation-Induced Brain Injuries Using In Vivo *Magnetic Resonance Imaging and Spectroscopy*

Preclinical mouse models enable the development of new approaches to investigate both neurogenic and nonneurogenic radio-induced effects. Magnetic resonance imaging (MRI) allows the whole brain, whether human or rodent, to be imaged noninvasively in an identical manner given the proper equipment for probing cerebral alterations. Compared to microscopic approaches, MRI allows us to consider the intact whole brain, preserving most of its complex and dynamic interactions (except for the anesthetic status). As a consequence, in vivo MRI studies can complement cellular and immunohistochemistry experiments, providing three-dimensional, or even four-dimensional (for longitudinal or functional studies), datasets.

2.5.1 General MRI Principles and MRI Contrasts

As the aim of this chapter is not to detail how MRI works, the reader is advised to look for more information in one of the many books or reviews referenced [67–69]. As illustrated in Fig. 11.3 (left panel), the basis of MRI is the nuclear magnetic resonance (NMR) phenomenon. This NMR phenomenon corresponds to the establishment of the macroscopic magnetization of hydrogen nuclei in tissues, mostly from water (H_2O) molecules when they are exposed to an intense static magnetic field. Using radiofrequency waves and gradients of a magnetic field, hydrogen nuclei and their magnetization can be manipulated, and an NMR signal can be measured whose characteristics (intensity, phase, resonance frequency, relaxation times) depend on various biophysical factors, such as the local density of water, the Brownian motion of the

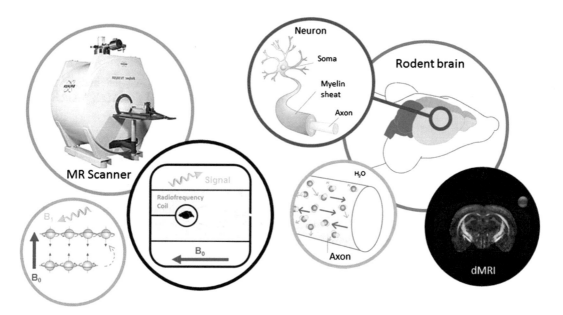

Fig. 11.3 MRI principle and applications in the mouse brain. (Left) To noninvasively study the mouse brain, mice are placed in a preclinical MR scanner (orange circle). The intense static magnetic field B0 generated by the MR scanner polarizes the hydrogen nuclei of water molecules ubiquitous in the mouse brain (gray circle) and establishes a macroscopic magnetic field along the B0 direction. Radiofrequency waves (B1) disturb the system of hydrogen nuclei, and the signal emitted by nuclear magnetic resonance is recorded as it returns to the equilibrium state (dark circle). (Right) Due to the influence of the anisotropic and tortuous diffusion of water molecules within and between neuronal and glial cells (red and green circles), diffusion-weighted MRI (dMRI) can be used to probe microstructural changes following brain irradiation. The color-coded image (full dark circle) reveals WM fiber orientation. Basically, red, blue, and green colors encode the left-right, anteroposterior, and superior to inferior fiber directions, respectively

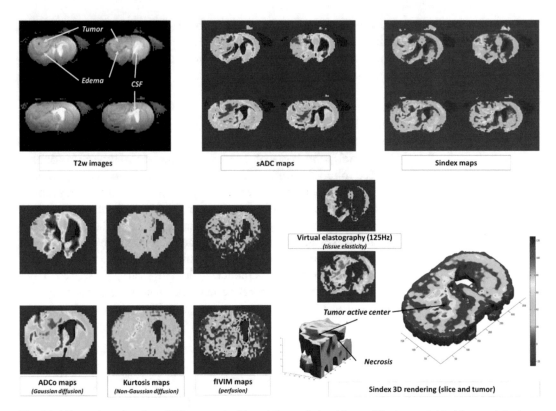

Fig. 11.4 Examples of various MRI contrasts (T_2 and those derived from diffusion weighted images) that can be obtained, here, in an implanted 9 L glioma tumor rat brain model (curtesy Dr. M. Iima et al., NeuroSpin)

molecules within and between cells, or the proximity of paramagnetic molecules (e.g., iron deposits or deoxyhemoglobin). A mathematical operation (often a two-dimensional Fourier transform) is necessary to reconstruct final 2D images from the measured NMR signal. As a consequence of the sensitivity of MRI to the biophysical factors mentioned above, various imaging contrasts can be generated depending on which morphological, microstructural, vascular, or metabolic biomarkers of brain function and organization are examined in normal and pathological conditions (Fig. 11.4) (e.g., Note 3).

Intrinsically, MRI provides images with contrast weighted by the "magnetic relaxation" properties of tissues, called T_1 (longitudinal relaxation time) and T_2 (transversal relaxation time). These properties are mainly related to the chemical content of the tissues. For instance, white matter (WM) has shorter T_1 and T_2 values than gray matter (GM), which results in a high level of contrast (inverted between T_{1w} and T_{2w} images), especially in T_{1w} images, which is further shortened by the lipids present in WM myelin. Globally, T_{1w} and T_{2w} images provide information about the overall brain anatomy, showing atrophy, ventricle dilatation, or the displacement of structures by lesions. T_{1w} images are often preferred because the GM/WM contrast is higher than that in T_{2w} images, and these images are also faster to acquire. Most lesions are detectable from surrounding tissue, as they increase both T_1 and T_2, especially in the presence of edema or necrosis (free water has higher T_1 and T_2 values than background tissue). However, to increase detectability, contrast agents, mainly paramagnetic agents such as gadolinium (Gd) chelates, can be injected into the vascular system to shorten the T_1 relaxation of tissues in their vicinity (molecular interactions). Hence, these contrast agents highlight tissues and lesions that have high vascularity and permeable vessels. Using appropriate models (dynamic contrast enhancement MRI, DCE-MRI [70]), the time course of the image contrast after contrast agent injection may provide information on perfusion and vessel permeability. There are, however, other MRI approaches that can depict the vasculature without the need for contrast agents (MR angiography, MRA) and perfusion, such as arterial spin labeling (ASL) [70, 71] or intravoxel incoherent motion (IVIM) MRI (e.g., Note 4) [72]. In these methods, the flow of blood in vessels is encoded through variations in the space of the magnetic field.

Beyond T_{1w} and T_{2w} images, which are qualitative (although T_1 and T_2 values can be calculated from a series of images), diffusion MRI (dMRI) provides quantitative maps of markers sensitive to tissue microstructure by encoding diffusion-driven molecular displacements in space and time (e.g., Note 5). The standard marker is the apparent diffusion coefficient (ADC) of water, which reflects the hindrance of molecular diffusion by tissue elements (mainly cell membranes, fibers) [73, 74]. The ADC is higher in GM than WM and generally increases in the presence of necrosis or edema and decreases in malignant lesions or proliferating tissues (e.g., Note 6). The sensitivity of dMRI to tissue hindrance effects (which render the distribution of diffusion displacements non-Gaussian) may be boosted by increasing the displacement encoding level in the images, and the hindrance can be quantified using specific markers (sADC, S-index [75]) or using models (e.g., the kurtosis model [76]). dMRI can also provide information on tissue perfusion without the need to inject tracers via the IVIM approach, which models blood flow in capillaries as a pseudorandom process.

In brain WM, water diffusion in axons, especially when myelinated, is highly anisotropic, as revealed by ADC variations when the diffusion-encoding direction is rotated. Using the diffusion tensor imaging (DTI) framework, these anisotropy effects can be quantified with markers, such as the mean diffusivity (MD) and the fractional anisotropy (FA) index [77]. Demyelination leads to an increase in MD and a decrease in FA. Furthermore, the direction in space of the WM tracts can be estimated from the directions of highest diffusivity to produce stunning 3D maps of brain structural connectivity (e.g., Note 7). A drawback of dMRI is its high sensitivity to head motion and geometric distortions in the images. These artifacts can be mitigated, to some extent, by ad hoc image processing methods.

T_{2w} images can also be made sensitive to local variations in magnetic susceptibility. The resulting T_2 values, called T_2^*, become shorter. For instance, the magnetic field in tissues surrounding vessels is sensitive to the balance between paramagnetic deoxyhemoglobin and diamagnetic oxyhemoglobin present in blood erythrocytes. A major application of this value is in functional MRI (fMRI), as the blood oxygenation status in vessels varies according to local blood flow and, thus, according to the underlying neuronal activity (blood-oxygen-level dependent fMRI, BOLD fMRI, [78]). In addition to showing brain loci activated by sensorimotor or cognitive stimuli, fMRI can also provide interesting information on the functional connections between brain areas in the resting state (rs-fMRI) in different physiological and pathological conditions (e.g., Note 8) [79].

In general, dMRI and fMRI images do not exhibit a high resolution or a high GM/WM contrast in T_{1w} or T_{2w} structural images. Hence, registration between MRI modalities is often performed to allow the precise labeling of structures or the comparison of results across subjects using anatomical templates and atlases (Fig. 11.5, left panel).

While MRI signals are related to water hydrogen nuclei, MR spectroscopy (MRS) [80] is a technique that allows other species – such as some energy metabolites or neurotransmitters, through their hydrogen, phosphorus, or carbon nuclei, or some ions (Na+) or molecules (O_2) – to be studied, as they have signature MR frequencies (chemical shifts) that depend on the nuclei and the immediate electronic environment of the molecules (e.g., Note 9). Because of their much lower concentration (a few millimoles/liter, at best) than that of water and the fact that natural isotopes are often not MR compatible (e.g., C^{12} and O^{16}), low abundance isotopes (C^{13} and O^{17}) must be used or even injected. As a result, signals remain very low, preventing high-resolution images from being obtained. Indeed, most MRS studies are performed using MR systems operating with high magnetic fields (7 T and higher).

2.5.2 MRI Studies of Radiation-Induced Brain Injuries

Morphological Damage

Following intense brain irradiation, abnormal signal intensities and volume changes can be observed directly using either T_{1w} or T_{2w} MRI. The signal intensity changes in the brain parenchyma are mostly related to a lengthening of the average T_2 and T_1 relaxation times due to endothelial damage, edema, ventricle compression, and inflammation followed by cellular disorganization, ventricle dilation, and, ultimately, necrosis in the case of acute high-dose brain irradiation. Some of those lesions, in particular, delayed WM necrosis, have been enhanced on T_{1w} MRI scans (hyperintensities) by the injection of gadolinium-based contrast agents [81]. While limited to preclinical studies due to the inherent toxicity of $MnCl_2$, manganese-enhanced T_{1w} MRI (MEMRI) [82] has been used to follow the accumulation of Mn^{2+} molecules within new neurons and their migration to study neurodevelopment alterations in irradiated juvenile mice [83, 84].

In addition, volume changes in various regions of interest (ROIs) can be assessed either with T_{1w} or T_{2w} MRI. While these anatomical features can be evaluated manually [85–88], more automatic and robust analysis methods are becoming increasingly available for studying brain-induced morphological damage using specific or generic brain templates (Fig. 11.5, left panel) to which individual brain scans can be registered to, allowing for voxel-based morphometry [83, 89] or atlas-based parcellation [84]. An example of brain parcellation for one mouse brain subject is given in the right panel of Fig. 11.5.

Structural Damage

Prior to the observation of morphometric changes that occur later, more subtle changes may occur at the cellular level in normal-appearing white and gray matter. The diffuse demyelination and progressive degradation of axons have been observed extensively in irradiated patients and rodents, leading to a reduction in fractional anisotropy (FA) [85, 90–92] due to a relative increase in the radial diffusivity of water molecules across the compromised myelin

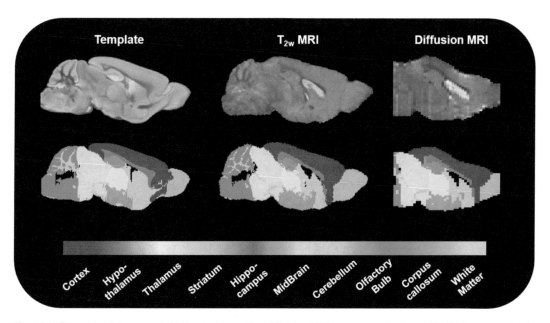

Fig. 11.5 Example of the coregistration between T_{2w}, diffusion MRI in a mouse brain, and both the mouse brain template and atlas. Anatomical and diffusion MRI can be registered to a generic or study-specific mouse brain template (top). Regions of interest (ROIs) in the anatomical or diffusion space of the individual mouse (middle and right bottom) can be generated from the previous registration applied on an atlas of a predefined ROI (right bottom)

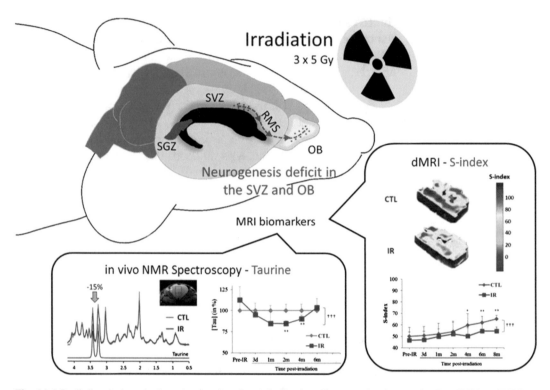

Fig. 11.6 Radiation-induced microstructural and metabolic alterations can be detected using dMRI and MRS. In the upper part, a scheme of the mouse brain and the areas implicated in neurogenesis are highlighted (from the left to the right: SGZ, SVZ, rostral migratory stream (RMS), OB). In the bottom part, Pérès et al. validated two relevant biomarkers in the OB of irradiated mice (3x5 Gy): S-index, a metric of cellular density (right panel) and taurine concentration (left panel) [88]. An example of averaged spectra of the irradiated (3x5 Gy) and unirradiated mice measured in the OB voxel (red box), as well as the time course of taurine concentrations, is shown in the bottom left panel. The time course of the S-index values in the OB for the irradiated and unirradiated groups, as well as the 3D rendering of the S-index maps, is plotted in the right bottom panel

sheath. ADC is a more sensitive biomarker than FA for the progressive stages of radiation-induced injury, particularly in GM. Indeed, increased ADC values have been reported following ionizing radiation exposure and were interpreted as the consequence of cell swelling and neuroinflammation [93], while decreased ADC values were observed at later stages due to necrosis and decreased cellular density [94]. In a recent study [88], we demonstrated the sensitivity of the S-index to microstructural changes underlying radiation-induced cognitive deficits in mice. In this longitudinal study, a decrease in the S-index was observed in the SVZ and the OB after 3x5 Gy whole brain irradiation (Fig. 11.6, right panel). These observations were further confirmed by immunohistochemistry as being the consequence of a decrease in the number of neural stem cells in the SVZ and in newborn neurons in the OB of irradiated animals.

Vascular Damage

Vascular malformations and blood-brain barrier (BBB) permeability following brain irradiation can be studied using MR angiography (MRA) and dynamic contrast-enhanced MRI (DCE-MRI). While challenging, arterial narrowing has been detected using MRA at 11.7 T in adult mice following in utero whole-body X-ray irradiation [95]. New malformed vessels surrounding the radionecrotic area have been assessed with DCE-MRI in cases of gamma knife irradiation [96].

Functional and Metabolic Alterations

A significant limitation of anatomical or structural MRI studies is that the main negative impact of brain irradiation is functional. Indeed, cognitive deficits of various degrees have been reported in most adult and pediatric brain-irradiated patients. This need explains the appeal of imaging techniques that focus on brain function or metabolism, such as resting-state functional MRI or NMR spectroscopy (MRS). While radiation-induced functional connectivity alterations have been reported in patients, their relevance remains to be demonstrated in irradiated rodents.

In other MRI modalities, an array of radiation-induced metabolic alterations has been reported using ^{1}H MRS, depending mostly on the irradiation protocol. Most of the studies have detected a decline in N-acetyl-aspartate (NAA), a biomarker of neuronal viability and density, indicating either transient neuronal distress or irreversible degeneration [86, 88, 97–100]. On a few occasions, this decline in NAA has been observed alongside a logical decline in neurotransmitter levels such as glutamate or GABA [88, 101] or an increase in total choline (tCho) levels, indicating probable glial activation [100]. Perturbed osmotic balance has also been observed via a decline in the level of taurine or myoinositol [98]. However, those metabolic changes can be rather ambiguous, as a decline in taurine levels (Fig. 11.6, bottom panel) could also be interpreted as a sign of impaired neurogenesis [88].

3 Conclusion

Exposing the brain to ionizing radiation provokes cognitive alterations resulting in perturbed neurogenesis in neurogenic niches both in the V-SVZ and SGZ in mice. The NSCs from various brain areas seem to be interesting targets involved in well-described hippocampal alterations, as well as many other brain structures. Methods are available to follow NSC cycling, differentiation, and migration after brain irradiation. As illustrated in Fig. 11.6, MRI and MRS have been very successful tools in assessing radiation-induced cerebral damage noninvasively in rodents. While there are other valuable in vivo imaging modalities for investigating the metabolic, molecular, or cellular effects of irradiation, such as positron emission tomography (PET) [99] and fluorescence-based

[102] or bioluminescence-based [103] optical imaging, none of these modalities can compare to MRI in terms of its versatility and potential for translation toward clinical applications.

Thoroughly described in preclinical animal studies, these findings have also been observed in patients after cancer treatment [27]. Persistent radiation-induced microglial inflammation is accompanied by the near-complete inhibition of neurogenesis after cancer treatment [27] in the irradiation field and beyond. These findings encourage the development of conformational radiation procedures that will avoid neurogenic regions, in particular the hippocampus. This approach would reduce the integral dose delivered to uninvolved normal brain tissues and may reduce late neurocognitive sequelae caused by cranial radiotherapy [104]. Cognitive outcomes in clinical studies are beginning to provide evidence of cognitive effects associated with hippocampal radiation doses and the cognitive benefits of sparing the hippocampus [105]. However, it is recommended that whole-brain radiation exposure be avoided for the treatment of brain metastasis because the neurocognitive side effects are not limited to verbal memory [106]. Animal models are of highest interest, since they could help researchers choose brain structures to protect and could help in the development of strategies to prevent or mitigate radiation-induced effects on normal brain cells.

4 Notes

Note 1: Confocal microscopy should be preferred for immunohistochemical studies of the V-SVZ because this method allows a better colocalization analysis and, therefore, better cell characterization.

Note 2: The dissection of the ventricular wall preparation is a critical step for flow cytometry analysis in V-SVZ cells. This procedure requires specific training [46]. The enzymatic (papain) digestion of brain tissues needs to be controlled and should not exceed 10 min at 37 °C in a water bath to avoid the loss of immunoreactivity. Controls to set the gates of positivity in flow cytometry experiments should be performed with cells incubated with all the fluorescent antibodies minus one (FMO) [46].

Note 3: For any study, the established protocol has to be applied in the same ordered steps for all the animals of the considered group. Regardless of the MRI protocol, caution must be taken both during animal preparation and data acquisition. Manipulations of the animals must be standardized to expose the mice to the same stress level. The position of the brain of the animals within the MRI scanner must be as similar as possible and centered on the magnet center (in left-right, anterior-posterior, rostro-caudal directions), usually with landmarks on the mouse head (eyes, ears).

The animal head is restrained with equipment such as a bite bar or ear bars. After the installation of the animal into the equipment, the brain position is verified using a quick anatomical MRI scout view. The field of view can then be adjusted, but it is best to have the mouse brain as straight and centered as possible beforehand.

During acquisition, physiological parameters must be monitored (respiration, temperature, heart rate, etc.). MRI calibration parameters (gain, shim, frequency) should be approximately the same for all animals. The MRI data quality depends on the signal-to-noise ratio (SNR), which implies the use of radiofrequency coils placed as close as possible to the animal's head. Surface coils provide very inhomogeneous signal sampling (higher close to the surface, lower far from the surface), which might lead to errors when analyzing images. Volume coils provide much more homogeneous signals but usually lower SNRs. The SNR can be increased by using high-field MRI systems (e.g., 11.7 T or higher), phased-array coils, or "cryo-probes," which are cooled to reduce electronic noise.

Note 4: The degree of diffusion weighting for each diffusion MRI image is defined using sets of "b values." At least two sets of images must be acquired: (1) one with the minimum allowed diffusion weighting for calibration (as close as possible to $b = 0$, but for technical reasons, it is usually higher than 20 or even 50 s/mm^2); (2) the other one with a b value approximately 1000 s/mm^2 (optimized value for brain tissue in vivo). With those two values, ADC maps can be calculated. However, higher sensitivity to tissue microstructure is obtained using high b values (e.g., higher than 1500 s/mm^2). The modeling of diffusion and IVIM (perfusion)-related parameters requires the acquisition of data sets with multiple b values covering lower ranges (e.g., 0–400 s/mm^2), medium ranges (e.g., 800–1200 s/mm^2), and higher ranges (e.g., 1800–3000 s/mm^2). Higher values (5000–20,000 s/mm^2) might also be used for tractography. A note of caution: the programmed b values may sometimes differ significantly from the effective b values, depending on the MRI scanner and its calibration or the MRI sequence. A quality control check using test objects is, therefore, recommended.

Note 5: dMRI is sensitive to motion. It is thus necessary to limit animal movement using anesthetics and restraint equipment (bite bar, ear bars). Any data (outliers) corrupted by motion must be discarded. Moderate amounts of movements can be corrected using ad hoc postprocessing steps (coregistration algorithms). Another source of artifacts is geometric distortions induced by air-bone interfaces and MRI gradient pulses used for diffusion MRI (especially with the EPI sequence). Those distortions could be large and require correction using dedicated MRI acquisitions (data are acquired twice with encoding in opposite directions) and the postprocessing method, which increase the overall acquisition time (with the issue of animal stability) and may corrupt signals.

Registration is also mandatory when group average images have to be obtained over several animals or when using a template atlas.

Note 6: ADC values are temperature dependent, and it is necessary to control and maintain the body temperature (37 °C in rodent) within acquisitions and across subjects.

Note 7: In the brain, water diffusion is anisotropic, especially in white matter fibers. This anisotropy effect can be exploited to generate fiber orientation maps and tractograms. In the presence of anisotropy, diffusion MRI acquisition must be performed in multiple directions (DTI framework), at least 6, but sometimes 12, 18, 30, or more, to extract relevant parameters correctly (e.g., fractional anisotropy, axial and radial diffusivities). Even if anisotropy effects are not considered, a geometric average of signals acquired along multiple directions is mandatory, or ADC values will vary substantially depending on diffusion-encoding directions. Noise effects must also be precisely taken into account when analyzing anisotropy effects, as noise results in pseudoanisotropy effects.

Note 8: Resting-state fMRI results obtained with anesthetized or sedated animals using the BOLD fMRI method (linked to neurovascular coupling) will highly depend on the nature of the anesthetic agent. Furthermore, residual pain or motion during acquisition will affect the results.

Note 9: Localized spectra acquired with MRS require shimming and water suppression. The spectral linewidth of the water signal must be consistent between subjects. Metabolite quantification considered with Cramér-Rao lower band (percentage of standard deviation) must be no higher than 20%.

References

1. Durand T, Bernier M-O, Léger I et al (2015) Cognitive outcome after radiotherapy in brain tumor. Curr Opin Oncol 27:510. https://doi.org/10.1097/CCO.0000000000000227
2. Soussain C, Ricard D, Fike JR et al (2009) CNS complications of radiotherapy and chemotherapy. Lancet 374:1639–1651
3. Mahajan A, Dong L, Prabhu S et al (2007) Application of deformable image registration to hippocampal doses and neurocognitive outcomes. Neuro-Oncology 9:538
4. Seibert TM, Karunamuni R, Bartsch H et al (2017) Radiation dose–dependent hippocampal atrophy detected with longitudinal volumetric magnetic resonance imaging. Int J Radiat Oncol 97:263–269. https://doi.org/10.1016/j.ijrobp.2016.10.035
5. Seibert TM, Karunamuni R, Kaifi S et al (2017) Cerebral cortex regions selectively vulnerable to radiation dose-dependent atrophy. Int J Radiat Oncol 97:910–918. https://doi.org/10.1016/j.ijrobp.2017.01.005
6. Omuro AMP, Ben-Porat LS, Panageas KS et al (2005) Delayed neurotoxicity in primary central nervous system lymphoma. Arch Neurol:62. https://doi.org/10.1001/archneur.62.10.1595
7. Connor M, Karunamuni R, McDonald C et al (2016) Dose-dependent white matter damage after brain radiotherapy. Radiother Oncol 121:209–216. https://doi.org/10.1016/j.radonc.2016.10.003
8. Connor M, Karunamuni R, McDonald C et al (2017) Regional susceptibility to dose-dependent white matter damage after brain radiotherapy. Radiother Oncol 123:209–217. https://doi.org/10.1016/j.radonc.2017.04.006
9. Doetsch F (2003) A niche for adult neural stem cells. Curr Opin Genet Dev 13:543–550. https://doi.org/10.1016/j.gde.2003.08.012
10. Capdevila C, Vázquez LR, Martí J (2017) Glioblastoma multiforme and adult neuro-

genesis in the ventricular-subventricular zone: a review. J Cell Physiol 232:1596–1601. https://doi.org/10.1002/jcp.25502

11. Gil-Perotin S, Marin-Husstege M, Li J et al (2006) Loss of p53 induces changes in the behavior of subventricular zone cells: implication for the genesis of glial tumors. J Neurosci 26:1107–1116. https://doi.org/10.1523/JNEUROSCI.3970-05.2006

12. Gupta T, Nair V, Paul SN et al (2012) Can irradiation of potential cancer stem-cell niche in the subventricular zone influence survival in patients with newly diagnosed glioblastoma? J Neuro-Oncol 109:195–203. https://doi.org/10.1007/s11060-012-0887-3

13. Khalifa J, Tensaouti F, Lusque A et al (2017) Subventricular zones: new key targets for glioblastoma treatment. Radiat Oncol 12. https://doi.org/10.1186/s13014-017-0791-2

14. Bompaire F, Lahutte M, Buffat S et al (2018) New insights in radiation-induced leukoencephalopathy: a prospective cross-sectional study. Support Care Cancer 26:4217–4226. https://doi.org/10.1007/s00520-018-4296-9

15. Doolittle ND, Korfel A, Lubow MA et al (2013) Long-term cognitive function, neuro-imaging, and quality of life in primary CNS lymphoma. Neurology 81:84–92. https://doi.org/10.1212/WNL.0b013e318297eeba

16. Ricard D, Idbaih A, Ducray F et al (2012) Primary brain tumours in adults. Lancet 379:1984–1996. https://doi.org/10.1016/S0140-6736(11)61346-9

17. Vigliani MC, Duyckaerts C, Hauw JJ et al (1999) Dementia following treatment of brain tumors with radiotherapy administered alone or in combination with nitrosourea-based chemotherapy: a clinical and pathological study. J Neuro-Oncol 41:137–149

18. Dropcho EJ (2010) Neurotoxicity of radiation therapy. Neurol Clin 28:217–234. https://doi.org/10.1016/j.ncl.2009.09.008

19. Ricard D, Soussain C, Psimaras D (2011) Neurotoxicity of the CNS: diagnosis, treatment and prevention. Rev Neurol (Paris) 167:737–745

20. Tofilon PJ, Fike JR (2000) The radioresponse of the central nervous system: a dynamic process. Radiat Res 153:357–370. https://doi.org/10.1667/0033-7587(2000)153[0357:TROTCN]2.0.CO;2

21. Fike JR (2011) Physiopathology of radiation-induced neurotoxicity. Rev Neurol (Paris) 167:746–750. https://doi.org/10.1016/j.neurol.2011.07.005

22. Lai R, Abrey LE, Rosenblum MK, DeAngelis LM (2004) Treatment-induced leukoenceph-alopathy in primary CNS lymphoma: a clinical and autopsy study. Neurology 62:451–456

23. El Waly B, Macchi M, Cayre M, Durbec P (2014) Oligodendrogenesis in the normal and pathological central nervous system. Front Neurosci 8:145. https://doi.org/10.3389/fnins.2014.00145

24. Hebb AO, Cusimano MD (2001) Idiopathic normal pressure hydrocephalus: a systematic review of diagnosis and outcome. Neurosurgery 49:1166–1184.; discussion 1184-1186. https://doi.org/10.1097/00006123-200111000-00028

25. Yoneoka Y, Satoh M, Akiyama K et al (1999) An experimental study of radiation-induced cognitive dysfunction in an adult rat model. Br J Radiol 72:1196–1201. https://doi.org/10.1259/bjr.72.864.10703477

26. Monje ML, Mizumatsu S, Fike JR, Palmer TD (2002) Irradiation induces neural precursor-cell dysfunction. Nat Med 8:955. https://doi.org/10.1038/nm749

27. Monje ML, Vogel H, Masek M et al (2007) Impaired human hippocampal neurogenesis after treatment for central nervous system malignancies. Ann Neurol 62:515–520. https://doi.org/10.1002/ana.21214

28. Obernier K, Alvarez-Buylla A (2019) Neural stem cells: origin, heterogeneity and regulation in the adult mammalian brain. Development 146:dev156059. https://doi.org/10.1242/dev.156059

29. Doetsch F, Caillé I, Lim DA et al (1999) Subventricular zone astrocytes are neural stem cells in the adult mammalian brain. Cell 97:703–716. https://doi.org/10.1016/S0092-8674(00)80783-7

30. Kempermann G, Gage FH, Aigner L et al (2018) Human adult neurogenesis: evidence and remaining questions. Cell Stem Cell 23:25–30. https://doi.org/10.1016/j.stem.2018.04.004

31. Sanai N, Nguyen T, Ihrie RA et al (2011) Corridors of migrating neurons in human brain and their decline during infancy. Nature 478:382–386. https://doi.org/10.1038/nature10487

32. Sanai N, Tramontin AD, Quiñones-Hinojosa A et al (2004) Unique astrocyte ribbon in adult human brain contains neural stem cells but lacks chain migration. Nature 427:740. https://doi.org/10.1038/nature02301

33. Fuentealba LC, Rompani SB, Parraguez JI et al (2015) Embryonic origin of postnatal neural stem cells. Cell 161:1644–1655. https://doi.org/10.1016/j.cell.2015.05.041

34. Furutachi S, Miya H, Watanabe T et al (2015) Slowly dividing neural progenitors are an embryonic origin of adult neural stem cells.

Nat Neurosci 18:657–665. https://doi.org/10.1038/nn.3989

35. Kippin TE, Martens DJ, van der Kooy D (2005) p21 loss compromises the relative quiescence of forebrain stem cell proliferation leading to exhaustion of their proliferation capacity. Genes Dev 19:756–767. https://doi.org/10.1101/gad.1272305

36. Molofsky AV, Pardal R, Iwashita T et al (2003) Bmi-1 dependence distinguishes neural stem cell self-renewal from progenitor proliferation. Nature 425:962–967. https://doi.org/10.1038/nature02060

37. Ottone C, Krusche B, Whitby A et al (2014) Direct cell-cell contact with the vascular niche maintains quiescent neural stem cells. Nat Cell Biol 16:1045–1056. https://doi.org/10.1038/ncb3045

38. Mira H, Andreu Z, Suh H et al (2010) Signaling through BMPR-IA regulates quiescence and long-term activity of neural stem cells in the adult hippocampus. Cell Stem Cell 7:78–89. https://doi.org/10.1016/j.stem.2010.04.016

39. Zawadzka M, Rivers LE, Fancy SPJ et al (2010) CNS-resident glial progenitor/stem cells produce Schwann cells as well as oligodendrocytes during repair of CNS demyelination. Cell Stem Cell 6:578–590. https://doi.org/10.1016/j.stem.2010.04.002

40. Picard-Riera N, Decker L, Delarasse C et al (2002) Experimental autoimmune encephalomyelitis mobilizes neural progenitors from the subventricular zone to undergo oligodendrogenesis in adult mice. Proc Natl Acad Sci U S A 99:13211–13216. https://doi.org/10.1073/pnas.192314199

41. Tepavčević V, Lazarini F, Alfaro-Cervello C et al (2011) Inflammation-induced subventricular zone dysfunction leads to olfactory deficits in a targeted mouse model of multiple sclerosis. J Clin Invest 121:4722–4734. https://doi.org/10.1172/JCI59145

42. Parras CM, Galli R, Britz O et al (2004) Mash1 specifies neurons and oligodendrocytes in the postnatal brain. EMBO J 23:4495–4505. https://doi.org/10.1038/sj.emboj.7600447

43. Codega P, Silva-Vargas V, Paul A et al (2014) Prospective identification and purification of quiescent adult neural stem cells from their in vivo niche. Neuron 82:545–559. https://doi.org/10.1016/j.neuron.2014.02.039

44. Beckervordersandforth R, Tripathi P, Ninkovic J et al (2010) In vivo fate mapping and expression analysis reveals molecular hallmarks of prospectively isolated adult neural stem cells. Cell Stem Cell 7:744–758. https://doi.org/10.1016/j.stem.2010.11.017

45. Mich JK, Signer RA, Nakada D et al (2014) Prospective identification of functionally distinct stem cells and neurosphere-initiating cells in adult mouse forebrain. elife 3. https://doi.org/10.7554/eLife.02669

46. Daynac M, Chicheportiche A, Pineda JR et al (2013) Quiescent neural stem cells exit dormancy upon alteration of GABAAR signaling following radiation damage. Stem Cell Res 11:516–528. https://doi.org/10.1016/j.scr.2013.02.008

47. Daynac M, Morizur L, Kortulewski T et al (2015) Cell sorting of neural stem and progenitor cells from the adult mouse subventricular zone and live-imaging of their cell cycle dynamics. J Vis Exp. https://doi.org/10.3791/53247

48. Morizur L, Chicheportiche A, Gauthier LR et al (2018) Distinct molecular signatures of quiescent and activated adult neural stem cells reveal specific interactions with their microenvironment. Stem Cell Rep 11:565–577. https://doi.org/10.1016/j.stemcr.2018.06.005

49. Daynac M, Morizur L, Chicheportiche A et al (2016) Age-related neurogenesis decline in the subventricular zone is associated with specific cell cycle regulation changes in activated neural stem cells. Sci Rep:6. https://doi.org/10.1038/srep21505

50. Daynac M, Pineda JR, Chicheportiche A et al (2014) TGFβ lengthens the G1 phase of stem cells in aged mouse brain. Stem Cells 32:3257–3265. https://doi.org/10.1002/stem.1815

51. Alam MJ, Kitamura T, Saitoh Y et al (2018) Adult neurogenesis conserves hippocampal memory capacity. J Neurosci 38:6854–6863. https://doi.org/10.1523/JNEUROSCI.2976-17.2018

52. Santarelli L, Saxe M, Gross C et al (2003) Requirement of hippocampal neurogenesis for the behavioral effects of antidepressants. Science 301:805–809. https://doi.org/10.1126/science.1083328

53. Lazarini F, Mouthon M-A, Gheusi G et al (2009) Cellular and behavioral effects of cranial irradiation of the subventricular zone in adult mice. PLoS One 4. https://doi.org/10.1371/journal.pone.0007017

54. Feierstein CE, Lazarini F, Wagner S et al (2010) Disruption of adult neurogenesis in the olfactory bulb affects social interaction but not maternal behavior. Front Behav Neurosci 4. https://doi.org/10.3389/fnbeh.2010.00176

55. Valley MT, Mullen TR, Schultz LC et al (2009) Ablation of mouse adult neurogenesis alters olfactory bulb structure and olfactory

fear conditioning. Front Neurosci 3. https://doi.org/10.3389/neuro.22.003.2009

56. Amano T, Inamura T, Wu C-M et al (2002) Effects of single low dose irradiation on subventricular zone cells in juvenile rat brain. Neurol Res 24:809–816. https://doi.org/10.1179/016164102101200771

57. Hellström NAK, Björk-Eriksson T, Blomgren K, Kuhn HG (2009) Differential recovery of neural stem cells in the subventricular zone and dentate gyrus after ionizing radiation. Stem Cells 27:634–641. https://doi.org/10.1634/stemcells.2008-0732

58. Fukuda H, Fukuda A, Zhu C et al (2004) Irradiation-induced progenitor cell death in the developing brain is resistant to erythropoietin treatment and caspase inhibition. Cell Death Differ 11:1166. https://doi.org/10.1038/sj.cdd.4401472

59. Osato K, Sato Y, Ochiishi T et al (2010) Apoptosis-inducing factor deficiency decreases the proliferation rate and protects the subventricular zone against ionizing radiation. Cell Death Dis 1:e84. https://doi.org/10.1038/cddis.2010.63

60. Baser A, Skabkin M, Kleber S et al (2019) Onset of differentiation is post-transcriptionally controlled in adult neural stem cells. Nature 566:100–104. https://doi.org/10.1038/s41586-019-0888-x

61. Achanta P, Capilla-Gonzalez V, Purger D et al (2012) Subventricular zone localized irradiation affects the generation of proliferating neural precursor cells and the migration of neuroblasts. Stem Cells 30:2548–2560. https://doi.org/10.1002/stem.1214

62. Capilla-Gonzalez V, Guerrero-Cazares H, Bonsu JM et al (2014) The subventricular zone is able to respond to a demyelinating lesion after localized radiation. Stem Cells 32:59–69. https://doi.org/10.1002/stem.1519

63. Panagiotakos G, Alshamy G, Chan B et al (2007) Long-term impact of radiation on the stem cell and oligodendrocyte precursors in the brain. PLoS One 2:e588. https://doi.org/10.1371/journal.pone.0000588

64. Begolly S, Olschowka JA, Love T et al (2018) Fractionation enhances acute oligodendrocyte progenitor cell radiation sensitivity and leads to long term depletion. Glia 66:846–861. https://doi.org/10.1002/glia.23288

65. Pineda JR, Daynac M, Chicheportiche A et al (2013) Vascular-derived TGF-β increases in the stem cell niche and perturbs neurogenesis during aging and following irradiation in the adult mouse brain. EMBO Mol Med 5:548–562. https://doi.org/10.1002/emmm.201202197

66. Monje ML, Palmer T (2003) Radiation injury and neurogenesis. Curr Opin Neurol 16:129–134. https://doi.org/10.1097/01.wco.0000063772.81810.b7

67. de Graaf RA (2019) In vivo NMR spectroscopy: principles and techniques, 3rd edn. John Wiley & Sons, New York

68. Le Bihan D (1995) Magnetic resonance imaging of diffusion and perfusion: applications to functional imaging. Raven Press, New York

69. Haacke EM (1999) Magnetic resonance imaging; physical principles and sequence design. Wiley, New York

70. Jahng G-H, Li K-L, Ostergaard L, Calamante F (2014) Perfusion magnetic resonance imaging: a comprehensive update on principles and techniques. Korean J Radiol 15:554. https://doi.org/10.3348/kjr.2014.15.5.554

71. Petcharunpaisan S, Ramalho J, Castillo M (2010) Arterial spin labeling in neuroimaging. World J Radiol 2:384–398. https://doi.org/10.4329/wjr.v2.i10.384

72. Le Bihan D (2019) What can we see with IVIM MRI? NeuroImage 187:56–67. https://doi.org/10.1016/j.neuroimage.2017.12.062

73. Le Bihan D, Breton E (1985) Imagerie de diffusion in-vivo par résonance magnétique nucléaire. Comptes-Rendus de l'Académie des Sciences 93:27–34

74. Le Bihan D (2014) Diffusion MRI: what water tells us about the brain. EMBO Mol Med 6:569–573

75. Iima M, Le Bihan D (2015) Clinical Intravoxel incoherent motion and diffusion MR imaging: past, present, and future. Radiology 278:13–32. https://doi.org/10.1148/radiol.2015150244

76. Jensen JH, Helpern JA (2010) MRI quantification of non-Gaussian water diffusion by kurtosis analysis. NMR Biomed 23:698–710. https://doi.org/10.1002/nbm.1518

77. Le Bihan D, Mangin J-F, Poupon C et al (2001) Diffusion tensor imaging: concepts and applications. J Magn Reson Imaging 13:534–546. https://doi.org/10.1002/jmri.1076

78. Ogawa S, Lee TM, Nayak AS, Glynn P (1990) Oxygenation-sensitive contrast in magnetic resonance image of rodent brain at high magnetic fields. Magn Reson Med 14:68–78

79. Keilholz SD, Pan W-J, Billings J et al (2017) Noise and non-neuronal contributions to the BOLD signal: applications to and insights from animal studies. NeuroImage 154:267–281. https://doi.org/10.1016/j.neuroimage.2016.12.019

80. Prost RW (2008) Magnetic resonance spectroscopy. Med Phys 35:4530–4544. https://doi.org/10.1118/1.2975225

81. Jiang X, Yuan L, Engelbach JA et al (2015) A gamma-knife-enabled mouse model of cerebral single-hemisphere delayed radiation necrosis. PLoS One 10:e0139596. https://doi.org/10.1371/journal.pone.0139596

82. Wadghiri YZ, Blind JA, Duan X et al (2004) Manganese-enhanced magnetic resonance imaging (MEMRI) of mouse brain development. NMR Biomed 17:613–619. https://doi.org/10.1002/nbm.932

83. Gazdzinski LM, Cormier K, Lu FG et al (2012) Radiation-induced alterations in mouse brain development characterized by magnetic resonance imaging. Int J Radiat Oncol Biol Phys 84:e631–e638. https://doi.org/10.1016/j.ijrobp.2012.06.053

84. Nieman BJ, de Guzman AE, Gazdzinski LM et al (2015) White and gray matter abnormalities after cranial radiation in children and mice. Int J Radiat Oncol Biol Phys 93:882–891. https://doi.org/10.1016/j.ijrobp.2015.07.2293

85. Trivedi R, Khan AR, Rana P et al (2012) Radiation-induced early changes in the brain and behavior: serial diffusion tensor imaging and behavioral evaluation after graded doses of radiation. J Neurosci Res 90:2009–2019. https://doi.org/10.1002/jnr.23073

86. Verreet T, Quintens R, Van Dam D et al (2015) A multidisciplinary approach unravels early and persistent effects of X-ray exposure at the onset of prenatal neurogenesis. J Neurodev Disord 7. https://doi.org/10.1186/1866-1955-7-3

87. Verreet T, Rangarajan JR, Quintens R et al (2016) Persistent impact of in utero irradiation on mouse brain structure and function characterized by MR imaging and Behavioral analysis. Front Behav Neurosci 10:83. https://doi.org/10.3389/fnbeh.2016.00083

88. Pérès EA, Etienne O, Grigis A et al (2018) Longitudinal study of irradiation-induced brain microstructural alterations with S-index, a diffusion MRI biomarker, and MR spectroscopy. Int J Radiat Oncol Biol Phys 102:1244–1254. https://doi.org/10.1016/j.ijrobp.2018.01.070

89. de Guzman AE, Gazdzinski LM, Alsop RJ et al (2015) Treatment age, dose and sex determine neuroanatomical outcome in irradiated juvenile mice. Radiat Res 183:541–549. https://doi.org/10.1667/RR13854.1

90. Kumar M, Haridas S, Trivedi R et al (2013) Early cognitive changes due to whole body γ-irradiation: a behavioral and diffusion tensor imaging study in mice. Exp Neurol 248:360–368. https://doi.org/10.1016/j.expneurol.2013.06.005

91. Gupta M, Mishra SK, Kumar BSH et al (2017) Early detection of whole body radiation induced microstructural and neuroinflammatory changes in hippocampus: a diffusion tensor imaging and gene expression study. J Neurosci Res 95:1067–1078. https://doi.org/10.1002/jnr.23833

92. Constanzo J, Dumont M, Lebel R et al (2018) Diffusion MRI monitoring of specific structures in the irradiated rat brain. Magn Reson Med 80:1614–1625. https://doi.org/10.1002/mrm.27112

93. Serduc R, van de Looij Y, Francony G et al (2008) Characterization and quantification of cerebral edema induced by synchrotron x-ray microbeam radiation therapy. Phys Med Biol 53:1153–1166. https://doi.org/10.1088/0031-9155/53/5/001

94. Watve A, Gupta M, Khushu S, Rana P (2018) Longitudinal changes in gray matter regions after cranial radiation and comparative analysis with whole body radiation: a DTI study. Int J Radiat Biol 94:532–541. https://doi.org/10.1080/09553002.2018.1466064

95. Saito S, Sawada K, Mori Y et al (2015) Brain and arterial abnormalities following prenatal X-ray irradiation in mice assessed by magnetic resonance imaging and angiography. Congenit Anom 55:103–106. https://doi.org/10.1111/cga.12101

96. Constanzo J, Masson-Côté L, Tremblay L et al (2017) Understanding the continuum of radionecrosis and vascular disorders in the brain following gamma knife irradiation: an MRI study. Magn Reson Med 78:1420–1431. https://doi.org/10.1002/mrm.26546

97. Herynek V, Burian M, Jirák D et al (2004) Metabolite and diffusion changes in the rat brain after Leksell gamma knife irradiation. Magn Reson Med 52:397–402. https://doi.org/10.1002/mrm.20150

98. Gupta M, Rana P, Trivedi R et al (2013) Comparative evaluation of brain neurometabolites and DTI indices following whole body and cranial irradiation: a magnetic resonance imaging and spectroscopy study. NMR Biomed 26:1733–1741. https://doi.org/10.1002/nbm.3010

99. Kovács N, Szigeti K, Hegedűs N et al (2018) Multimodal PET/MRI imaging results enable monitoring the side effects of radiation therapy. Contrast Media Mol Imaging, In. https://www.hindawi.com/journals/cmmi/2018/5906471/. Accessed 29 May 2019

100. Chan KC, Khong P-L, Cheung MM et al (2009) MRI of late microstructural and metabolic alterations in radiation-induced brain injuries. J Magn Reson Imaging 29:1013–1020. https://doi.org/10.1002/jmri.21736

101. Bálentová S, Hnilicová P, Kalenská D et al (2017) Effect of whole-brain irradiation

on the specific brain regions in a rat model: metabolic and histopathological changes. Neurotoxicology 60:70–81. https://doi.org/10.1016/j.neuro.2017.03.005

102. Yamaguchi M, Saito H, Suzuki M, Mori K (2000) Visualization of neurogenesis in the central nervous system using nestin promoter-GFP transgenic mice. NeuroReport 11 (9):1991-1996

103. Couillard-Despres S, Finkl R, Winner B et al (2008) In vivo optical imaging of neurogenesis: watching new neurons in the intact brain. Mol Imaging 7. https://doi.org/10.2310/7290.2008.0004

104. Leung HWC, Chan ALF, Chang MB (2016) Brain dose-sparing radiotherapy techniques for localized intracranial germinoma: Case report and literature review of modern irradiation. Cancer Radiother 20:210-216. https://doi.org/10.1016/j.canrad.2016.02.007

105. Kazda T, Jancalek R, Pospisil P et al (2014) Why and how to spare the hippocampus during brain radiotherapy: the developing role of hippocampal avoidance in cranial radiotherapy. Radiat Oncol 9:139. https://doi.org/10.1186/1748-717X-9-139

106. Welzel G, Fleckenstein K, Schaefer J et al (2008) Memory function before and after whole brain radiotherapy in patients with and without brain metastases. Int J Radiat Oncol Biol Phys 72:1311–1318. https://doi.org/10.1016/j.ijrobp.2008.03.009

Part IV

Clinical Imaging

Chapter 12

Mechano-Biological Features in a Patient-Specific Computational Model of Glioblastoma

Francesco Acerbi, Abramo Agosti, Jacopo Falco, Stefano Marchesi, Ignazio G. Vetrano, Francesco DiMeco, Alberto Bizzi, Paolo Ferroli, Giorgio Scita, and Pasquale Ciarletta

Abstract

Despite the technical-scientific developments of the last decades, the prognosis of patients affected by glioblastoma (GBM) still remains poor due to the extensive invasion and diffusion of tumor cells anywhere within the host brain tissue, which makes almost impossible the complete surgical removal. Furthermore, although the blood-brain barrier (BBB) is typically disrupted at the core of GBM, it can be structurally conserved at the tumor periphery, allowing infiltrative and spreading glioma cells to escape chemotherapy-induced death.

In this chapter, the authors present an innovative mechano-biological approach to study and describe the GBM progression taking into account not only biochemical factors but also mechanical interactions occurring between the local micro-environment and the tumor. This model is based upon the experimental evidence of a preferential diffusion of glioma cells among white matter fiber tracts, detectable by preoperative patient-specific diffusion tensor imaging. The possibility of predicting tumor patterns of recurrence could potentially modify the therapeutic strategies, by guiding resection including a supramaximal removal in the areas at higher risk of recurrence, and by improving the planning of radiation therapy, possibly increasing the patient survival.

Key words Glioblastoma, Mechano-biology, Neuroimaging, Biomathematics, Personalized medicine

1 Introduction

1.1 Glioblastoma Epidemiology

Brain tumors account for about 1–2% of all cancers, representing approximately the sixth-eighth neoplasia by frequency in adults and a major cause of cancer-related mortality and morbidity. The annual incidence of primary brain tumors is 16.5 per 100,000 population in the Western countries [1]. Particularly, the number of recorded glioblastoma multiforme (GBM) cases in Europe and North America is 3–5 per 100,000 adults each year [2]. GBM is

Giorgio Seano (ed.), *Brain Tumors*, Neuromethods, vol. 158,
https://doi.org/10.1007/978-1-0716-0856-2_12, © Springer Science+Business Media, LLC, part of Springer Nature 2021

more frequent in the sixth-eighth decades of life, especially at a median age of 64 years, but it can occur at any age [3].

Despite treatments with the current standard of care, the overall median progression-free (PFS) and overall survival (OS) times are approximately 7 and 15 months, respectively [4, 5].

Malignant gliomas arise in a multistep process involving sequential and cumulative genetic alterations resulting from intrinsic and environmental factors; the etiology of GBM still remains largely unknown. Established risk factors include genetic predisposition and exposure to ionizing radiation; there are instead inconclusive data for an association with occupational risk factors, exposure to mobile phones or to other kinds of electromagnetic fields, head trauma, foods containing N-nitroso compounds, aspartame, pesticide, smoking [6–14].

GBM can be divided into two categories, primary and secondary GBM, based on the different origin: primary GBM, which is the most frequent subtype, rapidly develops de novo, without clinical or histological evidence of a less malignant precursor lesion; otherwise, secondary GBM evolves through progression from a lower-grade glioma and usually affect younger patients [15]. Molecular alterations are specific to one of the two subtypes, as highlighted also by WHO 2016 histopathological classification [16]. Specifically, primary GBMs are genetically characterized by loss of heterozygosis 10q, *EGFR* amplification, *p16*INK4a deletion, and *PTEN* mutations whereas, in the pathway to secondary GBM, *TP53* mutations are the most frequent and earliest detectable genetic alteration and further mutations accumulate during progression to GBM. Isocitrate dehydrogenase (IDH) gene status is one of the most important molecular differences between primary and secondary GBM and represents a relevant diagnostic criterion with also a prognostic impact: it is frequently associated with secondary GBM, conferring favorable prognostic value and prediction of response to standard chemotherapy [17–19].

1.2 Clinical and Radiological Presentation of Glioblastomas

The clinical presentation can vary greatly according to tumor size, its location and eloquent brain regions involved. Headache is quite frequent, presenting in about 50% of patients at the diagnosis [20]. Patients often present other symptoms related to an increased intracranial pressure, such as dizziness, nausea and vomiting, lethargy, and papilledema [21]. Seizures occur only in a minority of patients newly diagnosed, approximately 25%, sometimes in the later stage of the disease [22]. Neurological disorders associated with GBM can vary widely, sometimes causing symptoms such as persistent weakness, numbness, hemiparesis, loss of vision, or alteration of language [23].

Computerized tomography (CT) and magnetic resonance imaging (MRI) are the routinely performed important imaging techniques. The firstly performed imaging technique, in particular in emergency setting, is usually a non-contrasted brain CT. Radiological hallmarks suggesting mass effect from the tumor, that appears hypodense, may comprise sulcal effacement, midline shift, compression of the ventricles, and intracranial brain herniation. MRI with contrast injection is a more accurate technique which is usually performed after CT scan: it shows GBM as a contrast-enhancing inhomogeneous mass, with a thickened rind of enhancement and a hypointense core, corresponding to necrotic areas. Tumor boundaries are irregular or poorly defined, with an infiltrative spread of the tumor along white matter tracts or trans-callosally into the opposite hemisphere; peritumoral edema is generally T2/FLAIR (fluid-attenuated inversion recovery) hyperintense signal abnormality which surrounds the main mass lesion [24]. MRI techniques give not only morphological information but may also quantify increased neoplastic vascularity: MR perfusion techniques (PWI) can estimate GBM angiogenesis and capillary permeability, which are important biologic markers of malignancy and tumor aggressiveness, related to a worst prognosis. Furthermore, GBM vascular network nourishes and supports the rapid proliferation of cancer cells, but it also constitutes a roadmap for cellular infiltration and spread [25]. Magnetic resonance spectroscopy (MRS), which is performed directly in a unitary volume of the patient brain, allows to measure levels of metabolites, as N-acetylaspartate, marker of neuronal integrity that decreases in tumors, total choline and total creatine, both markers of neoplastic proliferation. Single-voxel MRS has been used to compare levels of these metabolites in the suspected glioma, in adjacent infiltrated tissue and in normal brain parenchyma: the typical MRS profile of GBM shows elevated choline, decreased N-acetylaspartate with a higher choline-creatine ratio than low-grade gliomas [26]. In neuro-oncology, 18fluoro-D-glucose positron emission tomography (PET) has never truly obtained widespread acceptance because of the nonspecific uptake of glucose into glial cells; to obviate to this limits, PET with radioactive amino acids, such as ^{11}C-methionine and ^{18}F-flouroethyltyrosine, has been recently introduced, to obtain information about the metabolic state of infiltrated parenchyma and to discriminate recurrent/progressive glioma from pseudo-progression/treatment effect or radio-necrosis [27, 28].

1.3 Therapeutic Strategies and their Limitations

The standard therapy of GBM includes maximal safe resection, followed by chemoradiation therapy [29]. Due to the high degree of invasiveness, radical resection of the primary tumor mass is not curative and infiltrating glioma cells invariably are left within the surrounding brain, leading to later disease progression and recurrence. The extent of resection is the most important prognostic factor; in particular, a residual volume of less than 2 or 5 cm^3 predicts a significant survival benefit [30, 31]. Intraoperatively, GBM appears as an infiltrative mass, poorly delineated, bleeding, of increased consistency with peripheral greyish aspect and a central area of yellowish necrosis from myelin breakdown. In case of large, multifocal, deep located tumors or in patients with a poor performance status, a diagnostic and eventually cytoreductive biopsy should be performed, for histopathological and molecular characterization.

The need to improve tumor visualization and consequently the recognition of healthy margins has determined the development of many specific tools, such as neuronavigation [32], fluorophores application [33], intraoperative ultrasound [34, 35], and intraoperative MRI [36]. A fluorophore is a fluorescent chemical compound that can re-emit light upon light excitation: different fluorophores are excited by a specific wavelength and re-emit lights at a major wavelength, generally detected by specific filters; the final effect is to better discriminate between pathological and normal tissue. Fluorescent agents commonly used in neuro-oncological surgery are 5-aminolevulenic acid (5-ALA) and sodium fluorescein (SF). 5-ALA is based upon an enzymatic mechanism, since it is a precursor of hemoglobin which is converted by aminolevulinate dehydratase in fluorescent porphyrins which accumulate in malignant cells thus fluorescence is detectable specifically in tumors, determining a significant increase in the rate of complete resection and PFS [37], confirming that a gross-total resection has a positive impact in overall survival. On the other hand, SF is a small organic molecule that readily crosses capillaries, provides fluorescent contrast in the extracellular matrix, with a renal total clearance of 24–36 h after intravenous injection. SF has a nonspecific mechanism of action with the property to accumulate in brain areas with a damaged BBB, as it happens in GBM, in a similar way like the gadolinium in MRI. The accumulation of SF in tumor interstitial space is related to the grade of breakdown of the normal BBB, which becomes useful predominantly in the resection of aggressive tumors. The first multicentric phase II trial on fluorescein-guided resection of HGG [38] showed a complete resection in 82.6% of the cases, with a median residual tumor in the remaining cases of 0.1945 cm^3. An extent of resection >98% was obtained in 93.4% of the cases (Fig. 12.1).

After surgery, adjuvant radiotherapy combined with chemotherapy should be considered in all patients. The typical radiotherapy dose is 60 Gy divided into 30 fractions. Hypo-fractioned radiotherapy with higher fractions and a lower total dose can be appropriate in older patients [39]. The DNA alkylating agent temozolomide (TMZ) is administered orally, concomitantly with radiotherapy [29]. A post hoc tissue analysis suggested patients with tumors displaying promoter methylation of the DNA repair enzyme MGMT were more likely to benefit from the addition of TMZ to radiotherapy [40]. Optimal treatments for elderly patients (65 years of age and older) and for patients with poor KPS remain

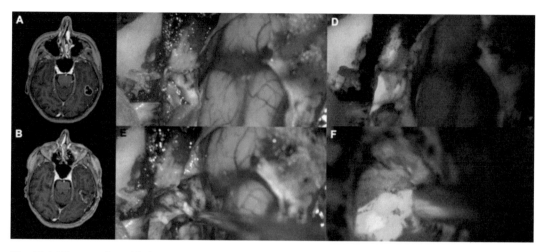

Fig. 12.1 [Fluorescein-guided resection of a left temporal glioblastoma]. Representative case of a left temporal GBM submitted to surgical resection under fluorescein visualization. (**a**) Preoperative post-contrast T1 MRI, showing a left irregular, contrast-enhancing lesion in the left temporal lobe. (**b**) Postoperative post-contrast T1 MRI, showing the surgical cavity, without macroscopic residual tissue. (**c–f**) Intraoperative pictures, demonstrating the increased visualization assured by the use of fluorescein (5 mg/kg injected i.v. at patient intubation), using a specific filter (Y560, Pentero Microscope, Carl Zeiss Meditec) on the surgical microscope (**d, f**), compared to white light illumination (**c, e**)

to be established; in elderly patients with methylated MGMT status, the addition of TMZ to short-course radiotherapy resulted in longer survival than short-course radiotherapy alone [41]. Lastly, in addition to standard treatment, a large number of targeted agents and novel approaches are being evaluated both in new onset and recurrent GBM [42].

1.4 Infiltrative Characteristics of Glioblastoma, and Mechano-Biological Feedbacks with the Extracellular Matrix

The most important hallmark of diffuse astrocytic tumors and, especially, of GBM is their invasive behavior and infiltrative spreading. At diagnosis, GBM is already largely disseminated; this hidden diffusion is the principal reason that makes gliomas difficult to be successfully removed and almost impossible to be cured. In fact, surgery and adjuvant therapies mainly focus on the highly proliferative, contrast-enhancing tumor mass, but local invasion generally leads to cancer relapse.

GBM is one of the tumors in which the role of extracellular matrix (ECM) has early being recognized as a key factor in its malignancy, actively concurring to its aggressiveness and dissemination. GBM highly motile cells are able to easily detach from the primary lesion or to migrate in collective patterns and disseminate locally in various districts of the central nervous system. Glioblastoma cells rarely metastasize to distal sites and organs, preferring, instead, to move along "already-tracked" routes such as vessels and white matter tracts [43–46].

Within this context, the physical and chemical properties of neuronal ECM have been shown to play an important role in GBM progression. Glioma cells have been shown to be able to form mature adhesions, migrate faster, and intensively proliferate on stiffer substrates [47]. Moreover, the increase in stiffness in tumor-associated ECM correlates with poor prognosis [48]. In turn, the perturbed mechanical properties in the tumor microenvironment have been shown to impair treatment efficacy, by compromising blood vessel integrity and decreasing the delivery of chemotherapeutics drugs [49–52].

In the adult brain, ECM occupies around 20% of the whole organ and is primarily composed of hyaluronic acid and tenascins [53–55]. Both molecules are frequently elevated in GBM and contribute to enhance ECM stiffness and invasive capacity of cancer cells [48, 56–58]. For example, tenascin-C (TNC) was recently found to be a part of tumor response to reduced oxygen availability and to actively participate in a signaling cascade, ultimately promoting GBM aggressiveness and resistance to chemotherapy. Indeed, TNC expression was shown to be directly induced by hypoxia-inducible factor-1α (HIF1α), a master transcription factor activated upon low oxygen availability. In this context, the mutational status of metabolic enzyme IDH1 was demonstrated to be crucial for the capacity of HIF1α-TNC axis to promote GBM aggressiveness [48]. Mutations of IDH1 previously associated with an increase in disease-free survival [19], including the R132H, have been shown to prevent the upregulation of HIF1α-TNC axis in response to low oxygen, thereby leading to impaired mechano-signaling pathways, ultimately interrupting the vicious cycles toward increased malignancy. However, an increase in ECM stiffness, as frequently scored in recurrent gliomas after surgical resections and chemotherapy, was shown to be able to bypass the protective effect of IDH1 mutation, reinstalling mechanical responses that ultimately promote tumor aggressiveness [48].

The acquisition of the mesenchymal phenotype by cancer cells was shown to correlate with a reduction in survival and a tendency to relapse. Activation of mechano-signaling pathways can promote a mesenchymal-like transcriptional program and an increased invasive capacity in primary GBM cells [59]. In addition, exogenous manipulation of glycocalyx-dependent signaling was found to directly control glioma aggressiveness and overall survival by enhancing integrin activation and mechano-response. Indeed, a sub-population of putative glioma stem cells was identified using glycoproteins as markers (such as CD44 and podoplanin). These cells were shown to display an elevated mechano-signaling and to be resistant to temozolomide treatment, which actually represents the "standard of care" in glioma chemotherapy. Conversely, reducing glycocalyx bulkiness or integrin-dependent mechano-signaling was demonstrated to be sufficient to lead to the sensitization of

CD44/podoplanin-positive cells to chemotherapeutic treatment [59].

Taken together, these experimental evidences argue for the vital importance of ECM-dependent signaling in sensing the mechanical properties of microenvironment and in the progression of glioblastoma. They further point to the notion that physical and chemical features of the environment surrounding tumor mass are overarching determinants of cancer progression, governing a myriad of biological processes such as proliferation, motility, invasion, and survival to therapy.

1.5 Aim of the Study

As widely described, GBM is a highly aggressive tumor with high mortality and morbidity and, mostly, at the current state of knowledge, with limited therapeutic possibilities. GBM recurrences generally occur within the FLAIR signal abnormality envelope but can develop also at distance from principal localization. Glioma cells motion is positively correlated with the white matter fiber tracts, since the tumor cells tend to migrate while attached to a solid support [60, 61]. During the past few decades, studying the evolution of gliomas from a mathematical point of view has become an issue of primary importance, mainly because, despite the impressive scientific, medical and technological advances, a cure for this disease has not yet been found. Here, we present an innovative approach combining in-vitro experiments and neuroimaging data to feed a personalized mathematical model that is used as an in silico tools to assist medical doctors in the clinical treatment of glioblastoma. The novelty of these efforts is taking into account the mechano-biological causes of cancer invasion, to aid in the understanding of experimental and clinical observations, and to help design new, targeted experiments and treatment strategies, in order to minimize patient suffering while maximizing treatment effectiveness [62, 63]. The possibility to predict in preoperative setting and after surgery tumor behavior and its pattern of regrowth thanks to mathematical modeling may tailor surgery, guide feasible local chemotherapies and personalize radiotherapy, both for the gross and the clinical target volume in order to better hit the infiltrating glioma cells to reduce and delay the risk of tumor relapse.

2 Methods

2.1 In Vitro Mechano-Biological Studies: The Spontaneous Aggregation of GBM Cells

Recent experimental works have highlighted that mechano-biological features play a key role in glioblastoma development.

In epithelial tissues, the progression from a primary neoplastic lesion to a metastasis is characterized by a number of steps, during which tumor cells acquire the capacity to migrate and detach from the bulk mass, degrade the basal membrane, invade surrounding areas, and disseminate to distant organs.

Along the various steps of this process, the transition from an epithelial to a mesenchymal identity (EMT) is thought to be essential to promote alteration of the architecture and apico-basolateral polarity of epithelial-derived carcinoma. Cells that have undergone EMT are typically characterized by the loss of epithelial-cadherin (E-cadherin) and weakening of cell-cell contacts, with the successive appearance of memsenchymal traits exemplified by the increased expression of prototypical mesenchymal markers, such as vimentin and N-cadherin [64–66].

EMT causes also a drastic alteration of tissue homeostasis, promoting the detachment of either single cells or clusters from the primary site. This transition, which is governed and accompanied by the rewiring of specific intracellular signaling pathways, also leads to profound changes in the mechanical properties of cancer cells. For its importance in cancer progression, numerous approaches and model systems have been developed with the intent to recapitulate the key aspect of the EMT process. A number of these approaches relied on simple in vitro culture models of epithelial cells, in which genetic manipulation of pivotal EMT factors such as SNAIL, ZEB, and TWIST was shown to be sufficient to promote the acquisition of mesenchymal molecular and morphological traits, thereby recapitulating the rewiring in cell identity program typically underlying EMT [67].

However, these in vitro single cell-based systems, albeit having been of tremendous help in dissecting molecular pathways to mesenchymal state, have intrinsic limitations. For example, they fail to mimic the complex set of biophysical supracellular alterations that accompany EMT and lead to the detachment of tumor clusters, rather than individual cells from the primary site and their subsequent dissemination in distal sites. In this regard, one of the simplest and attractive model systems is represented by the spontaneous formation of aggregates or spheroids arising from cell monolayers. This phenomenon has been documented in a number of cellular contexts, such as ovarian cancer cell lines and primary colon cancers [68, 69]. Spontaneous spheroid budding can occur in conditions where the interaction with the bottom substrate is weak or absent, allowing the detachment of 3D aggregates [70, 71]. Molecularly, the transition from a planar monolayer to a three-dimensional aggregate implies a profound alteration of the cell-substrate adhesive properties and, as a consequence, of the cell-cell adhesive interaction.

From the physical point of view, this transition shares several analogies with the process of wetting and de-wetting, i.e., the interaction between a fluid drop and the bottom solid surface where surface tension, cell-cell adhesion, and cell-substrate adhesion can be understood within the framework and laws governing the kinematic and behavior of fluids on surfaces. A recent paper, indeed, demonstrated that manipulation of cell-cell contacts by ectopic E-cadherin expression could influence the transition from monolayer spreading to aggregate budding [72].

In GBM, spontaneous budding of 3D aggregates was first reported in U-87 MG cell line and associated with the presence of a cancer stem cell subpopulation. These structures are formed in low serum-containing or serum-free media and express neural stem cell markers, such as CD133 and nestin [73–75].

Recently, we have studied the mechano-biological features of U-87 MG cells cultured in a monolayer adhering to a Petri dish at a room temperature of 20 °C and 5% CO_2 in Dulbecco's modified Eagle's medium plus fetal bovine serum (10% in volume), L-glutamine (2 mM), sodium pyruvate (1 mM), non-essential amino acids (1% in volume), and penicillin/streptomycin (100 mg/mL). At day 0, about 2.5 millions of U-87 MG cells were seeded into two dishes (considered as technical replicas) with a total medium volume of 10 mL/each.

We characterized the proliferation and invasion properties by phase-contrast microscopy. Images were acquired at 4X magnification using an EVOS-FL Imaging System (ThermoFisher Scientific), and the morphological features of the emerging clusters were quantified by an automated procedure using the Fiji software. These experimental observations have guided the quantification of the most relevant mechano-biological parameters. Thus, we have integrated these features into a mathematical model of tumor invasion, in which the tumor is considered as a mixture of tumor cells and healthy tissue, and growth is promoted by nutrient absorption from the culture medium. The tumor expands as a gradient flow with respect to the chemical potential, whose expression is calibrated to fit the experimental observation. In fact, the invasion characteristics of the in-vitro systems have been compared with the numerical results of the mathematical model. In particular, we found that the mechano-biological features, e.g., the competition between the friction with the substrate and the intercellular adhesion, play a key role in triggering the spontaneous formation and growth of three-dimensional cellular aggregates. Such clusters were found to obey a universality class following a peculiar dynamics, reaching a saturation regime characterized by a typical average size of the same order as the diffusion length of the nutrient in the medium (Figs. 12.2 and 12.3). On the basis of collected data, it is tempting to speculate that a similar process may occur in early or partial stages of EMT, independently of a full genetic and molecular rewiring of identity and further fostered by tumor heterogeneity. Alteration of cell-cell contacts seems, indeed, sufficient for the formation of collective clusters and their subsequent detachment from the primary tumor site, without an overt change in cell identity (i.e., a complete transition to mesenchymal state).

The in vitro experiments have been used to calibrate the mechano-biological parameters in a mathematical model [83]. In

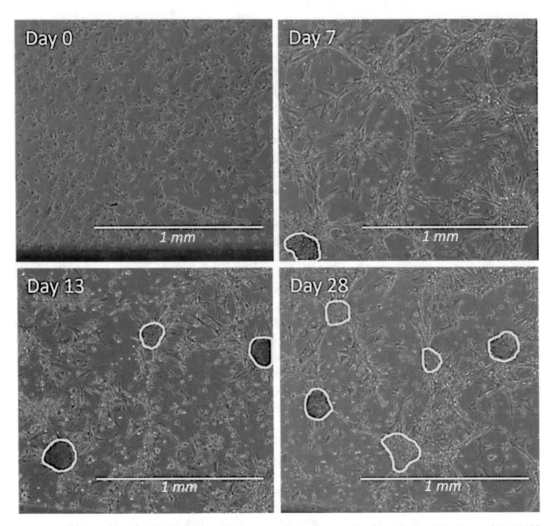

Fig. 12.2 [Tumor cell budding] Representative images of spontaneous budding in the 2D culture of U-87 MG cells, over time. Images were acquired with an EVOS wide-field microscope, using a 4X objective. The contour of the cellular aggregates has been highlighted with a yellow line. Scale-bar is 1 mm

the following, we further refine this model using clinical and radiological data from patients affected by GBM, developing a personalized computational tool that is able to predict its invasion and its response to treatments in a patient-specific manner.

2.2 Mechano-Biological Studies in GBM Patients

2.2.1 Clinical Protocol

We performed a prospective observational study in which patients were enrolled and evaluated in the context of the normal clinical practice. Patients suspected to have GBM, on the basis of radiological imaging, were considered for inclusion, with an estimated group of 30 patients in a period of 24 months. Inclusion criteria were: adult age (older than 18 years), suspected intracranial GBM, as suggested by MRI with a contrast agent, and surgical removal or biopsy of the lesions. Exclusion criteria were: (1) impossibility to

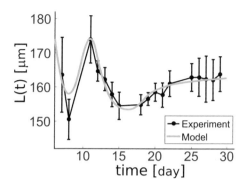

Fig. 12.3 [Budding dynamics] Evolution of U-87 MG spontaneous tumor budding. The chart shows the average size $L(t)$ of tumor aggregates over time. The black markers indicate experimental measures, obtained by tracing the clusters' contours as in Fig. 12.2. The green line depicts the prediction obtained by the numerical simulation of the mathematical model, showing a good quantitative agreement with experimental data

give consent due to cognitive deficits or language disorder; (2) women in their first trimester of pregnancy or lactation. The study period of the clinical part is of 36 months, including enrollment and follow up.

Screening visit was performed before surgery. Clinical and radiological evaluation included, as for routine institutional practice: physical examination, neurological examination, preoperative volumetric MRI including diffusion tensor imaging (DTI) (see below the radiological protocol), and recording of concomitant medications. Patients could be submitted to surgical removal of biopsy only. In both cases, the procedures were performed in a standard manner, with fluorescent visualization by 5-ALA or SF [76, 77]. The histological analysis of the biological samples taken during the surgery was performed according to the 2016 WHO classification [16].

Clinical and radiological postoperative examinations were carried out with the usual institutional practice. The early clinical evaluation, including a standard neurological and clinical examination, and the volumetric contrast-enhanced MRI for estimation of the extent of resection were made within 72 h from the surgical intervention. The protocol for early postoperative MRI was the same as performed in a setting without the DTI, which was excluded due to the possibility of artifacts caused by surgery; the following radiological examinations, performed every 2 months, were performed according to the same protocol of preoperative MRI with DTI.

All patients, upon confirmation of the histologic diagnosis of GBM, were offered adjuvant radio- and chemotherapy [4]; different dose fractioning and TMZ posology were established on the base of patient age, KPS, and methylation status of MGMT gene promoter, according to the EANO guideline [78].

2.2.2 *Radiological Protocol*

Radiological protocol included volumetric axial whole-brain T1-weighted MRI before and after intravenous gadolinium injection, useful to depict the structural anatomy of the patient's brain and to calculate the total volume of tumor extension after segmentation procedure; axial whole-brain 3D-FLAIR image at the high spatial resolution, useful to delineate the outline of the tumor and peri-tumor rim by suppressing signal from CSF.

DTI studies the diffusion of water molecules in tissues involving several directions of interrogation in order to provide information on anisotropic diffusion which can be used to derive numerous parameters. The DTI technique is nowadays the only noninvasive method to characterize the microstructural architecture of the brain bundles and to derive the preferential direction of water diffusion and of cell migration. For each patient, a set of 147 diffusion-weighted images DTI were acquired with different, specific parameters of acquisition; in particular, DTI is used to detect information regarding white tracts alignment and to estimate the parameters related to nutrients release and to the preferential direction of malignant cells motion (see below: *Integration of DTI data*).

2.2.3 *Integration of Clinical and Radiological Data in a Mechano-Biological Model*

We developed a computational platform to perform patient-specific simulations of tumor growth, surgical resection, recurrence, and response to adjuvant therapies by the means of an innovative integration of a mechano-biological mathematical model with neuro-imaging. The mathematical model considers that the brain tissue and the neoplasia are made of a cellular phase and a liquid phase, whose evolution over time is regulated by Newton's laws of motion. The cancer growth is coupled with the local availability of oxygen that diffuses anisotropically within the brain along the preferred principal directions extracted by DTI for each patient. We also assume that gliomas have a preferential growth direction given by the local orientation of white matter fibers, with a constitutive law depending on the eigenvalues extrapolated by the DTI data. Moreover, we include the inhibiting effects on GBM growth due to the application of radiotherapy and chemotherapy, following the clinical schedule of the Stupp protocol for each patient.

In the following, we resume the main feature of the developed computational platform that has been used to support the medical doctors in the clinical practice. Further discussion on the mathematical model with respect to the state-of-the-art in biomathematics of cancer can be found here [79–83].

The main steps are here summarized.

- *Medical segmentation.*

- We first used the MR sequences to create a computational grid reproducing the brain geometry of each patient at each clinical stage. We built an automated procedure for medical image segmentation, allowing to identify and to label the three brain tissues (i.e., gray matter, white matter, cerebrospinal (CSF) fluid) and the background. An illustrative example of the output is depicted in Fig. 12.4, showing the labeled tissues before and after surgery.

- *Mesh generation.*

- We later extract the boundary of the tumor and of the whole brain to generate the surface computational mesh, applying a preliminary smoothing as depicted in Fig. 12.5. This allows to create the computational tetrahedral mesh that is conveniently refined in the peri-tumor area. Each element of the mesh is then labeled using the previous medical segmentation of the brain tissues.

- Analyzing the MR data at each clinical stage, we quantify the corresponding area invaded by the tumor mass, registering the volumetric and morphological evolution over time.

- *Integration of DTI data.*

- We finally import the DTI tensor data using the same procedure used for tissue labeling to generate the six additional meshes corresponding to the six independent components of the diffusion tensor that will be used in the model for reconstructing the local diffusivity of the oxygen and the preferential direction of tumor growth. The DTI images are associated with the T1-MRI data in the process of constructing the computational mesh of the brain, so that the corresponding DTI data of each specific voxel are labeled in the computational mesh. An illustrative example is given in Fig. 12.6.

- *Numerical simulations.*

- The previous steps allowed to reconstruct the patient-specific brain geometry into a virtual environment that is used for an in-silico simulation of the mathematical growth model for assisting medical doctors in evaluating different approaches during the clinical treatment. The numerical computation is performed using the finite element method. The model parameters that could not be extracted by the patient's data are taken within the range extracted from the medical literature, using an iterative technique allowing to select the physiological value that better approaches the tumor growth rate observed clinically. When dealing with the simulation of tumor recurrence, we compare the T1-weighted and FLAIR MR images before and after surgery searching for residual masses of tumor infiltrated in the healthy tissue. We later simulate the recurrent

Pre Surgery

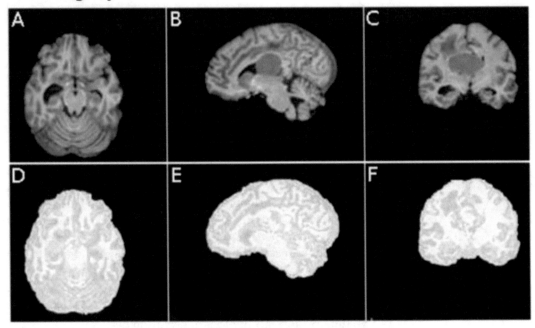

Post Surgery

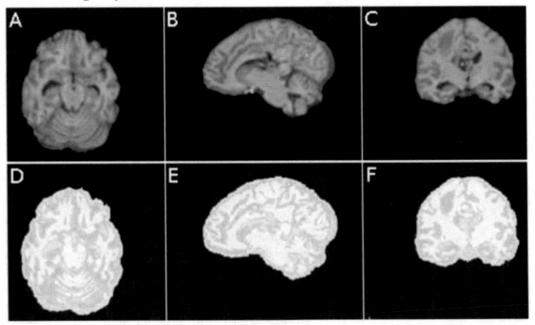

Fig. 12.4 [Segmentation of clinical images] An illustrative example of medical image segmentation inputs and outputs: T1-weighted MR images with segmented maps superposed at the before surgery (pre-surgery) and after surgery (post-surgery) events. Pre-surgery: axial (**a**), sagittal (**b**), and coronal (**c**) slices of T1-weighted MR images, together with the corresponding axial (**d**), sagittal (**e**), and coronal (**f**) segmented map of the brain tissues. Post-surgery: axial (**a**), sagittal (**b**), and coronal (**c**) slices of T1-weighted MR images, together with the corresponding axial (**d**), sagittal (**e**), and coronal (**f**) segmented map of the brain tissues. White matter, grey matter, and CSF are highlighted in white, grey, and blue colors respectively. In Figs. (**b**) and (**c**), the segmented map of the tumor mass before surgery is highlighted in red color

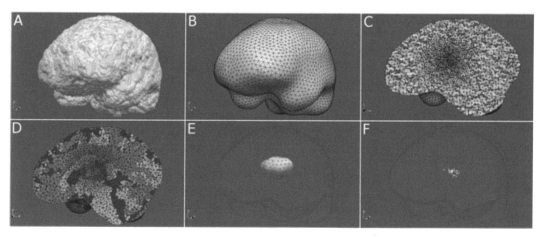

Fig. 12.5 [Computational brain] Representative images of the mesh generation process. (**a**) External brain surface extracted from the segmented map obtained by the medical images. (**b**) Smoothed and meshed external surface. (**c**) Computational tetrahedral mesh generated within the external brain surface, conveniently refined in the peritumoral area. (**d**) Labeled mesh obtained from the refined mesh, each element of which is labeled using the previous medical segmentation of the brain tissues: grey matter (grey), white matter (white), CSF (blue). (**e**) External brain (blue meshed surface) and tumor (light grey) surfaces obtained from the segmented maps before surgery. (**f**) External brain (blue meshed surface) and tumor (light grey) extension after surgery, obtained from the analysis of MR images as the residual mass of tumor infiltrated in the healthy tissue

growth and the response to therapy of these tumor aggregates. An illustrative simulation result of tumor recurrence is depicted in Fig. 12.7, showing the invading tumor mass over time after the surgery event, which is considered as day 0, and during the adjuvant therapy.

– *Benchmark results.*

– The results of the numerical simulations of the model have been found in very good qualitative and quantitative agreement with the clinical observation in benchmark clinical cases. We found that the local brain microstructure encoded in the MRI and DTI data is indeed correlated with the spatial invasive potential of GBM, as confirmed by the well-studied butterfly shape in tumor developed around the corpus callosum.

An illustrative example of the predictive ability of the simulated tumor growth patterns against clinical neuroimaging data acquired after adjuvant therapy is shown in Fig. 12.8. The simulated results not only capture the immediate inhibiting effect of therapies but also the long-term explosion of the tumor growth rate, as depicted in Fig. 12.9. The predictive ability of the model with respect to the tumor extension (expressed in terms of the Jaccard index, which measures the overall overlap between the tumor extensions from simulation and MRI data) is one of the highest found up to this date in biomathematical literature.

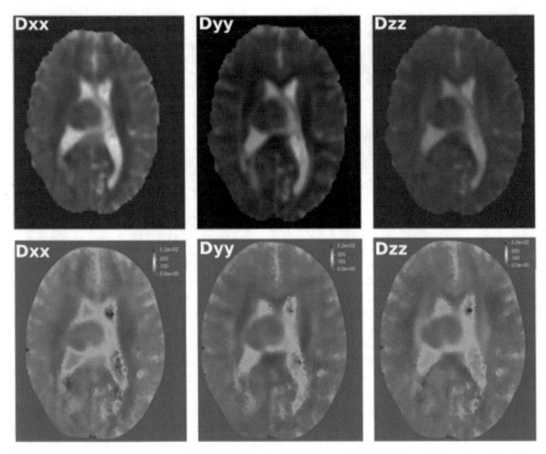

Fig. 12.6 [Computational DTI reconstruction] Representative images of the independent components of the diffusion tensor obtained from the DTI tensor data and of the additional meshes associated with the same components: axial slices of the diagonal components of the diffusion tensor, Dxx (**a**), Dyy (**b**), Dzz (**c**), with the corresponding meshes associated with the same components Dxx (**d**), Dyy (**e**), Dzz (**f**)

3   Conclusions and Future Perspectives

GBM is characterized by a highly complex pathology with a wide genotypic and phenotypic variability. Recently, many efforts have been aimed to improve the comprehension of the key mechano-biological features directing the inception and the development of such a disease. In this chapter, we applied a new model which represents a remarkable advancement with respect to the state-of-the-art GBM models because it takes into account the heterogeneity and the anisotropy of the brain tissues thanks to the introduction of a chemotactic flux of mass and the patient-specific diffusion tensor imaging data. In vitro experiments have been used to calibrate the mechano-biological parameters controlling the growth and invasion properties of the cell aggregates. MRI data were used in order to extract the information about the three principal tissues of the brain, white and grey matter and CSF, the shape of the brain

Time: 5.70   Days   Time: 25.65   Days   Time: 34.20   Days

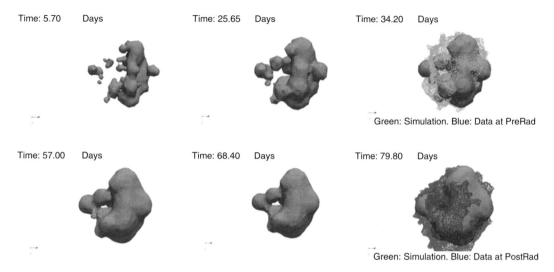

Green: Simulation. Blue: Data at PreRad

Time: 57.00   Days   Time: 68.40   Days   Time: 79.80   Days

Green: Simulation. Blue: Data at PostRad

Fig. 12.7 [Simulation results] Illustrative simulation results of tumor recurrence over time at different evolution stages after the surgery event (which is taken at day 0). The simulated tumor contour is highlighted in green color. The observed tumor boundaries from MRI data before radiotherapy (PreRad) and after radiotherapy (PostRad) are highlighted as blue and red meshed surfaces respectively

itself, and also to approximate position and shape of the GBM whereas the DTI clinical data were used to calculate the diffusion tensor of the nutrient and the tensor describing the preferential movement directions for the tumor cells. Therefore, the model could be ideally used to reproduce the recurrence of GBM after surgical resection and the effect of subsequent cycles of radiotherapy and chemotherapy treatments.

Preliminary data derived from the application of this study are suggestive in predicting the observed GBM behavior with a high degree of agreement (Figs. 12.7–12.9) showing the feasibility of the use of this mathematical model. From a translational point of view, this work purposes a method which will allow to know in advance the behavior of GBM, from its first diagnosis to its surgical removal and the subsequent adjuvant therapies.

For more than a decade, the Stupp regimen still remained the gold standard of GBM treatments, without any improvements in prognosis despite the large amount of pre-clinical and clinical researches in this field. Indeed, it is mandatory to look for the application of a personalized model in order to personalize the therapies: early knowledge of the timing and modality of tumor regrowth, in relation to the amount of surgical excision, could guide and influence some clinical choices, such as extending surgical resection, developing targeted local chemotherapies, and personalizing radiotherapy.

The principal limitation of this model is the lack of consideration of data deriving from molecular biology; the GBM is not a unique disease but it is characterized by a lot of genetic alterations,

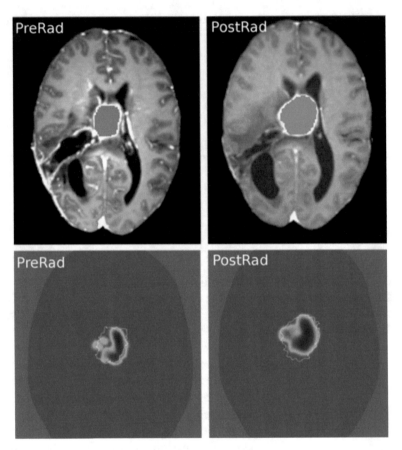

Fig. 12.8 [Simulation results versus neuroimaging] Representative images of the simulated tumor growth patterns against clinical neuroimaging data acquired before and after adjuvant therapy in a benchmark clinical case. Axial slices of the T1-weighted MR images at the before radiotherapy (PreRad) and after radiotherapy (PostRad) events are reported in the top panel, overlapped with the segmentation maps of the tumor extension, highlighted in blue (PreRad) and red (PostRad) colors. These medical data are compared to the corresponding axial slices of the tumor extension calculated from the simulations at the PreRad and PostRad events (bottom panels), overlapped with the contours (white lines) of the tumor extension from the corresponding MR images

molecular pathways, and peculiarities. Nevertheless, our model forecasts neither the possibility of varying prognostic factors nor modifying the estimated velocity of regrowth. Another important limitation is the lack of differentiation of the patient's age: in particular, the elderly constitutes about half of the GBM population and the incidence of HGG in the elderly is increasing; their median survival is 6 months regardless of the treatment, which is much shorter of the general GBM population.

As concerning future works, it will be necessary to include the bio-molecular data in the mathematical model in order to better correlate the different prognostic factors and genetic characteristics

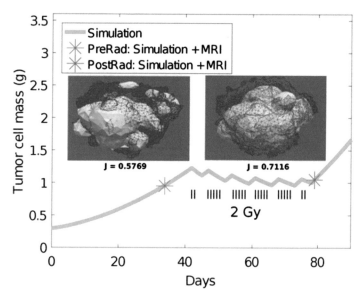

Fig. 12.9 [Simulated tumor dynamics] Time evolution of the tumor mass (green line) from the after surgery to the after radiotherapy (PostRad) events in a benchmark clinical case, in which a radiotherapy treatment 2 Gy/day, indicated by black segments, was applied. The insets display the simulated (light grey) and the observed tumor boundaries from MRI data (blue surface in before radiotherapy (PreRad) event, red surface in PostRad event). An indication of the Jaccard index J at the PreRad and PostRad events is also reported

to GBM growth simulation, also taking into account the patient's age. Furthermore, we will focus on refining the methods for fitting the model parameters through constrained optimization problems, and on the development of automated tumor segmentation tool, in order to avoid the manual segmentation that can introduce a bias and a human error into the process. Finally, whereas the proposed model considers a homogeneous distribution of blood vessels, through the nutrient supply term in the reaction-diffusion equation, thus neglecting the role of angiogenesis in GBM development, which is a hallmark of the disease, future refinements should consider a patient-specific nutrient supply term, considering both that physiologically blood vessels lie mostly in grey matter and the dynamic tumor neo-angiogenesis, elaborating data on brain perfusion and vessel location, from PWI techniques.

In conclusion, this model represents one more step toward the definition of a computational tool combining clinical data with a mathematical model, which is able to capture both chemical and mechanical phenomena driving GBM evolution. This multidisciplinary approach has the potential to help the clinicians in the definition of better therapeutic strategies in personalized oncology.

Acknowledgments

This work has been funded by the Associazione Italiana per la Ricerca sul Cancro (AIRC) through the MFAG grant 17412.

References

1. Ostrom QT, Gittleman H, Truitt G, Boscia A, Kruchko C, Barnholtz-Sloan JS (2018) CBTRUS statistical report: primary brain and other central nervous system tumors diagnosed in the United States in 2011–2015. Neuro-Oncology 20:iv1–iv86

2. Leece R, Xu J, Ostrom QT, Chen Y, Kruchko C, Barnholtz-Sloan JS (2017) Global incidence of malignant brain and other central nervous system tumors by histology, 2003–2007. Neuro-Oncology 19:1553–1564

3. Glaser SM, Dohopolski MJ, Balasubramani GK, Flickinger JC, Beriwal S (2017) Glioblastoma multiforme (GBM) in the elderly: initial treatment strategy and overall survival. J Neuro-Oncol 134:107–118

4. Stupp R, Mason WP, van den Bent MJ et al (2005) Radiotherapy plus concomitant and adjuvant temozolomide for glioblastoma. N Engl J Med 352:987–996

5. Wen PY, Macdonald DR, Reardon DA et al (2010) Updated response assessment criteria for high-grade gliomas: response assessment in neuro-oncology working group. J Clin Oncol 28:1963–1972

6. Cote DJ, Downer MK, Smith TR, Smith-Warner SA, Egan KM, Stampfer MJ (2018) Height, waist circumference, body mass index, and body somatotype across the life course and risk of glioma. Cancer Causes Control 29:707–719

7. Yang M, Guo W, Yang C, Tang J, Huang Q, Feng S, Jiang A, Xu X, Jiang G (2017) Mobile phone use and glioma risk: a systematic review and meta-analysis. PLoS One 12:e0175136

8. Wang Y, Guo X (2016) Meta-analysis of association between mobile phone use and glioma risk. J Cancer Res Ther 12:298

9. Yiin JH, Ruder AM, Stewart PA et al (2012) The upper Midwest health study: a case–control study of pesticide applicators and risk of glioma. Environ Health 11:39

10. Giles GG, McNeil JJ, Donnan G, Webley C, Staples MP, Ireland PD, Hurley SF, Salzberg M (1994) Dietary factors and the risk of glioma in adults: results of a case-control study in Melbourne, Australia. Int J Cancer 59:357–362

11. Ohgaki H, Kleihues P (2005) Population-based studies on incidence, survival rates, and genetic alterations in astrocytic and oligodendroglial gliomas. J Neuropathol Exp Neurol. https://doi.org/10.1093/jnen/64.6.479

12. Dubrow R, Darefsky AS, Park Y et al (2010) Dietary components related to N-Nitroso compound formation: a prospective study of adult glioma. Cancer Epidemiol Biomark Prev 19:1709–1722

13. Holick CN, Giovannucci EL, Rosner B, Stampfer MJ, Michaud DS (2007) Prospective study of cigarette smoking and adult glioma: dosage, duration, and latency. Neuro-Oncology 9:326–334

14. Silvera SAN, Miller AB, Rohan TE (2006) Cigarette smoking and risk of glioma: a prospective cohort study. Int J Cancer 118:1848–1851

15. Ohgaki H (2005) Genetic pathways to glioblastomas. Neuropathology 25:1–7

16. Louis DN, Perry A, Reifenberger G, von Deimling A, Figarella-Branger D, Cavenee WK, Ohgaki H, Wiestler OD, Kleihues P, Ellison DW (2016) The 2016 World Health Organization classification of tumors of the central nervous system: a summary. Acta Neuropathol. https://doi.org/10.1007/s00401-016-1545-1

17. Shirahata M, Ono T, Stichel D et al (2018) Novel, improved grading system(s) for IDH-mutant astrocytic gliomas. Acta Neuropathol 136:153–166.

18. Alifieris C, Trafalis DT (2015) Glioblastoma multiforme: pathogenesis and treatment. Pharmacol Ther. https://doi.org/10.1016/j.pharmthera.2015.05.005

19. Cancer Genome Atlas Research Network, Brat DJ, RGW V et al (2015) Comprehensive, integrative genomic analysis of diffuse lower-grade gliomas. N Engl J Med 372:2481–2498

20. Forsyth PA, Posner JB (1993) Headaches in patients with brain tumors: a study of 111 patients. Neurology. https://doi.org/10.1212/WNL.43.9.1678

21. Snyder H, Robinson K, Shah D, Brennan R, Handrigan M (1993) Signs and symptoms of patients with brain tumors presenting to the

emergency department. J Emerg Med. https://doi.org/10.1016/0736-4679(93)90042-6

22. Ertürk Çetin Ö, İşler C, Uzan M, Özkara Ç (2017) Epilepsy-related brain tumors. Seizure. https://doi.org/10.1016/j.seizure.2016.12.012

23. Alexander BM, Cloughesy TF (2017) Adult glioblastoma. J Clin Oncol. https://doi.org/10.1200/JCO.2017.73.0119

24. Mabray MC, Barajas RF, Cha S (2015) Modern brain tumor imaging. Brain Tumor Res Treat. https://doi.org/10.14791/btrt.2015.3.1.8

25. Cha S (2006) Update on brain tumor imaging: from anatomy to physiology. Am J Neuroradiol. https://doi.org/10.1007/s11910-005-0044-x

26. Horská A, Barker PB (2010) Imaging of brain tumors: MR spectroscopy and metabolic imaging. Neuroimaging Clin N Am. https://doi.org/10.1016/j.nic.2010.04.003

27. Chiang GC, Kovanlikaya I, Choi C, Ramakrishna R, Magge R, Shungu DC (2018) Magnetic resonance spectroscopy, positron emission tomography and radiogenomics-relevance to glioma. Front Neurol. https://doi.org/10.3389/fneur.2018.00033

28. Alongi P, Vetrano IG, Fiasconaro E, Alaimo V, Laudicella R, Bellavia M, Rubino F, Bagnato S, Galardi G (2018) Choline-PET/CT in the differential diagnosis between cystic glioblastoma and intraparenchymal hemorrhage. Curr Radiopharm. https://doi.org/10.2174/1874471011666180817122427

29. Stupp R, Mason W, van den Bent MJ et al (2005) Radiotherapy plus concomitant and adjuvant temozolomide for glioblastoma. N Engl J Med. https://doi.org/10.1056/NEJMoa043330

30. Chaichana KL, Jusue-Torres I, Navarro-Ramirez R et al (2014) Establishing percent resection and residual volume thresholds affecting survival and recurrence for patients with newly diagnosed intracranial glioblastoma. Neuro-Oncology. https://doi.org/10.1093/neuonc/not137

31. Grabowski MM, Recinos PF, Nowacki AS, Schroeder JL, Angelov L, Barnett GH, Vogelbaum MA (2014) Residual tumor volume versus extent of resection: predictors of survival after surgery for glioblastoma. J Neurosurg. https://doi.org/10.3171/2014.7.JNS132449

32. Schulz C, Waldeck S, Mauer UM (2012) Intraoperative image guidance in neurosurgery: development, current indications, and future trends. Radiol Res Pract. https://doi.org/10.1155/2012/197364

33. Cavallo C, De Laurentis C, Vetrano IG, Falco J, Broggi M, Schiariti M, Ferroli P, Acerbi F (2018) The utilization of fluorescein in brain tumor surgery: a systematic review. J Neurosurg Sci. https://doi.org/10.23736/S0390-5616.18.04480-6

34. Prada F, Perin A, Martegani A et al (2014) Intraoperative contrast-enhanced ultrasound for brain tumor surgery. Neurosurgery. https://doi.org/10.1227/NEU.0000000000000301

35. Prada F, Bene MD, Fornaro R et al (2016) Identification of residual tumor with intraoperative contrast-enhanced ultrasound during glioblastoma resection. Neurosurg Focus. https://doi.org/10.3171/2015.11.FOCUS15573

36. Senft C, Bink A, Franz K, Vatter H, Gasser T, Seifert V (2011) Intraoperative MRI guidance and extent of resection in glioma surgery: a randomised, controlled trial. Lancet Oncol. https://doi.org/10.1016/S1470-2045(11)70196-6

37. Stummer W, Pichlmeier U, Meinel T, Wiestler OD, Zanella F, Reulen HJ (2006) Fluorescence-guided surgery with 5-aminolevulinic acid for resection of malignant glioma: a randomised controlled multicentre phase III trial. Lancet Oncol. https://doi.org/10.1016/s1470-2045(06)70665-9

38. Acerbi F, Broggi M, Schebesch K-M et al (2018) Fluorescein-guided surgery for resection of high-grade gliomas: a multicentric prospective phase II study (FLUOGLIO). Clin Cancer Res 24:52–61

39. Beauchesne P, Bernier V, Carnin C et al (2010) Prolonged survival for patients with newly diagnosed, inoperable glioblastoma with 3-times daily ultrafractionated radiation therapy. Neuro-Oncology 12:595–602

40. Hegi ME, Diserens A-C, Gorlia T et al (2005) *MGMT* gene silencing and benefit from temozolomide in glioblastoma. N Engl J Med. https://doi.org/10.1056/NEJMoa043331

41. Perry JR, Laperriere N, O'Callaghan CJ et al (2017) Short-course radiation plus temozolomide in elderly patients with glioblastoma. N Engl J Med. https://doi.org/10.1056/NEJMoa1611977

42. Thomas AA, Brennan CW, DeAngelis LM, Omuro AM (2014) Emerging therapies for glioblastoma. JAMA Neurol. https://doi.org/10.1001/jamaneurol.2014.1701

43. Lugassy C, Haroun RI, Brem H, Tyler BM, Jones RV, Fernandez PM, Patierno SR, Kleinman HK, Barnhill RL (2002) Pericytic-like angiotropism of glioma and melanoma cells. Am J Dermatopathol 24:473–478

44. Winkler F, Kienast Y, Fuhrmann M, Von Baumgarten L, Burgold S, Mitteregger G, Kretzschmar H, Herms J (2009) Imaging glioma cell invasion in vivo reveals mechanisms of dissemination and peritumoral angiogenesis. Glia 57:1306–1315

45. Zagzag D, Amirnovin R, Greco MA, Yee H, Holash J, Wiegand SJ, Zabski S, Yancopoulos GD, Grumet M (2000) Vascular apoptosis and involution in gliomas precede neovascularization: a novel concept for glioma growth and angiogenesis. Lab Investig 80:837–849

46. Cuddapah VA, Robel S, Watkins S, Sontheimer H (2014) A neurocentric perspective on glioma invasion. Nat Rev Neurosci. https://doi.org/10.1038/nrn3765

47. Ulrich TA, de Juan Pardo EM, Kumar S (2009) The mechanical rigidity of the extracellular matrix regulates the structure, motility, and proliferation of glioma cells. Cancer Res 69:4167–4174

48. Miroshnikova YA, Mouw JK, Barnes JM et al (2016) Tissue mechanics promote IDH1-dependent HIF1α-tenascin C feedback to regulate glioblastoma aggression. Nat Cell Biol 18:1336–1345

49. Jain RK (1999) Transport of molecules, particles, and cells in solid Tumors. Annu Rev Biomed Eng 1:241–263

50. Netti PA, Berk DA, Swartz MA, Grodzinsky AJ, Jain RK (2000) Role of extracellular matrix assembly in interstitial transport in solid tumors. Cancer Res 60:2497–2503

51. Padera TP, Stoll BR, Tooredman JB, Capen D, di Tomaso E, Jain RK (2004) Pathology: cancer cells compress intratumour vessels. Nature 427:695

52. Pickup MW, Mouw JK, Weaver VM (2014) The extracellular matrix modulates the hallmarks of cancer. EMBO Rep 15:1243–1253

53. Nicholson C, Syková E (1998) Extracellular space structure revealed by diffusion analysis. Trends Neurosci 21:207–215

54. Ruoslahti E (1996) Brain extracellular matrix. Glycobiology 6:489–492

55. Zimmermann DR, Dours-Zimmermann MT (2008) Extracellular matrix of the central nervous system: from neglect to challenge. Histochem Cell Biol 130:635–653

56. Brösicke N, Faissner A (2015) Role of tenascins in the ECM of gliomas. Cell Adhes Migr 9:131–140

57. Kim Y, Kumar S (2014) CD44-mediated adhesion to hyaluronic acid contributes to mechanosensing and invasive motility. Mol Cancer Res 12:1416–1429

58. Knüpfer MM, Poppenborg H, Hotfilder M, Kühnel K, Wolff JE, Domula M (1999) CD44 expression and hyaluronic acid binding of malignant glioma cells. Clin Exp Metastasis 17:71–76

59. Barnes JM, Kaushik S, Bainer RO et al (2018) A tension-mediated glycocalyx-integrin feedback loop promotes mesenchymal-like glioblastoma. Nat Cell Biol 20:1203–1214

60. Zhang Y, Brady M, Smith S (2001) Segmentation of brain MR images through a hidden Markov random field model and the expectation-maximization algorithm. IEEE Trans Med Imaging. https://doi.org/10.1109/42.906424

61. Deisboeck TS, Berens ME, Kansal AR, Torquato S, Stemmer-Rachamimov AO, Chiocca EA (2001) Pattern of self-organization in tumour systems: complex growth dynamics in a novel brain tumour spheroid model. Cell Prolif. https://doi.org/10.1046/j.1365-2184.2001.00202.x

62. Cristini V, Lowengrub J (2010) Multiscale modeling of cancer: an integrated experimental and mathematical modeling approach. Multiscale Model Cancer An Integr Exp Math Model Approach doi: https://doi.org/10.1017/CBO9780511781452

63. Lowengrub J, Nie Q, Cristini V, Li X (2007) Nonlinear three-dimensional simulation of solid tumor growth. Discret Contin Dyn Syst – Ser B. https://doi.org/10.3934/dcdsb.2007.7.581

64. De Craene B, Berx G (2013) Regulatory networks defining EMT during cancer initiation and progression. Nat Rev Cancer 13:97–110

65. Przybyla L, Muncie JM, Weaver VM (2016) Mechanical control of epithelial-to-Mesenchymal transitions in development and cancer. Annu Rev Cell Dev Biol 32:527–554

66. Lambert AW, Pattabiraman DR, Weinberg RA (2017) Emerging biological principles of metastasis. Cell 168:670–691

67. Sánchez-Tilló E, Liu Y, de Barrios O, Siles L, Fanlo L, Cuatrecasas M, Darling DS, Dean DC, Castells A, Postigo A (2012) EMT-activating transcription factors in cancer: beyond EMT and tumor invasiveness. Cell Mol Life Sci 69:3429–3456

68. Pease JC, Brewer M, Tirnauer JS (2012) Spontaneous spheroid budding from monolayers: a potential contribution to ovarian cancer dissemination. Biol Open 1:622–628

69. Zajac O, Raingeaud J, Libanje F et al (2018) Tumour spheres with inverted polarity drive the formation of peritoneal metastases in

patients with hypermethylated colorectal carcinomas. Nat Cell Biol 20:296–306

70. Kunjithapatham R, Karthikeyan S, Geschwind J-F, Kieserman E, Lin M, Fu D-X, Ganapathy-Kanniappan S (2014) Reversal of anchorage-independent multicellular spheroid into a monolayer mimics a metastatic model. Sci Rep 4:6816

71. Shimada N, Saito M, Shukuri S, Kuroyanagi S, Kuboki T, Kidoaki S, Nagai T, Maruyama A (2016) Reversible monolayer/spheroid cell culture switching by UCST-type thermoresponsive ureido polymers. ACS Appl Mater Interfaces 8:31524–31529

72. Pérez-González C, Alert R, Blanch-Mercader C, Gómez-González M, Kolodziej T, Bazellieres E, Casademunt J, Trepat X (2019) Active wetting of epithelial tissues. Nat Phys 15(1):79-88

73. Ledur PF, Villodre ES, Paulus R, Cruz LA, Flores DG, Lenz G (2012) Extracellular ATP reduces tumor sphere growth and cancer stem cell population in glioblastoma cells. Purinergic Signal 8:39–48

74. Yu S, Ping Y, Yi L, Zhou Z, Chen J, Yao X, Gao L, Wang JM, Bian X (2008) Isolation and characterization of cancer stem cells from a human glioblastoma cell line U87. Cancer Lett 265:124–134

75. Zhou Y, Zhou Y, Shingu T, Feng L, Chen Z, Ogasawara M, Keating MJ, Kondo S, Huang P (2011) Metabolic alterations in highly tumorigenic glioblastoma cells: preference for hypoxia and high dependency on glycolysis. J Biol Chem 286:32843–32853

76. Stummer W, Stepp H, Wiestler OD, Pichlmeier U (2017) Randomized, prospective double-blinded study comparing 3 different doses of 5-aminolevulinic acid for fluorescence-guided resections of malignant gliomas. Neurosurgery. https://doi.org/10.1093/neuros/nyx074

77. Acerbi F, Broggi M, Broggi G, Ferroli P (2015) What is the best timing for fluorescein injection during surgical removal of high-grade gliomas? Acta Neurochir. https://doi.org/10.1007/s00701-015-2455-z

78. Weller M, van den Bent M, Tonn JC et al (2017) European Association for Neuro-Oncology (EANO) guideline on the diagnosis and treatment of adult astrocytic and oligodendroglial gliomas. Lancet Oncol 18:e315–e329

79. Colombo MC, Giverso C, Faggiano E, Boffano C, Acerbi F, Ciarletta P (2015) Towards the personalized treatment of glioblastoma: integrating patient-specific clinical data in a continuous mechanical model. PLoS One. https://doi.org/10.1371/journal.pone.0132887

80. Agosti A, Antonietti PF, Ciarletta P, Grasselli M, Verani M (2017) A Cahn-Hilliard-type equation with application to tumor growth dynamics. Math Methods Appl Sci 40:7598–7626

81. Agosti A, Cattaneo C, Giverso C, Ambrosi D, Ciarletta P (2018) A computational framework for the personalized clinical treatment of glioblastoma multiforme. ZAMM – J Appl Math Mech/Zeitschrift für Angew Math und Mech 98:2307–2327

82. Agosti A, Giverso C, Faggiano E, Stamm A, Ciarletta P, Agosti A, Giverso C, Faggiano E, Stamm A (2018) A personalized mathematical tool for neuro-oncology: a clinical case study. Int J Non Linear Mech 107:170–181

83. Agosti A, Marchesi S, Scita G, Ciarletta P (2020) Modelling cancer cell budding in-vitro as a self-organised, non-equilibrium growth process. J Theor Biol 492:110203

Chapter 13

Magnetic Resonance Imaging for Quantification of Brain Vascular Perfusion

Line Brennhaug Nilsen and Kyrre Eeg Emblem

Abstract

Perfusion refers to the biological process of blood flow through vascularized tissue and allows for sufficient delivery of vital nutrients to most organs in the body as well as removal of metabolic waste and heat. Thus, perfusion plays a critical role in determining physiological levels of oxygenation, bioenergetics status, and pH distributions. Neoplasms in the brain, including brain tumors, are typically characterized by irregular and insufficient perfusion from abnormal neo-vasculature. Consequently, brain tumors create a hostile microenvironment that promotes tumor aggressiveness and treatment resistance. Moreover, treatment-induced changes in physiological functions, such as perfusion, may occur rapidly and before any measurable reduction in tumor volume. Hence, spatial and temporal assessment of quantitative perfusion metrics is an ideal target for diagnosis and treatment response monitoring of brain tumors. While conventional magnetic resonance imaging (MRI) remains the gold standard for non-invasive characterization of tumors of the central nervous system (CNS), quantitative measures of perfusion by dynamic susceptibility contrast (DSC)-MRI and arterial spin labeling (ASL) have helped advance cancer imaging as a non-invasive diagnostic force in the fight against cancer. Here, we review DSC-MRI and ASL, and their current and potential use in the clinical management of brain tumors.

Key words Perfusion, Vascular function, Cerebral blood flow (CBF), Cerebral blood volume (CBV), Dynamic susceptibility contrast (DSC), Magnetic resonance imaging (MRI), Arterial spin labeling (ASL)

1 Introduction

Tumor angiogenesis, a long-recognized hallmark of cancer [1], is essential for tumor growth while at the same time contributing to the development of a dysfunctional, but highly aggressive vascular network in solid tumors. The tumor neo-vasculature is characterized by high density of immature, leaky, and disorganized vessels with increased inter-vessel distances [2, 3]. For brain tumors, increased tumor angiogenesis and disrupted blood-brain barrier (BBB) usually reflect higher tumor aggressiveness [4]. The abnormal vasculature leads to impaired perfusion and a dysfunctional

Giorgio Seano (ed.), *Brain Tumors*, Neuromethods, vol. 158,
https://doi.org/10.1007/978-1-0716-0856-2_13, © Springer Science+Business Media, LLC, part of Springer Nature 2021

microenvironment, characterized by severe acidity, high interstitial pressure, as well as chronic and acute hypoxia [2, 5]. These conditions contribute to reduced sensitivity for common anti-cancer treatments of brain tumors, including chemotherapy, post-operative chemo-radiation, stereotactic radiosurgery (SRS), as well as anti-angiogenic therapy and immunotherapy [6, 7]. Moreover, aggressive and metastatic brain tumors usually receive a combination of these treatments, leading to accelerated and complex changes in the tumor- and peri-tumoral vasculature. Therefore, frequent monitoring of vascular efficacy may be crucial for achieving optimal balance between treatment outcome and potential harmful side-effects [8].

Taken together, perfusion dictates structural and functional brain tumor characteristics. Optimal and individualized management of brain tumors, including tumor characterization and response monitoring, are therefore dependent on non-invasive methods that can capture both spatial and temporal changes in vascular function. By these means, quantitative perfusion metrics may also help develop agents targeting tumor vasculature [9].

1.1 MRI and Assessment of Brain Vascular Function

Conventional MRI, in particular contrast-enhanced T1-weighted and T2-weighted fluid-attenuated inversion recovery (FLAIR) MRI, play essential roles in pre- and post-treatment imaging of brain tumors [10]. The increased contrast enhancement of brain tumors on T1-weighted images following the injection of an external contrast agent stems from a compromised or defective BBB. Conventional MRI allows for superior morphological evaluation, including delineation of anatomical margins, and founds the basis for response assessment in neuro-oncological imaging. Initially, response criteria for solid brain tumors were solely based on changes in tumor size obtained from bidirectional size measures of contrast enhancement on T1-weighted images, but later refined to account for use of steroids and clinical status in 1990 [11]. However, during the past two decades, the development of novel vascular- and immune-targeting therapies have resulted in complex expressions of contrast-enhancement patterns on post-treatment MRI. For instance, effective immunotherapy may cause increased BBB disruption [12, 13]. This can lead to enhanced leakage of the contrast agent into the extravascular space, ultimately depicted as increased regions of contrast-enhancing tissue. This phenomenon can easily be misinterpreted as tumor progression. Therefore, the need for improved response criteria—beyond structural changes—has shifted research focus toward also including information on physiological function, such as perfusion [14]. Still, while including T2/FLAIR signal change, current consensus guidelines by the Response Assessment in Neuro-Oncology (RANO) working group for high-grade glioma [15] and brain metastases [16] do not yet incorporate metrics for assessment of physiological function. Therefore, the criteria are associated with several shortcomings, particularly related to the use of anti-angiogenic therapy and immunotherapy, alone or in combination with radiotherapy.

In MRI, three techniques are arguably referred to as perfusion-weighted imaging (PWI): dynamic contrast-enhanced (DCE)-MRI, DSC-MRI, and ASL. While DSC-MRI and ASL do provide measures of perfusion, DCE-MRI instead measures trans-capillary permeability by quantification of contrast agent based transfer constants and the volume of the extravascular extracellular space (EES) [17]. This is a separate process that may or may not depend on perfusion in a very complex manner [18]. Diffusion-weighted (DW)-MRI, which has gained a significant position in neuro-oncology imaging, is another MRI-based technique that may also provide estimation of micro-vessel perfusion by means of intra-voxel incoherent motion (IVIM) without the use of contrast agent [19]. Here we will focus on perfusion metrics that can be measured by DSC-MRI and ASL, and their current use and potential clinical use in brain tumors.

2   Technical Background

Here we provide an overview of the physical principles behind DSC-MRI and ASL, together with the use of tracer kinetic modeling[1] for quantification of perfusion-related metrics. A more in-depth presentation of the physical principles of the techniques is outside the scope of this chapter, and can be found elsewhere for DSC-MRI [20–24] and ASL [25–28].

2.1   Dynamic Susceptibility Contrast (DSC) Magnetic Resonance Imaging (MRI)

2.1.1   Physical Principles

In DSC-MRI, an externally injected paramagnetic contrast agent is used to induce local magnetic susceptibility[2] effects, leading to shortening of T2/T2* relaxation times and transiently reduced MR signal [23, 29, 30]. Clinically approved contrast agents are typically gadolinium-based, and are injected intravenously and rapidly to produce a compact bolus of the tracer through the organ of interest. The transport of the contrast agent bolus is in order of seconds, thus rapid T2 or T2* Echo Planar Imaging (EPI) sequences are acquired before, during, and after the contrast agent injection. In the brain, the EPI technique provides sufficient speed

---

[1] Tracer kinetic models are based on the tracking of a bolus of tracer through a region of interest which produces changes in tissue relaxations times, ultimately altering the signal intensity. The measured alteration in signal intensity depends on the transient presence of the tracer.

[2] Magnetic susceptibility is a dimensionless proportionality constant that indicates the degree of magnetization of a material in response to an applied magnetic field.

for adequate coverage while being sensitive to the accelerated T2/ T2* relaxation times of the contrast agent. To obtain image contrast from T2 effects, the spin-echo (SE) readout scheme with its 180° refocusing pulse is used, whereas the gradient-echo (GE) scheme is used to image corresponding T2* effects. Gradient-echo imaging provides considerably higher signal-to-noise ratio (SNR) than spin-echo and is therefore commonly preferred in clinical practice.

2.1.2 Kinetic Modeling— Quantitative Perfusion Metrics

The traditional kinetic models used to generate perfusion-based images from DSC-MRI assume an inverse linear relationship between the increase in the concentration of the injected tracer and the corresponding drop in the MR signal (the indicator-dilution theory) [20, 31]. Under the assumption of a mono-exponential signal decay with increasing relaxation rates $\Delta R2$ for SE (and $\Delta R2^*$ for GE), the MR signal as a function of time, $S(t)$, is converted into a new signal curve depicting the concentration as a function of time, $C(t)$;

$$C(t) = -\frac{k}{\text{TE}} \ln\left(\frac{S(t)}{S(0)}\right) \propto \Delta R_2(t) \tag{1}$$

where k is an unknown proportionality constant that for simplicity is set independent of vessel size, vascular geometry, and capillary density [22], and TE is the echo time (image sampling time). The function $S(0)$ represents the initial MR signal at baseline, i.e., before the contrast agent bolus arrives the tissue of interest. The tracer is also assumed to stay within the intravascular space, an assumption that is violated when the BBB is disrupted. Moreover, with the central volume principle, we can derive measures of cerebral blood volume (CBV), cerebral blood flow (CBF), mean transit time (MTT), and time to peak (TTP) [32]. The central volume principle postulates that the level of tissue blood volume (CBV) is equal to its flow (CBF) multiplied by the time the tracer uses to pass through the tissue (MTT);

$$\text{CBV} = \text{CBF} \cdot \text{MTT} \tag{2}$$

For clinical applications, a qualitative measure of CBV is set equal to the area under the curve $C(t)$ in Eq. (1), and obtained by assuming the contrast agent is injected as a tight bolus with zero duration. The corresponding values of MTT and CBF can thus be calculated from the so-called Zierler's area-height relationship and the central volume principle, respectively [33]. This simplification also implies that the blood flow into the feeding artery is constant [29]. In turn, this makes qualitative DSC-MRI limited to visual inspection of regional differences, or, for relative intra- and inter-patient comparisons by normalizing the vascular maps to a standardized mean value of selected tissue (such as white matter).

To derive a more realistic estimate of tissue perfusion by DCE-MRI, the analysis needs to account for the circulatory transport throughout the body from the site of injection, as well as the properties of a blood tracer in a magnetic field. In practice, the contrast agent bolus will not be in the form of an indefinitely short injection when arriving at the organ at interest, but rather susceptible to regional delays and dispersion [34]. This error can be corrected for by tracking the corresponding contrast agent concentration in a feeding artery as a function of time. This curve is known as the arterial input function (AIF) and a more accurate representation of CBF and MTT can thus be estimated by deconvolution of the AIF and the tissue concentration curves [29].

As pointed out in the guidelines for clinical DSC-MRI published by the American Society for Functional Neuroradiology (ASFNR) [35], CBV is commonly denoted "rCBV" in the MRI-perfusion literature, and referring to both "relative CBV" and "regional CBV." Generally, "relative CBV" means that CBV has been determined qualitatively without devolution of an AIF. Often the "relative CBV" is normalized to the contralateral normal-appearing white matter, recommended to be denoted "normalized rCBV or nrCBV," but is commonly referred to as "rCBV.". "Regional CBV" refers to absolute quantification of CBV in a particular anatomic region, requiring the measurement of an AIF, and the guidelines recommend abandoning the use of "rCBV" for regional CBV. In this chapter, rCBV will therefore exclusively refer to "relative CBV".

The assumption of a linear relationship between contrast agent concentration and the apparent signal change is fairly accurate when the contrast agent is distributed in the intravascular space [36]. The situation is, however, dramatically changed in areas of BBB breakdown and permeable vessels. Here, the extravasation of contrast agent through permeable vessels is in violation of the indicator-dilution theory, and may lead to substantial underestimation of CBV (from T1-dominated leakage), overestimation of CBV (from T2-dominated leakage), or probably a combination of the two [37]. Extravasation of contrast agent may be compensated for by saturating the tissue with an initial "pre-dose" injection of contrast agent and by rigorous post-processing algorithms [37–42]. Still, a complete understanding of the influence of contrast agent leakage on DSC-MRI, as well as consensus strategies to correct for this phenomenon, are lacking.

2.1.3 Vessel Size Imaging and the Road Ahead

Another intriguing concept of DSC-MR is vessel caliber imaging or vessel size MRI [43]. Requiring a minimum of a GE- and a SE-EPI readout, the highly susceptibility-sensitive GE signal is sensitive to blood vessels of all calibers, whereas SE is predominantly sensitive to microscopic vessels or capillaries (radius $< 10 \mu m$) [21, 44]. By computation, this allows for parametric measures of the mean vessel diameter and density by the ratio of GE to SE signal, and vice versa, respectively. In addition, an apparent measure of the vessel size index may also be estimated by accounting for levels of water diffusion and the blood volume fraction [45].

For traditional vessel size MRI, a DSC-MRI acquisition using GE- or SE-EPI readout is set with a repetition time of ~1.5 s, resulting in a total imaging time of a few minutes over +100 volumes [24, 42]. However, recent advances in scanner hardware combined with new MRI acquisition techniques now dramatically push the limits of conventional DSC-MRI. Multiple image echo readouts have long been available for clinical MRI [46], but are now combined with so-called accelerated or parallel imaging methods [47]. Several approaches are currently in clinical review, of which simultaneous multi-slice (SMS) EPI has received much attention [48]. In SMS-EPI, multiple slice planes are excited simultaneously to gain higher spatial and temporal image resolution and brain coverage [49].

An additional, potential feature of these time-accelerated imaging methods is rapid and reliable estimates of R2 and R2*, that in turn may improve the estimates of tissue perfusion and permeability [50]. Moreover, the use of multiple GE or SE readouts may further improve the selection of the AIF [51], or even help eliminate the need for an AIF altogether in certain clinical situations [52].

2.2 Arterial Spin Labeling (ASL)

Whereas DSC-MRI relies on the use of an exogenous contrast agent, ASL is truly non-invasive in nature by utilizing magnetically labeled blood water protons as an endogenous tracer. In light of the recently increased concern regarding the safety and toxicity of gadolinium-based contrast agents, efforts to reduce the need for gadolinium-based tracers in MRI while maintaining adequate image quality at clinical magnetic field strengths are warranted [53, 54].

2.2.1 Physical Principles

In ASL, images are acquired with- "labeled arterial spins" (labeled images), and without labeling (control images). In brain ASL, labeling of arterial spins is performed by applying radiofrequency (RF) pulses on upstream blood water protons in the pre-cerebral arteries of the neck. The RF pulses invert the magnetization of the blood protons for a definite period of time, called the labeling duration. After the labeling duration there is a critical period of waiting time before the labeled images are acquired. This time is called the post labeling delay (PLD) or inversion time, and reflects the inflow time before the labeled blood water protons reach the brain parenchyma. The acquired images contain signal from the magnetically labeled blood protons as well as signal from brain tissue protons. By subtracting the control images from the labeled images, most of the signal from the brain tissue is removed. The remaining signal is proportional to the amount of magnetization inverted and delivered to the tissue, and scales directly with tissue perfusion. Because the signal difference of one control-label pair is in the order of 1–2% only, the labeling process must be repeated several times to obtain SNR for robust estimation of tissue perfusion. Owing to this subtraction of images, ASL is inherently sensitive to motion, and optimal suppression of static background signal in the post labeling period is therefore critical to obtain sufficient image quality. This is usually compensated by 3D volumetric readouts [25]. Another benefit of 3D readouts is higher SNR and less susceptibility artifacts compared to that of a 2D readout. Several different approaches for the labeling scheme exist, of which pseudo-continuous ASL (pCASL) is the recommended labeling strategy for most clinical and research applications [25].

2.2.2 Kinetic Modeling—Quantitative Perfusion Metrics

The magnetically labeled blood water protons in ASL can be well approximated as freely diffusible tracers, which rapidly distributes to all tissue compartments in the brain. In contrast to DSC-MRI, ASL most commonly measures the effect of the tracer in the brain at a single time point and consequently does not provide information about the dynamic effect of the tracer. The same kinetic modeling theory used for DSC-MRI still applies, but CBV and MTT cannot be estimated from the single time point measure. The selection of PLD, at which the effect of the tracer is measured, is critical for obtaining appropriate measures of CBF. It must be sufficiently long for the labeled blood water protons to reach the brain parenchyma and not stay in the macrovasculature producing unwanted "arterial noise," while at the same time being short enough to avoid too much influence from T1 relaxation of the labeled blood water protons and thus loss of signal [27]. Generally, PLD has to match the arterial transit time, which increases with age and may also vary with different pathologies, such as cervical and intracranial vascular conditions, as well as in cases of reduced cardiac output [26]. Contrary to DSC-MRI, ASL provides quantification of CBF without the need for an AIF because the CBF is proportional to the signal difference of the label and control images. For pCASL [55];

$$\text{CBF} = \frac{6000 \lambda \left(S_{\text{control}} - S_{\text{label}} \right) e^{\frac{\text{PLD}}{T1_{\text{blood}}}}}{2 \alpha T1_{\text{blood}} S_{\text{PD}} \left(1 - e^{-\frac{\tau}{T1_{\text{blood}}}} \right)} \left[\text{mL} / 100 \, \text{g} / \min \right] \qquad (3)$$

where λ is the brain/blood partition coefficient in mL/g, S_{control} and S_{label} are the time-averaged MR signals in the control and labeled images, respectively, $T1_{\text{blood}}$ is the longitudinal relaxation time of blood measured in seconds, α is the labeling efficiency, S_{PD} is the MR signal of a proton density-weighted image, τ is the label duration, and PLD is the post labeling delay [25]. The factor of 6000 converts the units from mL/g/s to mL/100 g/min. Several of the parameters needed for CBF quantification in Eq. (3) may however be difficult to measure and influence the accuracy of the measured CBF. In the 2015 ASL White Paper [25], a recommended ASL sequence (3D pCASL), labeling parameter values, and CBF parameters at different physiologies and diseases are suggested. Although single time point measures are most common, approaches using multiple time points (multi-PLD) do exist where dynamic measures of the effect of the tracer can be measured [25].

2.3 Recommended Brain Perfusion Protocols

Characteristics of the two perfusion techniques are summarized in Table 13.1, including current recommendations of key imaging parameters for DSC-MRI [56] and of ASL [25], and assumptions and considerations for kinetic modeling [57].

In short, both GE-EPI and SE-EPI readouts can be used for DSC-MRI, but generally GE-EPI is recommended in clinical routine as it provides higher SNR due to increased contrast agent sensitivity. Moreover, the linearity between the contrast agent concentration and change in R2* for GE applies to a broader range of vessel sizes compared to the change in R2 for SE [56]. On the other hand, signal loss and geometric and intensity distortions in regions with large variations in magnetic susceptibility, typically in interfaces between tissue and bone or air, are more prominent for GE-EPI than SE-EPI. The use of SE-EPI may thus be advantageous at higher field strengths to reduce the influence from susceptibility-dependent artifacts. In vessel size imaging, the benefits of both the GE- and SE-EPI are utilized, but typically at the expense of a slightly more complex image acquisition process. Both 2D and 3D readouts can be used for DSC-MRI, but currently 2D is more commonly used due to better spatial resolution and characterization of the contrast agent bolus passage [56].

In DSC-MRI, a well-defined first-pass bolus of contrast agent is important for obtaining accurate perfusion estimates, and it is recommended using a minimum injection rate of 3 mL/s (range; 3–5 mL/s) immediately followed by a 25 mL (range; 10–30 mL) saline flush at the same injection rate. The current recommended dose of Gd-based contrast agent is 0.1 mmol/kg of body weight, although efforts to obtain high-quality data while reducing this dose even further are ongoing. Also, to reduce the effect of T1-shortening due to contrast agent extravasation, some institutions recommend using a pre-load dose to saturate the tissue signal and thereby getting more accurate perfusion metrics [39]. This is particularly relevant in permeable tissue and if higher flip angles are used (>60°). The pre-load dose is recommended to be approximately 25% of the total dose, and to be administered 5–10 min prior to the DSC-MRI acquisition [35]. As an alternative, by using a combination of lower flip angle, longer TR, and longer TE in the

Table 13.1
Overview of key characteristic of DSC-MRI and ASL

Parameter	DSC-MRI	ASL
Type of method	Exogenous	Endogenous
	Intravenous bolus injection of contrast agent	No contrast agent (Magnetic labeling of blood H_2O)
	Increased susceptibility → change in T2/T2*-relaxation → decreased MR signal	Magnetic labeling of blood → magnetization exchange from labeled H_2O → decreased MR signal
	First pass of intravascular contrast agent through regional circulation	Continuous arrival and diffusion of labeled H_2O into cells and interstitium
Acquisition—Recommendations	ASFN	ISMRM and European ASL in Dementia Consortium
Tracer/labeling	Gd-based contrast agent: 0.1 mmol/kg body weight at minimum rate of 3 mL/s	Pseudo-continuous labeling (pCASL): τ = 1800 ms, labeling duration = 1800 ms, PLD = 1500–2000
Field strength	1.5 or 3 T	3 T (if 1.5 T; lower spatial resolution)
Sequence	GE-EPI (T2*-weighted) or SE-EPI (T2-weighed)	3D RARE stack of spiral or 3D GRASE
Total scan time	Short: < 2 min	Long: ~ 4 min
TR	≤ 1.5 s (or *as short as possible*)	3–4 s
TE	1.5 T; 40–50 ms 3 T; 20–35 ms	~15 ms (preserve MR signal from T2/T2* decay)
Flip angle	60°–70° (with pre-dose), 30° (without pre-dose)	90°
Turbo factor		8–12
Temporal resolution	90–120 time points (2–3 min)	
Minimum matrix size	128 × 128 or higher	64 × 64
FOV	20 × 20 cm (range; 16 × 16–24 × 24 cm)	~25 × 25 cm
In-plane resolution	~1.5 mm (range; 1.5–2.0 mm)	3–4 mm
Slice thickness	3–5 mm	4–8 mm
No. of slices	15–20	~11
No. of dynamic frames	90–120 (2–3 min)	

(continued)

Table 13.1 (continued)

Parameter	DSC-MRI	ASL
No. of pre-contrast images	10	
Kinetic modeling	Central volume principle	
Physiological quantitative parameters	• Absolute and relative CBF • Absolute and relative CBV • Absolute and relative MTT	• Absolute CBF
Assumptions/considerations	– Intact BBB – Stable flow – Negligible change in T1 – No recirculation of contrast agent – No dispersion and delay of bolus for AIF estimation – Linear relationship: Contrast agent concentration and ΔR2* (GE)/ΔR2 (SE)	– Inversion efficiency ($a = 0-1$) – T1blood, T1tissue – Inflow and outflow effects – Blood-tissue partition coefficient of water – Arterial transit time – Vascular signal contamination

Parameters that are particularly favorable for one of the perfusion techniques are highlighted with thumps up (pros) as opposed to thumb down (cons)

AIF arterial input function, *ASL* arterial spin labeling, *ASFN* American Society for Functional Neuroradiology, *BBB* blood-brain barrier, *CBF* cerebral blood flow, *CBV* cerebral blood volume, *DSC* dynamic susceptibility contrast, *EPI* echo planer imaging, *Gd* gadolinium, *GE* gradient echo, *GRASE* gradient and spin echo, *ISMRM* International Society for Magnetic Resonance in Medicine, *MTT* mean transit time, *PLD* post labeling delay, *RARE* rapid acquisition with refocusing echoes, *SE* spin echo, *TE* echo time, *TR* repetition time, *τ* labeling duration

EPI readout, reduced T1 effects can be achieved without the use of a pre-load dose [37].

Currently, pCASL with a single PLD is the recommended labeling approach of ASL, and should include as short as possible RF pulse spacing (1 ms between the centers of two consecutive RF pulses) [25]. The selection of optimal label duration is a compromise between increased SNR and greater power deposition, and T1-sensitivity. Longer label duration increases the signal, but is reduced with labeling durations longer than the T1 relaxation time of the tissue. Also, increased label duration implies a prolonged TR with subsequent reduction of the number of averages that can be obtained per unit time. Currently, the label duration is recommended to be 1800 ms [25]. The selection of PLD is important for obtaining accurate estimates of CBF. Ideally, the PLD should be just above the longest arterial transit time of the tissue. In tissues with unusually long arterial transit time, artificially low CBF may therefore be observed [25].

Currently, segmented 3D readouts with a single excitation per TR are recommended for ASL, providing optimal background signal suppression while being fairly insensitive to field inhomogeneities. It also provides shorter scan times compared to 2D readouts. Suggested 3D methods include multi-echo RARE (rapid acquisition with refocusing echoes) or GRASE (gradient and spin echo). However, multi-slice 2D EPI or spiral read-outs with minimum TE is suggested as a fair alternative because these techniques are less sensitive to motion artifacts and available on most scanner systems [25].

2.3.1 Pros and Cons of DSC-MRI Versus ASL

Currently, DSC-MRI provides whole-brain coverage with shorter acquisition time (~2 versus ~4 min), and better in-plane resolution than ASL (~1.5 versus ~3 mm) (Table 13.1). Furthermore, DSC-MRI can be performed on both 1.5 T and 3 T, with 3 T providing better SNR at the expense of increased potential for susceptibility artifacts compared to 1.5 T. For ASL, 3 T is recommended because of the inherently low SNR of this technique. However, sufficient SNR can be obtained at 1.5 T, but with the cost of reduced spatial resolution. In contrast to ASL, DSC-MRI provides a measure of blood volume in addition to blood flow. However, absolute quantification of CBF in DSC-MRI is more challenging because of the need for an AIF [31, 56]. ASL on the other hand does not require similar rigorous processing steps, and has the advantage of using an endogenous tracer. ASL is therefore particularly well suited for repeated imaging exams, as well as measurements in patients where gadolinium-based contrast agent should be limited, including children and patients with reduced kidney function [57]. Figure 13.1 shows perfusion (CBF) maps obtained from DSC-MRI and ASL in a patient with a glioblastoma, respectively. To this end, these techniques may work well in symphony to cover most clinical applications and needs, and serve as independent validations of perfusion.

3   Clinical Applications

Glioblastoma multiforme, a type of high-grade glioma, is the most common primary brain tumor in adults. According to the 2016 World Health Organization (WHO) grading standard, gliomas are graded from I to IV [58], where high-grade gliomas include grade III–IV, low-grade gliomas have grade II, and grade I are generally benign tumors. While the prognosis for high-grade gliomas is dismal with median overall survival of 15 months [59], patients with low-grade gliomas can live for several years (median overall survival of 13 years with aggressive treatment [60]). The longstanding treatment option for newly diagnosed high-grade gliomas consists of surgical resection followed by adjuvant chemo-radiotherapy, i.e., concomitant chemotherapy (usually, temozolomide) and fractionated radiotherapy. In contrast, low-grade gliomas can be treated much less aggressively, and depending on tumor location

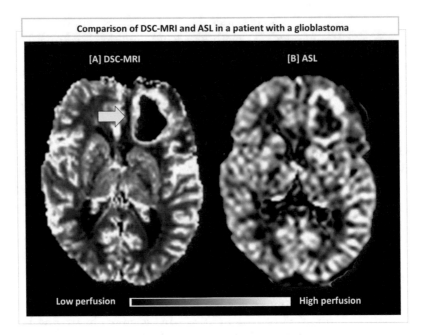

Fig. 13.1 Comparison of DSC-MRI (**a**) and ASL (**b**) in a patient with glioblastoma. The images show axial projections of perfusion (CBF) acquired during the same MRI exam. Areas of abnormal blood flow are observed in viable tumor tissue (yellow arrow) surrounding a necrotic core. Note the heterogeneous signal in the active tumor, with patches of both high and low perfusion. For a standard clinical image protocol and compared to ASL, DSC-MRI tends to have better spatial resolution and SNR over a similar acquisition period. However, the processing steps for acquiring biologically reasonable values of perfusion are considered easier and more robust with ASL

and the clinical status of the patient, sometimes a watch-and-wait approach is preferred.

While conventional treatment response criteria focus on changes in the tumor size from contrast agent enhancement, DSC-MRI is neither dependent on nor limited by, leakage of the contrast agent into the extravascular space in order to characterize glioma status and prognosis [61–63]. Intriguingly, DSC-MRI can therefore detect malignant transformations in gliomas up to a year before changes are visible on conventional MRI [64]. Also, by providing absolute quantification of blood flow without the use of an exogenous contrast agent, the research activity of ASL is growing rapidly.

3.1 Discriminative Power

3.1.1 Pre-Surgical Characterization in Gliomas—Grading, Predicting Prognosis and Genetic Profiling

As a complement to histopathological analyses, conventional contrast-enhanced T1-weighted MRI is used to characterize glioma grade and prognosis by the presence of contrast agent leakage into the extravascular space. Typically, high-grade gliomas show increased enhancement compared to low-grade gliomas, owing to the higher density of permeable vessels associated with increased tumor angiogenesis and BBB breakdown [65]. However, it has been shown that up to 1/3 of malignant gliomas are non-enhancing [66].

Moreover, increased tumor angiogenesis and high micro-vessel density in high-grade gliomas are generally reflected by elevated levels of rCBV [67, 68]. Early on, DSC-MRI studies demonstrated the usefulness of rCBV for pre-surgical character-ization of malignancy in gliomas [68, 69]. Since then, these find-ings have been confirmed in a large number of studies [70]. In contrast, only a limited number of studies have assessed ASL for brain tumor characterization, but they are unambiguously dem-onstrating elevated regional or global tumor CBF values in high-grade gliomas compared to that of low grades. This reinforces the reported strong correlation between perfusion metrics from ASL and DSC-MRI [71, 72]. Still, a recent published meta-analysis of 15 ASL studies concluded that ASL can provide excellent diag-nostic accuracy for tumor grading; it also emphasized the need for more studies with larger sample sizes and use of uniform tumor WHO classification [73].

Elevated pre-treatment levels of rCBV in gliomas have been associated with shorter survival [74–76], whereas lower rCBV translates into prolonged time to progression [77]. Moreover, rCBV has been found to be associated with the genetic profile of gliomas. Molecular and genetic characteristics have long been rec-ognized as important factors of tumor aggressiveness and progno-sis, and are integrated in the recent 2016 WHO classification standard for CNS tumors [58]. In particular, elevated levels of rCBV in high-grade gliomas may be associated with increased amplification of the epidermal growth factor receptor (EGFR) [15], un-methylated 6-methylguanine DNA methyltransferase (MGMT) status [78], and increased percentage of wild-type isoci-trate dehydrogenase (IDH), which is indirectly associated with hypoxia-dependent angiogenesis [79, 80]. Similar associations between CBF and IDH mutation status, more specifically the IDH1 gene, are also observed with ASL [81].

While perfusion-based MRI may aid in the patients' treatment decision-making process [82], a major challenge still remains owing to inherent overlap in the parametric estimations between different methodological approaches and physiologic conditions. Oligodendroglial tumors typically have elevated levels of rCBV regardless of glioma grade, and the diagnostic accuracy for grading may be substantially reduced when including this glioma type [83, 41]. Moreover, and regardless of efforts to normalize the rCBV measures, variability in reported values occurs owing to the use of different scanners, acquisition protocols, and post-processing rou-tines including the choice of leakage correction methods [84, 41]. Furthermore, normalized rCBV has traditionally been obtained by manual placement of a region of interest (ROI) in rCBV "hot-spots" of the tumor and typically in contralateral non-affected white matter [85, 67]. The normalized rCBV is thus prone to incorrect placement of the ROIs, potentially hampering reliable determination of a cut-off/threshold value. In particular, large blood vessels and non-cerebral structures should be avoided from the tumor and normal-appearing brain matter ROIs, respectively [35, 41, 85]. Promising approaches to overcome these limitations include histogram analysis [85–88] and automatic removal of macro-vessels that may induce artificially high perfusion values [88, 89]. Ultimately, the clinical applicability of tumor characterization by DSC-MRI will greatly benefit from standardized acquisition and approaches for quantification, and future studies should strive to follow consensus guidelines [56].

Improved discriminative power of DSC-MRI may be obtained by applying a multi-parametric approach, where other functional MRI techniques provide complementary information to that of perfusion. For example, the combined use of DSC-MRI and DW-MRI has shown to improve the diagnostic performance of tumor characterization [82]. Moreover, increased vessel permeability by DCE-MRI may also help distinguish tumor grades [62, 90–92].

3.1.2 Differential Diagnosis

While depicting similar contrast-enhancing pattern on conventional contrast-enhanced MRI, different brain tumor types may show unique vascular and perfusion characteristics. Similar to high-grade gliomas, the blood vessels of a primary CNS lymphoma are highly permeable, leading to increased contrast enhancement on conventional MR. However, the degree of neo-vascularization in primary CNS lymphomas is much lower, with reduced blood volume and perfusion as indicated by lower rCBV from DSC-MRI [93, 94]. Interestingly, parametric leakage correction may actually reduce the differentiation between high-grade gliomas and primary CNS lymphomas. This apparent contradiction is in part attributed to higher permeability in primary CNS lymphomas [93]. Finally, similar to tumor characterization, the combined use of functional MRI techniques; DSC- and DCE-MRI [95] or DSC- and DW-MRI [96] has been shown to improve the differential diagnoses of high-grade glioma and primary CNS lymphoma.

Although gliomas may be differentiated from brain metastases on conventional MRI, large solitary brain metastases with central necrosis may mimic a high-grade glioma [97]. Similarly, a high-grade glioma with small enhancing satellites may be mistaken for a brain metastasis. However, as illustrated in Fig. 13.2, the peri-tumoral T2 hyperintense region of high-grade gliomas are, in contrast to brain metastases, characterized by diffuse tumor infiltration resulting in higher rCBV [98, 99]. Consequently, the discriminating power of rCBV is reported stronger in the peri-tumoral region [100]. This finding of the peri-tumoral region is also corroborated by measures of CBF from ASL [101, 102]. Contrary to DSC-MRI,

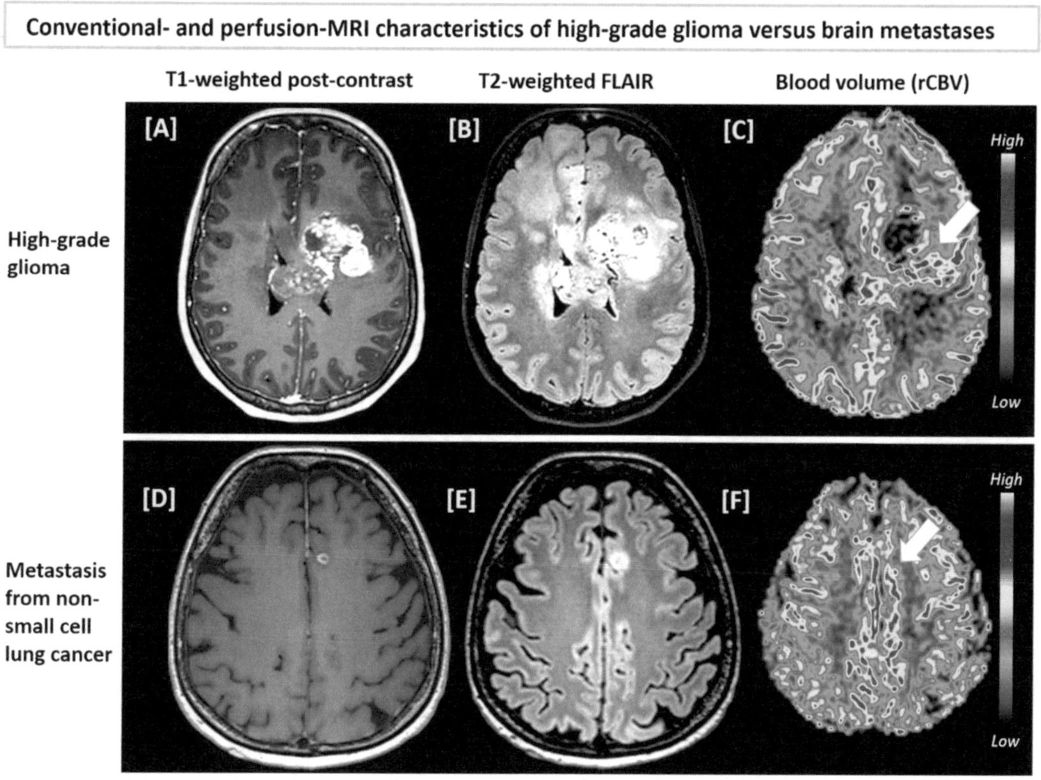

Conventional- and perfusion-MRI characteristics of high-grade glioma versus brain metastases

Fig. 13.2 Conventional- and perfusion-MRI characteristics of high-grade glioma versus brain metastasis. Representative axial pre-treatment MR images of a patient with glioblastoma (**a–c**) and a patient with a metastasis to the brain from non-small cell lung cancer (**d–f**). High-grade gliomas such as glioblastomas tend to have a complex shape with central necrosis, and on conventional T1-weighted images the contrast enhancement is often diffuse in peri-tumoral region (**a**). In contrast, brain metastases are often spherical in shape, and have well-defined borders to the surrounding brain parenchyma, and without central necrosis for small metastases (**d**). Both high-grade gliomas and brain metastases are typically surrounded by vasogenic edema depicted by T2 hyperintense signal on fluid-attenuated images (FLAIR) [B, E], often pronounced for high-grade gliomas (**b**) and for brain metastases having received immunotherapy (not the case for this patient) (**d**). Because of diffuse tumor infiltration in the peri-tumoral region of high-grade gliomas [**c**; white arrow], the blood volume (rCBV) is characteristically elevated compared to the peri-tumoral region of brain metastases [**f**; white arrow]

intratumoral measures by ASL have shown significantly higher CBF in high-grade gliomas compared to brain metastases [101], reflecting higher microvascular density as assessed by histopathology [103]. Intratumoral rCBV measures are however dominated by overlapping rCBV of high-grade glioma and brain metastases, in part owing to complex extravasation patterns of the contrast agent. Furthermore, brain metastases of different origins may have a broad range of vascular signatures. For instance, the average rCBV levels in metastases from malignant melanoma and renal metastases are found to be elevated compared to metastases from primary lung cancer [104].

3.2 Monitoring Treatment Response

A major conundrum for MRI-based treatment response monitoring is that similar contrast-enhancement patterns may represent very different treatment responses and hence clinical endpoints of survival. Among others, an increase in contrast enhancement from treatment-induced changes may be indistinguishable from that of an active tumor. Furthermore, with the vascular remodeling of anti-angiogenic therapies, absence or decrease of contrast enhancement on conventional MRI may not necessarily represent true anti-tumor effect [15]. Research efforts are thus focused on improving image-based response criteria by including assessment of vascular and also metabolic activity [105, 106].

3.2.1 Chemo-Radiation and Stereotactic Radiosurgery (SRS)

Similar to primary CNS tumors, brain metastases may also be removed by surgery in selected patients, and followed by adjuvant chemo-radiotherapy. Alternatively, patients with metastases to the brain usually receive radiotherapy with whole-brain radiation and/or SRS where a high radiation dose of >15 Gy is given in one to three fractions. Chemo-radiotherapy and SRS however may lead to pseudoprogression.[3] Pseudoprogression is a term describing non-tumoral radiation-induced effects causing transient increase in contrast enhancement and/or increased areas of peri-tumoral edema on conventional MR, without further evidence of any viable tumor [107]. For patients with high-grade gliomas and brain metastases, pseudoprogression is typically observed in one-third of the patients receiving chemo-radiotherapy and SRS [108–110]. Pseudoprogression is a reversible form of radionecrosis that also manifests by increased contrast enhancement [111]. Differentiation of true tumor progression or recurrence from pseudoprogression and/or radionecrosis therefore poses a significant challenge for robust radiographic response assessment in both high-grade gliomas and brain metastases. While the pathological features of pseudoprogression are not yet fully understood, it is related to transient changes to the vascular endothelium and the BBB, as well as treatment-induced hypoxia [112]. True radionecrosis is characterized by coagulative necrosis of white matter, wall thickening and hyalinization of vessels as well as capillary obliteration [83, 113]. Contrasting true tumor progression, contrast enhancement from pseudoprogression is typically reversed after 3–6 months, but the window of observation may range from a few weeks up to more than a year after chemo-radiotherapy or SRS. Moreover, compared to no apparent contrast enhancement, development of pseudoprogression after chemo-radiotherapy has been associated with prolonged survival for patients with high-grade gliomas [114]. To

---

[3]The term pseudoprogression has traditionally been used in relation to response assessment of chemoradiotherapy in patients with high-grade gliomas, but is now also widely used to describe transient effects of SRS treatment in patients with metastases to the brain.

reliably confirm pseudoprogression on traditional contrast-enhanced MRI, long follow-up times are therefore required [115].

Although the literature is characterized by variable as well as interchangeable use of pseudoprogression and radionecrosis, a large body of evidence points toward lower levels of normalized rCBV in brain tissue with non-tumoral radiation-induced changes compared to recurrent tissue in high-grade glioma patients. Indeed, several studies have confirmed reduced levels of normalized rCBV in treatment-related changes compared to the recurrent tumor by histopathology [116, 117]. Though not without limitations, histopathology is considered the gold standard for distinguishing pseudoprogression, radionecrosis, and viable tumor tissue. The higher normalized rCBV in recurrent high-grade gliomas is believed to reflect increased tumor neo-vasculature as opposed to reduced or absent neo-vasculature in pseudoprogression and radionecrosis [118]. These DSC-MRI results have been substantiated by recent findings using ASL, showing increased CBF in recurrent high-grade gliomas compared to pseudoprogression and radionecrosis [119, 120]. However, there is large variability in reported normalized rCBV thresholds for distinguishing recurrent high-grade glioma from non-tumoral treatment-induced changes, ranging from 0.71 to 2.0 [117, 121]. In a study by Hu et al. a threshold of 0.71 provided sensitivity 92% and specificity of 100%, while Prager et al. reported sensitivity of 87% and specificity 84% for a threshold of 1.27 [116]. In a similar pilot study on ASL, high sensitivity of 94% combined with low specificity of 50% was reported [122]. The overlap in normalized rCBV may in part be due to confounding effects of dissimilar contrast agent extravagation, i.e., underestimation of rCBV from high degree of BBB disruption. Other contributing factors include different definitions for treatment-induced changes, various time points for post-treatment assessment, and lack of standardized acquisition and post-processing routines of DSC-MRI [118]. Based on studies differentiating true radionecrosis from pseudoprogression, tumor recurrence is easier separated from radionecrosis compared to pseudoprogression. The higher variability associated with pseudoprogression is probably owing to the coexistence of tumor and necrosis, ultimately resulting in a broad range of vascular volumes and thus also rCBV [123]. This hurdle may in tumor be addressed by the alternative use of histogram analyses of the DSC-MRI data as demonstrated by Baek et al. 2012 [124]. Furthermore, improved differentiation of treatment-related necrosis and recurrent high-grade neoplasms [125], as well as improved prediction of overall survival [126] has been achieved by the use of estimated fractional tumor burden post-treatment (fraction of enhancing mass above a certain threshold of normalized rCBV).

In addition to post-treatment normalized rCBV, the relative change in normalized rCBV before and after chemo-radiotherapy has been shown to increase in high-grade glioma patients with disease progression, while decreasing in patients developing pseudoprogression. In a study by Manlge et al., including 36 high-grade glioma patients, the mean increase in normalized rCBV 1 month after chemo-radiotherapy was 12% in patients with true tumor progression, whereas pseudoprogressing patients had a mean decrease of 41% [127]. Moreover, while the change in normalized rCBV correlated with overall survival, a change in tumor size did not. Compared to post-treatment assessment only, the use of relative change is less prone to variations in DSC-MRI acquisitions and post-processing methods.

DSC-MRI for response monitoring of brain metastases treated with SRS provides similar results as reported for high-grade gliomas, where metastases have increased levels of normalized rCBV, as well as CBF from ASL, compared to non-tumor tissue [115, 128, 129]. Also, reduced normalized rCBV 1 month post-SRS is shown to be associated with therapeutic benefit [130]. With SRS, while very high doses are given to the tumor region (often 25 Gy in a single fraction), healthy brain tissue receives significantly lower doses due to high conformity and steep dose-fall-off. Nevertheless, alterations in the vasculature of healthy brain tissue after SRS are reported [131]. In particular, rCBV as well as rCBF may increase 6 weeks after SRS in healthy white matter receiving 5–10 Gy. Improved understanding of the underlying mechanisms of responses to radiotherapy in healthy brain tissue and potential dose-response relationships may help minimize toxicity and neurological complications. Finally, for brain metastases receiving SRS, a recent finding suggests the vascular signature of the lesion as assessed by DSC-MRI prior to SRS is suggestive of the risk for subsequent development of pseudoprogression [132]. In particular, metastases with abnormal peri-tumoral vascular function including reduced perfusion and high fractions of micro-vessel rCBV were especially prone to the development of pseudoprogression.

3.2.2 *Anti-Angiogenic Therapy*

Angiogenesis was early identified as a potent target for anti-cancer treatment of solid tumors, and in particular high-grade gliomas, owing to their characteristic high level of tumor angiogenesis. Anti-angiogenic therapy includes agents targeting angiogenic factors to prevent abnormal angiogenesis as well as agents targeting preexisting tumor vasculature [133]. Although the concept of anti-angiogenic therapy has been discussed for centuries, the exact mechanisms by which anti-angiogenic drugs work in humans are still not fully understood, nor the direct microvascular response to treatment [105, 134, 135].

With bevacizumab in particular, a humanized monoclonal antibody that targets vascular endothelial-derived growth factor (VEGF), DSC-MRI already has a long history of use. Collectively, results indicate a reduction, or a gradient toward normalization, of abnormal blood volume levels (by rCBV) in malignant gliomas on follow-up MRIs after treatment [136]. In a multicenter, randomized, Phase II trial of bevacizumab with irinotecan or temozolomide in high-grade gliomas, decreases in rCBV at week 2 and week 16, but not week 8, were predictive of improved survival [137]. Other works suggest a corresponding, favorable pre-therapeutic vascular status and window in high-grade gliomas, resulting in prolonged survival after bevacizumab [138–140]. A similar, optimal vascular signature was not identified in patients treated with traditional chemotherapy. Interestingly, these findings may indicate pruning of abnormal vessels as one of the dominant mechanisms of action of anti-angiogenic therapies [141, 134], whereupon a favorable response is a further contingent on the vascular network staying above the limits in which it maintains the long-range connectivity required for sufficient communication and metabolic supply. This phenomenon is termed the percolation threshold [142].

The mechanisms of other anti-angiogenic agents such as cediranib, an oral pan-VEGF receptor kinase inhibitor, have also been the target of DSC-MRI. Focusing rather on perfusion (by rCBF), metabolic status and vessel architecture, a series of studies showed that cediranib-treated patients with glioblastoma whose subnormal tumor perfusion increased, as well as abnormal vessel calibers and permeability decreased, survived longer compared to patients' tumors who did not show a similar coherent normalization [141, 143, 144] (Andronesi et al. 2017). Vascular normalization by DSC-MRI is also shown using tipifarnib, a farnesyl transferase inhibitor, and radiotherapy, where abnormal rCBV levels subsequently decreased and subnormal rCBV levels increased [145]. A somewhat similar normalization was seen with tivozanib, another VEGF inhibitor hitting all three VEGF receptors [146]. However, intriguingly, there was no apparent vessel pruning with tivozanib, or change in the enhancing tumor volume, indicating the drug had limited anti-tumor activity and no survival benefit. Another interesting observation with anti-angiogenic therapies is the apparent reduction in the incidence of pseudoprogression [105, 147]. The therapy-induced remodeling of the vessel wall limits vascular permeability, and reduces contrast agent enhancement as well as edema on anatomical MRI [147]. This effect further complicates the use of the RANO criteria, while strengthening the argument for complementary techniques such as DSC-MRI [148, 149].

3.2.3 Immunotherapy

For decades it has been known that intact immune surveillance plays a key role in controlling outgrowth of neoplastic transformations. Immunotherapy is a promising therapy option that is now finally coming of age with astonishing results of prolonged patient survival, especially for melanomas [150]. For brain tumors however, the role of immunotherapeutic agents has not been adequately explored and why only some patients respond to these immunotherapies remains unclear [151]. Moreover, there is still a complex, poorly understood interaction between the immune system and blood vessels, especially in the setting of MRI of cancer [135, 152]. For metastases to the brain in particular, misleading regions of immunotherapy-induced contrast agent enhancement can be observed not only in regions receiving concomitant radiotherapy, but also in distant regions because of radiation-induced unmasking of tumor antigens (the *abscopal effect*) [151, 153, 154]. The increase in enhancement is associated with the apparent inflammatory response of immunotherapy.

With the increasing use of immunotherapy, the importance of incorporating immune-related mechanisms in the response assessment of immunotherapy has been highlighted by the RANO working group through the iRANO criteria [155]. Here, the emphasis is put on the importance of including neurological status together with sufficient follow-up time for detection of tumor progression, avoiding abruption of treatment in patients showing image-based, but not neurological decline [156].

So far, a very limited number of studies on DSC-MRI and ASL in the assessment of immunotherapy of brain tumors are available. A small, retrospective study suggests elevated rCBV levels may separate real progression from a therapy-induced inflammatory immune response in high-grade patients receiving autologous mature dendritic cells treatment [157]. In another small study, including eight patients with high-grade glioma of whom six received immunotherapy in addition to surgery, lack of normalized rCBV within post-operatively contrast-enhancing regions was consistent with aggressive recurrent tumor and not reactive inflammatory changes [158]. For brain metastases, preliminary data show increased vessel calibers in the peri-tumoral region in patients treated with SRS and immunotherapy compared to SRS only, indicating that vascular MRI may help show the added value of immunotherapy in combination with radiotherapy [159]. Figure 13.3 illustrates increased vessel calibers and decreased blood flow 3 and 6 months after SRS in a patient with brain metastases from malignant melanoma treated with additional immunotherapy. Clearly, the usefulness DSC-MRI and ASL in this context is still in its infancy and further research is warranted.

Conventional- and perfusional-MRI responses to combined SRS and immunotherapy

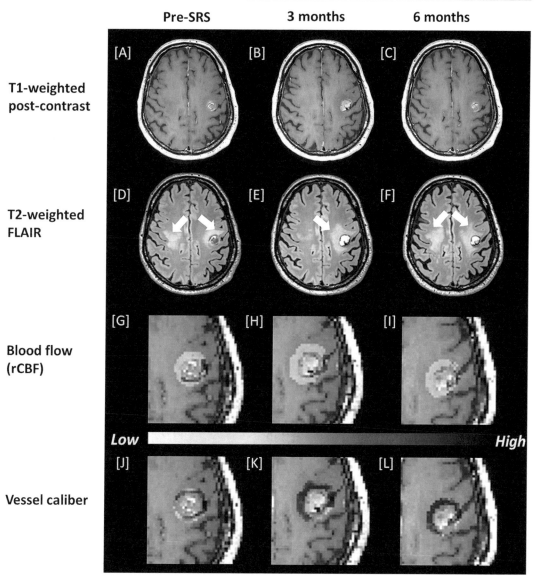

Fig. 13.3 Conventional- and perfusional-MRI responses to combined SRS and immunotherapy. MR images and perfusion metric maps of a patient with metastases to the brain from malignant melanoma treated with ste-reotactic radiosurgery (SRS) and concomitant immunotherapy (monoclonal antibody ipilimumab). Conventional T1-weighted post-contrast images (**a–c**), T2-weighted fluid-attenuated inversion recovery (FLAIR) (**d–e**), peri-tumoral relative blood flow (rCBF) (**g–i**), and peri-tumoral relative vessel caliber (**j–l**) before SRS (left column), 3 months (middle column), and 6 months after SRS (right column). Large regions of edema are typically observed in patients treated with immunotherapy (white arrows in FLAIR images [**d–f**]). For this patient, while the blood flow is decreased [**g–i**], especially in the peri-tumoral region, the vessel caliber parameter shows increased values at 3 months (**k**) and at 6 months (**l**) after SRS indicating the often complex response of brain tumor patients to combination therapy

4   Conclusions and Future Directions

A large body of evidence substantiates the importance of including tumor physiological function, such as perfusion, for optimal management of both primary and secondary brain tumors. Perfusion metrics obtained from DSC-MRI and ASL, in particular normalized rCBV and CBF, have to a great extent shown their utility for tumor characterization, prognosis, differential diagnosis, and for distinguishing non-tumoral treatment-induced changes. Still, the availability and usefulness of DSC-MRI is in part challenged by complicated imaging and post-processing routines that historically has limited the clinical impact of the technique outside the major research-oriented hospitals and institutions [160]. Efforts to provide much-needed consensus guidelines for clinical imaging are ongoing by the Quantitative Imaging Biomarker Alliance (QIBA), as well as guidelines by the clinical practice committee of the ASFNR [56].

In conjunction with ongoing efforts to improve the clinical utility of DSC-MRI by implementing consensus guidelines, we look to research to speculate where the field is heading. For CNS cancers, DSC-MRI holds a golden ticket in the ability to measure hemodynamic properties beyond the visible tumor margins, i.e., enhancing regions. Researchers are increasingly appreciating the importance of the peri-tumoral zone to understand the heterogeneity of the disease and how its interactions extend into normal-appearing tissue [161, 162]. Moreover, while a new therapeutic drug may target a specific pathway with great success, most treatments do not account for other components of the microenvironment that also regulate tumorigenesis [61, 163, 164]. We hypothesize that measuring how mechanical forces of the tumor microenvironment restrict perfusion and promotes treatment resistance will be critical for patient care.

With the advents of new imaging techniques and increased computational power, another application of DSC-MRI with high potential is the combination of image observations with artificial intelligence and vascular model systems. Coined as vascular fingerprinting among others, advanced computer models are used to describe microvascular networks and vessel architecture in a range of disease scenarios, and subsequently applied to quantify the corresponding MRI findings [165–168]. With time, this approach may also include features of tissue cytoarchitecture to derive a unique signature of the disease [169].

Acknowledgements

We thank Wibeke Nordhøy and Alte Bjørnerud from the Department of Diagnostic Physics, Oslo University Hospital, Oslo, Norway, for critical reading of this text and for providing figure images, respectively. This work has received funding from the European Research Council (ERC) under the European Union's Horizon 2020 research and innovation program (ERC Grant Agreement No. 758657-ImPRESS), as well as the South-Eastern Norway Regional Health Authority Grants 2016102 and 2013069 and The Norwegian Cancer Society Grant 6817564.

References

1. Hanahan D, Weinberg RA (2000) The hallmarks of cancer. Cell 100(1):57–70
2. Horsman MR, Vaupel P (2016) Pathophysiological basis for the formation of the tumor microenvironment. Front Oncol 6:66. https://doi.org/10.3389/fonc.2016.00066
3. Jain RK, di Tomaso E, Duda DG, Loeffler JS, Sorensen AG, Batchelor TT (2007) Angiogenesis in brain tumours. Nat Rev Neurosci 8(8):610–622. https://doi.org/10.1038/nrn2175
4. Folkman J (2002) Role of angiogenesis in tumor growth and metastasis. Semin Oncol 29(6 Suppl 16):15–18. https://doi.org/10.1053/sonc.2002.37263
5. Kato Y, Ozawa S, Miyamoto C, Maehata Y, Suzuki A, Maeda T, Baba Y (2013) Acidic extracellular microenvironment and cancer. Cancer Cell Int 13(1):89. https://doi.org/10.1186/1475-2867-13-89
6. Wen PY, Kesari S (2008) Malignant gliomas in adults. N Engl J Med 359(5):492–507. https://doi.org/10.1056/NEJMra0708126
7. Gabani P, Fischer-Valuck BW, Johanns TM, Hernandez-Aya LF, Keller JW, Rich KM, Kim AH, Dunn GP, Robinson CG, Chicoine MR, Huang J, Abraham CD (2018) Stereotactic radiosurgery and immunotherapy in melanoma brain metastases: patterns of care and treatment outcomes. Radiother Oncol 128(2):266–273. https://doi.org/10.1016/j.radonc.2018.06.017
8. Huang Y, Stylianopoulos T, Duda DG, Fukumura D, Jain RK (2013) Benefits of vascular normalization are dose and time dependent–letter. Cancer Res 73(23):7144–7146. https://doi.org/10.1158/0008-5472.can-13-1989
9. Pahernik S, Griebel J, Botzlar A, Gneiting T, Brandl M, Dellian M, Goetz AE (2001) Quantitative imaging of tumour blood flow by contrast-enhanced magnetic resonance imaging. Br J Cancer 85(11):1655–1663. https://doi.org/10.1054/bjoc.2001.2157
10. Kessler AT, Bhatt AA (2018) Brain tumour post-treatment imaging and treatment-related complications. Insights Imaging. https://doi.org/10.1007/s13244-018-0661-y
11. Macdonald DR, Cascino TL, Schold SC Jr, Cairncross JG (1990) Response criteria for phase II studies of supratentorial malignant glioma. J Clin Oncol 8(7):1277–1280. https://doi.org/10.1200/JCO.1990.8.7.1277
12. Chiou VL, Burotto M (2015) Pseudoprogression and immune-related response in solid tumors. J Clin Oncol 33(31):3541–3543. https://doi.org/10.1200/jco.2015.61.6870
13. Okada H, Kohanbash G, Zhu X, Kastenhuber ER, Hoji A, Ueda R, Fujita M (2009) Immunotherapeutic approaches for glioma. Crit Rev Immunol 29(1):1–42
14. Sharma M, Juthani RG, Vogelbaum MA (2017) Updated response assessment criteria for high-grade glioma: beyond the MacDonald criteria. Chin Clin Oncol 6(4):37. https://doi.org/10.21037/cco.2017.06.26
15. Wen PY, Macdonald DR, Reardon DA, Cloughesy TF, Sorensen AG, Galanis E, Degroot J, Wick W, Gilbert MR, Lassman AB, Tsien C, Mikkelsen T, Wong ET, Chamberlain MC, Stupp R, Lamborn KR, Vogelbaum MA, van den Bent MJ, Chang SM (2010) Updated response assessment criteria for high-grade gliomas: response assessment in neuro-oncology working group. J Clin Oncol 28(11):1963–1972
16. Lin NU, Lee EQ, Aoyama H, Barani IJ, Barboriak DP, Baumert BG, Bendszus M, Brown PD, Camidge DR, Chang SM, Dancey J, de Vries EG, Gaspar LE, Harris GJ, Hodi FS, Kalkanis SN, Linskey ME, Macdonald DR, Margolin K, Mehta MP, Schiff D, Soffietti R, Suh JH, van den Bent MJ,

Vogelbaum MA, Wen PY (2015) Response assessment criteria for brain metastases: proposal from the RANO group. Lancet Oncol 16(6):e270–e278. https://doi.org/10.1016/S1470-2045(15)70057-4

17. Sourbron SP, Buckley DL (2012) Tracer kinetic modelling in MRI: estimating perfusion and capillary permeability. Phys Med Biol 57(2):R1–33. https://doi.org/10.1088/0031-9155/57/2/R1

18. Sourbron SP, Buckley DL (2011) On the scope and interpretation of the Tofts models for DCE-MRI. Magn Reson Med 66(3):735–745. https://doi.org/10.1002/mrm.22861

19. Paschoal AM, Leoni RF, Dos Santos AC, Paiva FF (2018) Intravoxel incoherent motion MRI in neurological and cerebrovascular diseases. Neuroimage Clin 20:705–714. https://doi.org/10.1016/j.nicl.2018.08.030

20. Barbier EL (2013) T2-*weighted perfusion MRI. Diagn Interv Imaging 94(12):1205–1209. https://doi.org/10.1016/j.diii.2013.06.007

21. Kiselev VG, Novikov DS (2018) Transverse NMR relaxation in biological tissues. NeuroImage 182:149–168. https://doi.org/10.1016/j.neuroimage.2018.06.002

22. Knutsson L, Stahlberg F, Wirestam R (2010) Absolute quantification of perfusion using dynamic susceptibility contrast MRI: pitfalls and possibilities. MAGMA 23(1):1–21. https://doi.org/10.1007/s10334-009-0190-2

23. Rosen BR, Belliveau JW, Vevea JM, Brady TJ (1990) Perfusion imaging with NMR contrast agents. Magn Reson Med 14(2):249–265

24. Zaharchuk G (2007) Theoretical basis of hemodynamic MR imaging techniques to measure cerebral blood volume, cerebral blood flow, and permeability. AJNR Am J Neuroradiol 28(10):1850–1858. https://doi.org/10.3174/ajnr.A0831

25. Alsop DC, Detre JA, Golay X, Gunther M, Hendrikse J, Hernandez-Garcia L, Lu H, MacIntosh BJ, Parkes LM, Smits M, van Osch MJ, Wang DJ, Wong EC, Zaharchuk G (2015) Recommended implementation of arterial spin-labeled perfusion MRI for clinical applications: a consensus of the ISMRM perfusion study group and the European consortium for ASL in dementia. Magn Reson Med 73(1):102–116. https://doi.org/10.1002/mrm.25197

26. Grade M, Hernandez Tamames JA, Pizzini FB, Achten E, Golay X, Smits M (2015) A neuroradiologist's guide to arterial spin labeling MRI in clinical practice. Neuroradiology 57(12):1181–1202. https://doi.org/10.1007/s00234-015-1571-z

27. MacIntosh BJ, Lindsay AC, Kylintireas I, Kuker W, Gunther M, Robson MD, Kennedy J, Choudhury RP, Jezzard P (2010) Multiple inflow pulsed arterial spin-labeling reveals delays in the arterial arrival time in minor stroke and transient ischemic attack. AJNR Am J Neuroradiol 31(10):1892–1894. https://doi.org/10.3174/ajnr.A2008

28. Williams DS, Detre JA, Leigh JS, Koretsky AP (1992) Magnetic resonance imaging of perfusion using spin inversion of arterial water. Proc Natl Acad Sci U S A 89(1):212–216

29. Weisskoff RM, Zuo CS, Boxerman JL, Rosen BR (1994) Microscopic susceptibility variation and transverse relaxation: theory and experiment. Magn Reson Med 31(6):601–610

30. Barbier EL, Lamalle L, Decorps M (2001) Methodology of brain perfusion imaging. J Magn Reson Imaging 13(4):496–520

31. Covarrubias DJ, Rosen BR, Lev MH (2004) Dynamic magnetic resonance perfusion imaging of brain tumors. Oncologist 9(5):528–537. https://doi.org/10.1634/theoncologist.9-5-528

32. Meier P, Zierler KL (1954) On the theory of the indicator-dilution method for measurement of blood flow and volume. J Appl Physiol 6(12):731–744

33. Zierler KL (1965) Equations for measuring blood flow by external monitoring of radio-isotopes. Circ Res 16:309–321

34. Calamante F (2005) Bolus dispersion issues related to the quantification of perfusion MRI data. J Magn Reson Imaging 22(6):718–722. https://doi.org/10.1002/jmri.20454

35. Welker K, Boxerman J, Kalnin A, Kaufmann T, Shiroishi M, Wintermark M, American Society of Functional Neuroradiology, M.R.P.S., Practice Subcommittee of the, A.C.P.C (2015) ASFNR recommendations for clinical performance of MR dynamic susceptibility contrast perfusion imaging of the brain. AJNR Am J Neuroradiol 36(6):E41–E51. https://doi.org/10.3174/ajnr.A4341

36. Knutsson L, Lindgren E, Ahlgren A, van Osch MJ, Bloch KM, Surova Y, Stahlberg F, van Westen D, Wirestam R (2014) Dynamic susceptibility contrast MRI with a prebolus contrast agent administration design for improved absolute quantification of perfusion. Magn Reson Med 72(4):996–1006. https://doi.org/10.1002/mrm.25006

37. Bjornerud A, Sorensen AG, Mouridsen K, Emblem KE (2011) T(1)- and T(2)(*)-dominant extravasation correction in DSC-MRI: Part I-theoretical considerations and implications for assessment of tumor hemodynamic properties. J. Cereb. Blood Flow Metab 31(10):2041-2053. https://doi.org/10.1038/jcbfm.2011.52

38. Leu K, Boxerman JL, Cloughesy TF, Lai A, Nghiemphu PL, Liau LM, Pope WB, Ellingson BM (2016) Improved leakage correction for single-echo dynamic susceptibility contrast perfusion MRI estimates of relative cerebral blood volume in high-grade gliomas by accounting for bidirectional contrast agent exchange. AJNR Am J Neuroradiol 37(8):1440–1446. https://doi.org/10.3174/ajnr.A4759

39. Leu K, Boxerman JL, Ellingson BM (2017) Effects of MRI protocol parameters, preload injection dose, fractionation strategies, and leakage correction algorithms on the Fidelity of dynamic-susceptibility contrast MRI estimates of relative cerebral blood volume in gliomas. AJNR Am J Neuroradiol 38(3):478–484. https://doi.org/10.3174/ajnr.A5027

40. Skinner JT, Moots PL, Ayers GD, Quarles CC (2016) On the use of DSC-MRI for measuring vascular permeability. AJNR Am J Neuroradiol 37(1):80–87. https://doi.org/10.3174/ajnr.A4478

41. Boxerman JL, Schmainda KM, Weisskoff RM (2006) Relative cerebral blood volume maps corrected for contrast agent extravasation significantly correlate with glioma tumor grade, whereas uncorrected maps do not. AJNR Am J Neuroradiol 27(4):859–867

42. Paulson ES, Schmainda KM (2008) Comparison of dynamic susceptibility-weighted contrast-enhanced MR methods: recommendations for measuring relative cerebral blood volume in brain tumors. Radiology 249(2):601–613. https://doi.org/10.1148/radiol.2492071659

43. Kiselev VG, Strecker R, Ziyeh S, Speck O, Hennig J (2005) Vessel size imaging in humans. Magn Reson Med 53(3):553–563

44. Boxerman JL, Hamberg LM, Rosen BR, Weisskoff RM (1995) MR contrast due to intravascular magnetic susceptibility perturbations. Magn Reson Med 34(4):555–566

45. Tropres I, Grimault S, Vaeth A, Grillon E, Julien C, Payen JF, Lamalle L, Decorps M (2001) Vessel size imaging. Magn Reson Med 45(3):397–408

46. Schmainda KM, Rand SD, Joseph AM, Lund R, Ward BD, Pathak AP, Ulmer JL, Badruddoja MA, Krouwer HG (2004) Characterization of a first-pass gradient-echo spin-echo method to predict brain tumor grade and angiogenesis. AJNR Am J Neuroradiol 25(9):1524–1532

47. Skinner JT, Robison RK, Elder CP, Newton AT, Damon BM, Quarles CC (2014) Evaluation of a multiple spin- and gradient-echo (SAGE) EPI acquisition with SENSE acceleration: applications for perfusion imaging in and outside the brain. Magn Reson Imaging 32(10):1171–1180. https://doi.org/10.1016/j.mri.2014.08.032

48. Eichner C, Jafari-Khouzani K, Cauley S, Bhat H, Polaskova P, Andronesi OC, Rapalino O, Turner R, Wald LL, Stufflebeam S, Setsompop K (2014) Slice accelerated gradient-echo spin-echo dynamic susceptibility contrast imaging with blipped CAIPI for increased slice coverage. Magn Reson Med 72(3):770–778. https://doi.org/10.1002/mrm.24960

49. Chakhoyan A, Leu K, Pope WB, Cloughesy TF, Ellingson BM (2018) Improved spatio-temporal resolution of dynamic susceptibility contrast perfusion MRI in brain tumors using simultaneous multi-slice echo-planar imaging. AJNR Am J Neuroradiol 39(1):43–45. https://doi.org/10.3174/ajnr.A5433

50. Stokes AM, Skinner JT, Yankeelov T, Quarles CC (2016) Assessment of a simplified spin and gradient echo (sSAGE) approach for human brain tumor perfusion imaging. Magn Reson Imaging 34(9):1248–1255. https://doi.org/10.1016/j.mri.2016.07.004

51. Newton AT, Pruthi S, Stokes AM, Skinner JT, Quarles CC (2016) Improving perfusion measurement in DSC-MR imaging with multiecho information for arterial input function determination. AJNR Am J Neuroradiol 37(7):1237–1243. https://doi.org/10.3174/ajnr.A4700

52. Nasel C, Boubela R, Kalcher K, Moser E (2017) Normalised time-to-peak-distribution curves correlate with cerebral white matter hyperintensities—could this improve early diagnosis? J Cereb Blood Flow Metab 37(2):444–455. https://doi.org/10.1177/0271678x16629485

53. Kanda T, Oba H, Toyoda K, Furui S (2016) Macrocyclic gadolinium-based contrast agents do not cause hyperintensity in the dentate nucleus. AJNR Am J Neuroradiol 37(5):E41. https://doi.org/10.3174/ajnr.A4710

54. Bjornerud A, Vatnehol SAS, Larsson C, Due-Tonnessen P, Hol PK, Groote IR (2017) Signal enhancement of the dentate nucleus at unenhanced MR imaging after very high cumulative doses of the macrocyclic gadolinium-based contrast agent gadobutrol: an observational study. Radiology 285(2):434–444. https://doi.org/10.1148/radiol.2017170391

55. Buxton RB, Frank LR, Wong EC, Siewert B, Warach S, Edelman RR (1998) A general kinetic model for quantitative perfusion imaging with arterial spin labeling. Magn Reson Med 40(3):383–396

56. Welker K, Boxerman J, Kalnin A, Kaufmann T, Shiroishi M, Wintermark M (2015) ASFNR recommendations for clinical performance of MR dynamic susceptibility contrast

perfusion imaging of the brain. AJNR Am J Neuroradiol 36(6):E41–E51. https://doi.org/10.3174/ajnr.A4341

57. Jahng GH, Li KL, Ostergaard L, Calamante F (2014) Perfusion magnetic resonance imaging: a comprehensive update on principles and techniques. Korean J Radiol 15(5):554–577. https://doi.org/10.3348/kjr.2014.15.5.554

58. Louis DN, Perry A, Reifenberger G, von Deimling A, Figarella-Branger D, Cavenee WK, Ohgaki H, Wiestler OD, Kleihues P, Ellison DW (2016) The 2016 World Health Organization classification of tumors of the central nervous system: a summary. Acta Neuropathol 131(6):803–820. https://doi.org/10.1007/s00401-016-1545-1

59. Ameratunga M, Pavlakis N, Wheeler H, Grant R, Simes J, Khasraw M (2018) Anti-angiogenic therapy for high-grade glioma. Cochrane Database Syst Rev 11:Cd008218. https://doi.org/10.1002/14651858.CD008218.pub4

60. Buckner JC, Shaw EG, Pugh SL, Chakravarti A, Gilbert MR, Barger GR, Coons S, Ricci P, Bullard D, Brown PD, Stelzer K, Brachman D, Suh JH, Schultz CJ, Bahary JP, Fisher BJ, Kim H, Murtha AD, Bell EH, Won M, Mehta MP, Curran WJ Jr (2016) Radiation plus procarbazine, CCNU, and vincristine in low-grade glioma. N Engl J Med 374(14):1344–1355. https://doi.org/10.1056/NEJMoa1500925

61. Jain R, Poisson LM, Gutman D, Scarpace L, Hwang SN, Holder CA, Wintermark M, Rao A, Colen RR, Kirby J, Freymann J, Jaffe CC, Mikkelsen T, Flanders A (2014) Outcome prediction in patients with glioblastoma by using imaging, clinical, and genomic biomarkers: focus on the nonenhancing component of the tumor. Radiology 272(2):484–493. https://doi.org/10.1148/radiol.14131691

62. Santarosa C, Castellano A, Conte GM, Cadioli M, Iadanza A, Terreni MR, Franzin A, Bello L, Caulo M, Falini A, Anzalone N (2016) Dynamic contrast-enhanced and dynamic susceptibility contrast perfusion MR imaging for glioma grading: preliminary comparison of vessel compartment and permeability parameters using hotspot and histogram analysis. Eur J Radiol 85(6):1147–1156. https://doi.org/10.1016/j.ejrad.2016.03.020. doi:S0720-048X(16)30095-X [pii]

63. Shiroishi MS, Castellazzi G, Boxerman JL, D'Amore F, Essig M, Nguyen TB, Provenzale JM, Enterline DS, Anzalone N, Dorfler A, Rovira A, Wintermark M, Law M (2015) Principles of T2 *-weighted dynamic susceptibility contrast MRI technique in brain tumor imaging. J Magn Reson Imaging 41(2):296–313. https://doi.org/10.1002/jmri.24648

64. Danchaivijitr N, Waldman AD, Tozer DJ, Benton CE, Brasil Caseiras G, Tofts PS, Rees JH, Jager HR (2008) Low-grade gliomas: do changes in rCBV measurements at longitudinal perfusion-weighted MR imaging predict malignant transformation? Radiology 247(1):170–178. https://doi.org/10.1148/radiol.2471062089

65. Maia AC Jr, Malheiros SM, da Rocha AJ, da Silva CJ, Gabbai AA, Ferraz FA, Stavale JN (2005) MR cerebral blood volume maps correlated with vascular endothelial growth factor expression and tumor grade in nonenhancing gliomas. AJNR Am J Neuroradiol 26(4):777–783

66. Scott JN, Brasher PM, Sevick RJ, Rewcastle NB, Forsyth PA (2002) How often are nonenhancing supratentorial gliomas malignant? A population study. Neurology 59(6):947–949

67. Lev MH, Ozsunar Y, Henson JW, Rasheed AA, Barest GD, Harsh GR 4th, Fitzek MM, Chiocca EA, Rabinov JD, Csavoy AN, Rosen BR, Hochberg FH, Schaefer PW, Gonzalez RG (2004) Glial tumor grading and outcome prediction using dynamic spin-echo MR susceptibility mapping compared with conventional contrast-enhanced MR: confounding effect of elevated rCBV of oligodendrogliomas [corrected]. AJNR Am J Neuroradiol 25(2):214–221

68. Aronen HJ, Gazit IE, Louis DN, Buchbinder BR, Pardo FS, Weisskoff RM, Harsh GR, Cosgrove GR, Halpern EF, Hochberg FH et al (1994) Cerebral blood volume maps of gliomas: comparison with tumor grade and histologic findings. Radiology 191(1):41–51. https://doi.org/10.1148/radiology.191.1.8134596

69. Rosen BR, Belliveau JW, Buchbinder BR, McKinstry RC, Porkka LM, Kennedy DN, Neuder MS, Fisel CR, Aronen HJ, Kwong KK (1991) Contrast agents and cerebral hemodynamics. Magn Reson Med 19(2):285–292

70. Zhang J, Liu H, Tong H, Wang S, Yang Y, Liu G, Zhang W (2017) Clinical applications of contrast-enhanced perfusion MRI techniques in Gliomas: recent advances and current challenges. Contrast Media Mol Imaging 2017:7064120. https://doi.org/10.1155/2017/7064120

71. Järnum H, Steffensen EG, Knutsson L, Fründ E-T, Simonsen CW, Lundbye-Christensen S, Shankaranarayanan A, Alsop DC, Jensen FT, Larsson E-M (2009) Perfusion MRI of brain tumours: a comparative study of pseudo-continuous arterial spin labelling and dynamic susceptibility contrast imaging.

Neuroradiology 52(4):307–317. https://doi.org/10.1007/s00234-009-0616-6

72. Hirai T, Kitajima M, Nakamura H, Okuda T, Sasao A, Shigematsu Y, Utsunomiya D, Oda S, Uetani H, Morioka M, Yamashita Y (2011) Quantitative blood flow measurements in gliomas using arterial spin-labeling at 3T: intermodality agreement and inter- and intraobserver reproducibility study. Am J Neuroradiol 32(11):2073–2079. https://doi.org/10.3174/ajnr.A2725

73. Falk Delgado A, De Luca F, van Westen D, Falk Delgado A (2018) Arterial spin labeling MR imaging for differentiation between high- and low-grade glioma-a meta-analysis. Neuro-Oncology 20(11):1450–1461. https://doi.org/10.1093/neuonc/noy095

74. Burth S, Kickingereder P, Eidel O, Tichy D, Bonekamp D, Weberling L, Wick A, Low S, Hertenstein A, Nowosielski M, Schlemmer HP, Wick W, Bendszus M, Radbruch A (2016) Clinical parameters outweigh diffusion- and perfusion-derived MRI parameters in predicting survival in newly diagnosed glioblastoma. Neuro-Oncology 18(12):1673–1679. https://doi.org/10.1093/neuonc/now122

75. Emblem KE, Due-Tonnessen P, Hald JK, Bjornerud A, Pinho MC, Scheie D, Schad LR, Meling TR, Zoellner FG (2014) Machine learning in preoperative glioma MRI: survival associations by perfusion-based support vector machine outperforms traditional MRI. J Magn Reson Imaging 40(1):47–54. https://doi.org/10.1002/jmri.24390

76. Lee J, Jain R, Khalil K, Griffith B, Bosca R, Rao G, Rao A (2016) Texture feature ratios from relative CBV maps of perfusion MRI are associated with patient survival in glioblastoma. AJNR Am J Neuroradiol 37(1):37–43. https://doi.org/10.3174/ajnr.A4534

77. Law M, Young RJ, Babb JS, Peccerelli N, Chheang S, Gruber ML, Miller DC, Golfinos JG, Zagzag D, Johnson G (2008) Gliomas: predicting time to progression or survival with cerebral blood volume measurements at dynamic susceptibility-weighted contrast-enhanced perfusion MR imaging. Radiology 247(2):490–498. https://doi.org/10.1148/radiol.2472070898

78. Ryoo I, Choi SH, Kim JH, Sohn CH, Kim SC, Shin HS, Yeom JA, Jung SC, Lee AL, Yun TJ, Park CK, Park SH (2013) Cerebral blood volume calculated by dynamic susceptibility contrast-enhanced perfusion MR imaging: preliminary correlation study with glioblastoma genetic profiles. PLoS One 8(8):e71704. https://doi.org/10.1371/journal.pone.0071704

79. Lee S, Choi SH, Ryoo I, Yoon TJ, Kim TM, Lee SH, Park CK, Kim JH, Sohn CH, Park SH, Kim IH (2015) Evaluation of the microenvironmental heterogeneity in high-grade gliomas with IDH1/2 gene mutation using histogram analysis of diffusion-weighted imaging and dynamic-susceptibility contrast perfusion imaging. J Neuro-Oncol 121(1):141–150. 10.1007/s11060-014-1614-z

80. Kickingereder P, Sahm F, Radbruch A, Wick W, Heiland S, Deimling A, Bendszus M, Wiestler B (2015) IDH mutation status is associated with a distinct hypoxia/angiogenesis transcriptome signature which is noninvasively predictable with rCBV imaging in human glioma. Sci Rep 5:16238. https://doi.org/10.1038/srep16238

81. Yamashita K, Hiwatashi A, Togao O, Kikuchi K, Hatae R, Yoshimoto K, Mizoguchi M, Suzuki SO, Yoshiura T, Honda H (2016) MR imaging-based analysis of glioblastoma multiforme: estimation of IDH1 mutation status. AJNR Am J Neuroradiol 37(1):58–65. https://doi.org/10.3174/ajnr.A4491

82. Hilario A, Sepulveda JM, Perez-Nunez A, Salvador E, Millan JM, Hernandez-Lain A, Rodriguez-Gonzalez V, Lagares A, Ramos A (2014) A prognostic model based on preoperative MRI predicts overall survival in patients with diffuse gliomas. AJNR Am J Neuroradiol 35(6):1096–1102. https://doi.org/10.3174/ajnr.A3837

83. Delgado AF, Delgado AF (2017) Discrimination between glioma grades II and III using dynamic susceptibility perfusion MRI: a meta-analysis. AJNR Am J Neuroradiol 38(7):1348–1355. https://doi.org/10.3174/ajnr.A5218

84. Abrigo JM, Fountain DM, Provenzale JM, Law EK, Kwong JS, Hart MG, Tam WWS (2018) Magnetic resonance perfusion for differentiating low-grade from high-grade gliomas at first presentation. Cochrane Database Syst Rev 1:Cd011551. https://doi.org/10.1002/14651858.CD011551.pub2

85. Emblem KE, Nedregaard B, Nome T, Due-Tonnessen P, Hald JK, Scheie D, Borota OC, Cvancarova M, Bjornerud A (2008) Glioma grading by using histogram analysis of blood volume heterogeneity from MR-derived cerebral blood volume maps. Radiology 247(3):808–817. https://doi.org/10.1148/radiol.2473070571

86. Law M, Young R, Babb J, Pollack E, Johnson G (2007) Histogram analysis versus region of interest analysis of dynamic susceptibility contrast perfusion MR imaging data in the grading of cerebral gliomas. AJNR Am J Neuroradiol 28(4):761–766

87. Young R, Babb J, Law M, Pollack E, Johnson G (2007) Comparison of region-of-interest analysis with three different histogram analysis methods in the determination of perfusion metrics in patients with brain gliomas. J Magn Reson Imaging 26(4):1053–1063. https://doi.org/10.1002/jmri.21064

88. Arisawa A, Watanabe Y, Tanaka H, Takahashi H, Matsuo C, Fujiwara T, Fujimoto Y, Yamamoto K, Tomiyama N (2017) Vessel-masked perfusion magnetic resonance imaging with histogram analysis improves diagnostic accuracy for the grading of glioma. J Comput Assist Tomogr 41(6):910–915. https://doi.org/10.1097/RCT.0000000000000614

89. Emblem KE, Due-Tonnessen P, Hald JK, Bjornerud A (2009) Automatic vessel removal in gliomas from dynamic susceptibility contrast imaging. Magn Reson Med 61(5):1210–1217. https://doi.org/10.1002/mrm.21944

90. Roberts HC, Roberts TP, Brasch RC, Dillon WP (2000) Quantitative measurement of microvascular permeability in human brain tumors achieved using dynamic contrast-enhanced MR imaging: correlation with histologic grade. AJNR Am J Neuroradiol 21(5):891–899

91. Jung SC, Yeom JA, Kim JH, Ryoo I, Kim SC, Shin H, Lee AL, Yun TJ, Park CK, Sohn CH, Park SH, Choi SH (2014) Glioma: application of histogram analysis of pharmacokinetic parameters from T1-weighted dynamic contrast-enhanced MR imaging to tumor grading. AJNR Am J Neuroradiol 35(6):1103–1110. https://doi.org/10.3174/ajnr.A3825

92. Nguyen TB, Cron GO, Mercier JF, Foottit C, Torres CH, Chakraborty S, Woulfe J, Jansen GH, Caudrelier JM, Sinclair J, Hogan MJ, Thornhill RE, Cameron IG (2015) Preoperative prognostic value of dynamic contrast-enhanced MRI-derived contrast transfer coefficient and plasma volume in patients with cerebral gliomas. AJNR Am J Neuroradiol 36(1):63–69. https://doi.org/10.3174/ajnr.A4006

93. Toh CH, Wei KC, Chang CN, Ng SH, Wong HF, Lin CP (2014) Differentiation of brain abscesses from glioblastomas and metastatic brain tumors: comparisons of diagnostic performance of dynamic susceptibility contrast-enhanced perfusion MR imaging before and after mathematic contrast leakage correction. PLoS One 9(10):e109172. https://doi.org/10.1371/journal.pone.0109172

94. Neska-Matuszewska M, Bladowska J, Sasiadek M, Zimny A (2018) Differentiation of glioblastoma multiforme, metastases and primary central nervous system lymphomas using multiparametric perfusion and diffusion MR imaging of a tumor core and a peritumoral zone-searching for a practical approach. PLoS One 13(1):e0191341. https://doi.org/10.1371/journal.pone.0191341

95. Murayama K, Nishiyama Y, Hirose Y, Abe M, Ohyu S, Ninomiya A, Fukuba T, Katada K, Toyama H (2018) Differentiating between central nervous system lymphoma and high-grade glioma using dynamic susceptibility contrast and dynamic contrast-enhanced MR imaging with histogram analysis. Magn Reson Med Sci 17(1):42–49. https://doi.org/10.2463/mrms.mp.2016-0113

96. Kickingereder P, Sahm F, Wiestler B, Roethke M, Heiland S, Schlemmer HP, Wick W, von Deimling A, Bendszus M, Radbruch A (2014) Evaluation of microvascular permeability with dynamic contrast-enhanced MRI for the differentiation of primary CNS lymphoma and glioblastoma: radiologic-pathologic correlation. AJNR Am J Neuroradiol 35(8):1503–1508. https://doi.org/10.3174/ajnr.A3915

97. Law M, Cha S, Knopp EA, Johnson G, Arnett J, Litt AW (2002) High-grade gliomas and solitary metastases: differentiation by using perfusion and proton spectroscopic MR imaging. Radiology 222(3):715–721. https://doi.org/10.1148/radiol.2223010558

98. Sparacia G, Gadde JA, Iaia A, Sparacia B, Midiri M (2016) Usefulness of quantitative peritumoural perfusion and proton spectroscopic magnetic resonance imaging evaluation in differentiating brain gliomas from solitary brain metastases. Neuroradiol J 29(3):160–167. https://doi.org/10.1177/1971400916638358

99. Bauer AH, Erly W, Moser FG, Maya M, Nael K (2015) Differentiation of solitary brain metastasis from glioblastoma multiforme: a predictive multiparametric approach using combined MR diffusion and perfusion. Neuroradiology 57(7):697–703. https://doi.org/10.1007/s00234-015-1524-6

100. Server A, Orheim TE, Graff BA, Josefsen R, Kumar T, Nakstad PH (2011) Diagnostic examination performance by using microvascular leakage, cerebral blood volume, and blood flow derived from 3-T dynamic susceptibility-weighted contrast-enhanced perfusion MR imaging in the differentiation of glioblastoma multiforme and brain metastasis. Neuroradiology 53(5):319–330. https://doi.org/10.1007/s00234-010-0740-3

101. Sunwoo L, Yun TJ, You SH, Yoo RE, Kang KM, Choi SH, Kim JH, Sohn CH, Park SW, Jung C, Park CK (2016) Differentiation of glioblastoma from brain metastasis: qualitative and quantitative analysis using arterial spin labeling MR imaging. PLoS

One   11(11):e0166662.   https://doi.org/10.1371/journal.pone.0166662

102. Lin L, Xue Y, Duan Q, Sun B, Lin H, Huang X, Chen X (2016) The role of cerebral blood flow gradient in peritumoral edema for differentiation of glioblastomas from solitary metastatic lesions. Oncotarget 7(42):69051–69059.   https://doi.org/10.18632/oncotarget.12053

103. Weber MA, Zoubaa S, Schlieter M, Juttler E, Huttner HB, Geletneky K, Ittrich C, Lichy MP, Kroll A, Debus J, Giesel FL, Hartmann M, Essig M (2006) Diagnostic performance of spectroscopic and perfusion MRI for distinction of brain tumors. Neurology 66(12):1899–1906.   https://doi.org/10.1212/01.wnl.0000219767.49705.9c

104. Kremer S, Grand S, Berger F, Hoffmann D, Pasquier B, Remy C, Benabid AL, Bas JF (2003) Dynamic contrast-enhanced MRI: differentiating melanoma and renal carcinoma metastases from high-grade astrocytomas and other metastases. Neuroradiology 45(1):44–49.   https://doi.org/10.1007/s00234-002-0886-8

105. Boxerman JL, Ellingson BM (2015) Response assessment and magnetic resonance imaging issues for clinical trials involving high-grade gliomas. Top Magn Reson Imaging 24(3):127–136.   https://doi.org/10.1097/RMR.0000000000000054

106. Patel P, Baradaran H, Delgado D, Askin G, Christos P, John Tsiouris A, Gupta A (2017) MR perfusion-weighted imaging in the evaluation of high-grade gliomas after treatment: a systematic review and meta-analysis. Neuro-Oncology   19(1):118–127.   https://doi.org/10.1093/neuonc/now148

107. de Wit MC, de Bruin HG, Eijkenboom W, Sillevis Smitt PA, van den Bent MJ (2004) Immediate post-radiotherapy changes in malignant glioma can mimic tumor progression. Neurology 63(3):535–537

108. Abbasi AW, Westerlaan HE, Holtman GA, Aden KM, van Laar PJ, van der Hoorn A (2018) Incidence of tumour progression and pseudoprogression in high-grade gliomas: a systematic review and meta-analysis. Clin Neuroradiol 28(3):401–411.   https://doi.org/10.1007/s00062-017-0584-x

109. Tsao M, Xu W, Sahgal A (2012) A meta-analysis evaluating stereotactic radiosurgery, whole-brain radiotherapy, or both for patients presenting with a limited number of brain metastases. Cancer 118(9):2486–2493.   https://doi.org/10.1002/cncr.26515

110. Patel TR, McHugh BJ, Bi WL, Minja FJ, Knisely JP, Chiang VL (2011) A comprehensive review of MR imaging changes following radiosurgery to 500 brain metastases.

AJNR Am J Neuroradiol 32(10):1885–1892.   https://doi.org/10.3174/ajnr.A2668

111. Thust SC, van den Bent MJ, Smits M (2018) Pseudoprogression of brain tumors. J Magn Reson Imaging. https://doi.org/10.1002/jmri.26171

112. Brandsma D, Stalpers L, Taal W, Sminia P, van den Bent MJ (2008) Clinical features, mechanisms, and management of pseudoprogression in malignant gliomas. Lancet Oncol 9(5):453–461.   https://doi.org/10.1016/S1470-2045(08)70125-6

113. Miyatake S, Nonoguchi N, Furuse M, Yoritsune E, Miyata T, Kawabata S, Kuroiwa T (2015) Pathophysiology, diagnosis, and treatment of radiation necrosis in the brain. Neurol Med Chir 55(1):50–59. https://doi.org/10.2176/nmc.ra.2014-0188

114. Taal W, Brandsma D, de Bruin HG, Bromberg JE, Swaak-Kragten AT, Smitt PA, van Es CA, van den Bent MJ (2008) Incidence of early pseudo-progression in a cohort of malignant glioma patients treated with chemoirradiation with temozolomide. Cancer 113(2):405–410. https://doi.org/10.1002/cncr.23562

115. Hoefnagels FW, Lagerwaard FJ, Sanchez E, Haasbeek CJ, Knol DL, Slotman BJ, Vandertop WP (2009) Radiological progression of cerebral metastases after radiosurgery: assessment of perfusion MRI for differentiating between necrosis and recurrence. J Neurol 256(6):878–887. https://doi.org/10.1007/s00415-009-5034-5

116. Prager AJ, Martinez N, Beal K, Omuro A, Zhang Z, Young RJ (2015) Diffusion and perfusion MRI to differentiate treatment-related changes including pseudoprogression from recurrent tumors in high-grade gliomas with histopathologic evidence. AJNR Am J Neuroradiol 36(5):877–885. https://doi.org/10.3174/ajnr.A4218

117. Hu LS, Baxter LC, Smith KA, Feuerstein BG, Karis JP, Eschbacher JM, Coons SW, Nakaji P, Yeh RF, Debbins J, Heiserman JE (2009) Relative cerebral blood volume values to differentiate high-grade glioma recurrence from posttreatment radiation effect: direct correlation between image-guided tissue histopathology and localized dynamic susceptibility-weighted   contrast-enhanced perfusion MR imaging measurements. AJNR Am J Neuroradiol 30(3):552–558. https://doi.org/10.3174/ajnr.A1377

118. Thust SC, Heiland S, Falini A, Jager HR, Waldman AD, Sundgren PC, Godi C, Katsaros VK, Ramos A, Bargallo N, Vernooij MW, Yousry T, Bendszus M, Smits M (2018) Glioma imaging in Europe: a survey of 220 centres and recommendations for best clinical practice. Eur Radiol 28(8):3306–

3317. https://doi.org/10.1007/s00330-018-5314-5

119. Ye J, Bhagat SK, Li H, Luo X, Wang B, Liu L, Yang G (2016) Differentiation between recurrent gliomas and radiation necrosis using arterial spin labeling perfusion imaging. Exp Ther Med 11(6):2432–2436. https://doi.org/10.3892/etm.2016.3225

120. Xu Q, Liu Q, Ge H, Ge X, Wu J, Qu J, Xu K (2017) Tumor recurrence versus treatment effects in glioma: a comparative study of three dimensional pseudo-continuous arterial spin labeling and dynamic susceptibility contrast imaging. Medicine 96(50):e9332. https://doi.org/10.1097/md.0000000000009332

121. Ellingson BM, Chung C, Pope WB, Boxerman JL, Kaufmann TJ (2017) Pseudoprogression, radionecrosis, inflammation or true tumor progression? Challenges associated with glioblastoma response assessment in an evolving therapeutic landscape. J Neuro-Oncol 134(3):495–504. https://doi.org/10.1007/s11060-017-2375-2

122. Ozsunar Y, Mullins ME, Kwong K, Hochberg FH, Ament C, Schaefer PW, Gonzalez RG, Lev MH (2010) Glioma recurrence versus radiation necrosis? A pilot comparison of arterial spin-labeled, dynamic susceptibility contrast enhanced MRI, and FDG-PET imaging. Acad Radiol 17(3):282–290. https://doi.org/10.1016/j.acra.2009.10.024

123. Boxerman JL, Shiroishi MS, Ellingson BM, Pope WB (2016) Dynamic susceptibility contrast MR imaging in glioma: review of current clinical practice. Magn Reson Imaging Clin N Am 24(4):649–670. https://doi.org/10.1016/j.mric.2016.06.005

124. Baek HJ, Kim HS, Kim N, Choi YJ, Kim YJ (2012) Percent change of perfusion skewness and kurtosis: a potential imaging biomarker for early treatment response in patients with newly diagnosed glioblastomas. Radiology 264(3):834–843. https://doi.org/10.1148/radiol.12112120

125. Gasparetto EL, Pawlak MA, Patel SH, Huse J, Woo JH, Krejza J, Rosenfeld MR, O'Rourke DM, Lustig R, Melhem ER, Wolf RL (2009) Posttreatment recurrence of malignant brain neoplasm: accuracy of relative cerebral blood volume fraction in discriminating low from high malignant histologic volume fraction. Radiology 250(3):887–896. https://doi.org/10.1148/radiol.2502071444

126. Hu LS, Eschbacher JM, Heiserman JE, Dueck AC, Shapiro WR, Liu S, Karis JP, Smith KA, Coons SW, Nakaji P, Spetzler RF, Feuerstein BG, Debbins J, Baxter LC (2012) Reevaluating the imaging definition of tumor progression: perfusion MRI quantifies recurrent glioblastoma tumor fraction, pseudo-progression, and radiation necrosis to predict survival. Neuro-Oncology 14(7):919–930. https://doi.org/10.1093/neuonc/nos112

127. Mangla R, Singh G, Ziegelitz D, Milano MT, Korones DN, Zhong J, Ekholm SE (2010) Changes in relative cerebral blood volume 1 month after radiation-temozolomide therapy can help predict overall survival in patients with glioblastoma. Radiology 256(2):575–584. https://doi.org/10.1148/radiol.10091440

128. Barajas RF, Chang JS, Sneed PK, Segal MR, McDermott MW, Cha S (2009) Distinguishing recurrent intra-axial metastatic tumor from radiation necrosis following gamma knife radiosurgery using dynamic susceptibility-weighted contrast-enhanced perfusion MR imaging. AJNR Am J Neuroradiol 30(2):367–372. https://doi.org/10.3174/ajnr.A1362

129. Mitsuya K, Nakasu Y, Horiguchi S, Harada H, Nishimura T, Bando E, Okawa H, Furukawa Y, Hirai T, Endo M (2010) Perfusion weighted magnetic resonance imaging to distinguish the recurrence of metastatic brain tumors from radiation necrosis after stereotactic radiosurgery. J Neuro-Oncol 99(1):81–88. https://doi.org/10.1007/s11060-009-0106-z

130. Jakubovic R, Sahgal A, Soliman H, Milwid R, Zhang L, Eilaghi A, Aviv RI (2014) Magnetic resonance imaging-based tumour perfusion parameters are biomarkers predicting response after radiation to brain metastases. Clin Oncol (R Coll Radiol) 26(11):704–712. https://doi.org/10.1016/j.clon.2014.06.010

131. Jakubovic R, Sahgal A, Ruschin M, Pejovic-Milic A, Milwid R, Aviv RI (2015) Non tumor perfusion changes following stereotactic radiosurgery to brain metastases. Technol Cancer Res Treat 14(4):497–503. https://doi.org/10.1177/1533034614600279. doi:14/4/497 [pii]

132. Digernes I, Grovik E, Nilsen LB, Saxhaug C, Geier O, Reitan E, Saetre DO, Breivik B, Reese T, Jacobsen KD, Helland A, Emblem KE (2018) Brain metastases with poor vascular function are susceptible to pseudoprogression after stereotactic radiation surgery. Adv Radiat Oncol 3(4):559–567. https://doi.org/10.1016/j.adro.2018.05.005

133. Carmeliet P, Jain RK (2011) Molecular mechanisms and clinical applications of angiogenesis. Nature 473(7347):298–307. https://doi.org/10.1038/nature10144

134. Lu-Emerson C, Duda DG, Emblem KE, Taylor JW, Gerstner ER, Loeffler JS, Batchelor TT, Jain RK (2015) Lessons from anti-vascular endothelial growth factor and anti-vascular endothelial growth factor receptor trials in patients with glioblastoma. J Clin

Oncol 33(10):1197–1213. https://doi.org/10.1200/JCO.2014.55.9575

135. Huang RY, Neagu MR, Reardon DA, Wen PY (2015) Pitfalls in the neuroimaging of glioblastoma in the era of antiangiogenic and immuno/targeted therapy—detecting illusive disease, defining response. Front Neurol 6, 33. https://doi.org/10.3389/fneur.2015.00033

136. Choi SH, Jung SC, Kim KW, Lee JY, Choi Y, Park SH, Kim HS (2016) Perfusion MRI as the predictive/prognostic and pharmacodynamic biomarkers in recurrent malignant glioma treated with bevacizumab: a systematic review and a time-to-event meta-analysis. J Neuro-Oncol 128(2):185–194. https://doi.org/10.1007/s11060-016-2102-4

137. Schmainda KM, Zhang Z, Prah M, Snyder BS, Gilbert MR, Sorensen AG, Barboriak DP, Boxerman JL (2015) Dynamic susceptibility contrast MRI measures of relative cerebral blood volume as a prognostic marker for overall survival in recurrent glioblastoma: results from the ACRIN 6677/RTOG 0625 multicenter trial. Neuro-Oncology 17(8):1148–1156. https://doi.org/10.1093/neuonc/nou364

138. Bonekamp D, Deike K, Wiestler B, Wick W, Bendszus M, Radbruch A, Heiland S (2015) Association of overall survival in patients with newly diagnosed glioblastoma with contrast-enhanced perfusion MRI: comparison of intraindividually matched T1- and T2 (*) -based bolus techniques. J Magn Reson Imaging 42(1):87–96. https://doi.org/10.1002/jmri.24756

139. Kickingereder P, Wiestler B, Burth S, Wick A, Nowosielski M, Heiland S, Schlemmer HP, Wick W, Bendszus M, Radbruch A (2015) Relative cerebral blood volume is a potential predictive imaging biomarker of bevacizumab efficacy in recurrent glioblastoma. Neuro-Oncology 17(8):1139–1147. https://doi.org/10.1093/neuonc/nov028

140. Kickingereder P, Radbruch A, Burth S, Wick A, Heiland S, Schlemmer HP, Wick W, Bendszus M, Bonekamp D (2016) MR perfusion-derived hemodynamic parametric response mapping of bevacizumab efficacy in recurrent glioblastoma. Radiology 279(2):542–552. https://doi.org/10.1148/radiol.2015151172

141. Emblem KE, Mouridsen K, Bjornerud A, Farrar CT, Jennings DL, Borra RJ, Wen PY, Ivy P, Batchelor TT, Rosen BR, Jain RK, Sorensen AG (2013) Vessel architectural imaging identifies cancer patient responders to anti-angiogenic therapy. Proc American Association for Cancer Research (AACR) Annual Meeting LB-297: Minisymposium—Biomarkers of Clinical Response

142. Baish JW, Stylianopoulos T, Lanning RM, Kamoun WS, Fukumura D, Munn LL, Jain RK (2011) Scaling rules for diffusive drug delivery in tumor and normal tissues. Proc Natl Acad Sci U S A 108(5):1799–1803. https://doi.org/10.1073/pnas.1018154108

143. Batchelor TT, Gerstner ER, Emblem KE, Duda DG, Kalpathy-Cramer J, Snuderl M, Ancukiewicz M, Polaskova P, Pinho MC, Jennings D, Plotkin SR, Chi AS, Eichler AF, Dietrich J, Hochberg FH, Lu-Emerson C, Iafrate AJ, Ivy SP, Rosen BR, Loeffler JS, Wen PY, Sorensen AG, Jain RK (2013) Improved tumor oxygenation and survival in glioblastoma patients who show increased blood perfusion after cediranib and chemoradiation. Proc Natl Acad Sci U S A 110(47):19059–19064. https://doi.org/10.1073/pnas.1318022110

144. Sorensen AG, Emblem KE, Polaskova P, Jennings D, Kim H, Ancukiewicz M, Wang M, Wen PY, Ivy P, Batchelor TT, Jain RK (2012) Increased survival of glioblastoma patients who respond to antiangiogenic therapy with elevated blood perfusion. Cancer Res 72(2):402–407. https://doi.org/10.1158/0008-5472.CAN-11-2464

145. Ken S, Deviers A, Filleron T, Catalaa I, Lotterie JA, Khalifa J, Lubrano V, Berry I, Peran P, Celsis P, Moyal EC, Laprie A (2015) Voxel-based evidence of perfusion normalization in glioblastoma patients included in a phase I-II trial of radiotherapy/tipifarnib combination. J Neuro-Oncol 124(3):465–473. https://doi.org/10.1007/s11060-015-1860-8

146. Kalpathy-Cramer J, Chandra V, Da X, Ou Y, Emblem KE, Muzikansky A, Cai X, Douw L, Evans JG, Dietrich J, Chi AS, Wen PY, Stufflebeam S, Rosen B, Duda DG, Jain RK, Batchelor TT, Gerstner ER (2017) Phase II study of tivozanib, an oral VEGFR inhibitor, in patients with recurrent glioblastoma. J Neuro-Oncol 131(3):603–610. https://doi.org/10.1007/s11060-016-2332-5

147. Pinho MC, Polaskova P, Kalpathy-Cramer J, Jennings D, Emblem KE, Jain RK, Rosen BR, Wen PY, Sorensen AG, Batchelor TT, Gerstner ER (2014) Low incidence of pseudoprogression by imaging in newly diagnosed glioblastoma patients treated with cediranib in combination with chemoradiation. Oncologist 19(1):75–81. https://doi.org/10.1634/theoncologist.2013-0101

148. Artzi M, Blumenthal DT, Bokstein F, Nadav G, Liberman G, Aizenstein O, Ben Bashat D (2015) Classification of tumor area using combined DCE and DSC MRI in patients with glioblastoma. J Neuro-Oncol 121(2):349–357. https://doi.org/10.1007/s11060-014-1639-3

149. Netto JP, Schwartz D, Varallyay C, Fu R, Hamilton B, Neuwelt EA (2016) Misleading early blood volume changes obtained using ferumoxytol-based magnetic resonance imaging perfusion in high grade glial neoplasms treated with bevacizumab. Fluids Barriers CNS 13(1):23. https://doi.org/10.1186/s12987-016-0047-9

150. Wolchok JD, Chiarion-Sileni V, Gonzalez R, Rutkowski P, Grob JJ, Cowey CL, Lao CD, Wagstaff J, Schadendorf D, Ferrucci PF, Smylie M, Dummer R, Hill A, Hogg D, Haanen J, Carlino MS, Bechter O, Maio M, Marquez-Rodas I, Guidoboni M, McArthur G, Lebbe C, Ascierto PA, Long GV, Cebon J, Sosman J, Postow MA, Callahan MK, Walker D, Rollin L, Bhore R, Hodi FS, Larkin J (2017) Overall survival with combined Nivolumab and Ipilimumab in advanced melanoma. N Engl J Med 377(14):1345–1356. https://doi.org/10.1056/NEJMoa1709684

151. Margolin K, Ernstoff MS, Hamid O, Lawrence D, McDermott D, Puzanov I, Wolchok JD, Clark JI, Sznol M, Logan TF, Richards J, Michener T, Balogh A, Heller KN, Hodi FS (2012) Ipilimumab in patients with melanoma and brain metastases: an open-label, phase 2 trial. Lancet Oncol 13(5):459–465. https://doi.org/10.1016/s1470-2045(12)70090-6

152. Wong CS, Van der Kogel AJ (2004) Mechanisms of radiation injury to the central nervous system: implications for neuroprotection. Mol Interv 4(5):273–284. https://doi.org/10.1124/mi.4.5.7

153. Owonikoko TK, Arbiser J, Zelnak A, Shu HK, Shim H, Robin AM, Kalkanis SN, Whitsett TG, Salhia B, Tran NL, Ryken T, Moore MK, Egan KM, Olson JJ (2014) Current approaches to the treatment of metastatic brain tumours. Nat Rev Clin Oncol 11(4):203–222. https://doi.org/10.1038/nrclinonc.2014.25

154. Rosenberg SA (2014) Decade in review-cancer immunotherapy: entering the mainstream of cancer treatment. Nat Rev Clin Oncol 11(11):630–632. https://doi.org/10.1038/nrclinonc.2014.174

155. Okada H, Weller M, Huang R, Finocchiaro G, Gilbert MR, Wick W, Ellingson BM, Hashimoto N, Pollack IF, Brandes AA, Franceschi E, Herold-Mende C, Nayak L, Panigrahy A, Pope WB, Prins R, Sampson JH, Wen PY, Reardon DA (2015) Immunotherapy response assessment in neuro-oncology: a report of the RANO working group. Lancet Oncol 16(15):e534–e542. https://doi.org/10.1016/S1470-2045(15)00088-1

156. Aquino D, Gioppo A, Finocchiaro G, Bruzzone MG, Cuccarini V (2017, 2017) MRI in glioma immunotherapy: evidence, pitfalls, and perspectives. J Immunol Res:5813951. https://doi.org/10.1155/2017/5813951

157. Vrabec M, Van Cauter S, Himmelreich U, Van Gool SW, Sunaert S, De Vleeschouwer S, Suput D, Demaerel P (2011) MR perfusion and diffusion imaging in the follow-up of recurrent glioblastoma treated with dendritic cell immunotherapy: a pilot study. Neuroradiology 53(10):721–731. https://doi.org/10.1007/s00234-010-0802-6

158. Stenberg L, Englund E, Wirestam R, Siesjo P, Salford LG, Larsson EM (2006) Dynamic susceptibility contrast-enhanced perfusion magnetic resonance (MR) imaging combined with contrast-enhanced MR imaging in the follow-up of immunogene-treated glioblastoma multiforme. Acta Radiol (Stockholm, Sweden: 1987) 47(8):852–861. https://doi.org/10.1080/02841850600815341

159. Digernes I, Grovik E, Nilsen LB, Saxhaug C, Geier O, Saetre DO, Breivik B, Jacobsen KD, Helland A, Emblem K. Vascular Responses to Pembrolizumab and Ipilimumab in Patients with Metastases to the Brain Receiving Stereotactic Radiosurgery. In: 2nd Special Conference European Association for Cancer Research American Association for Cancer Research, Florence, Italy, 24–27 June 2017

160. Willats L, Calamante F (2013) The 39 steps: evading error and deciphering the secrets for accurate dynamic susceptibility contrast MRI. NMR Biomed 26(8):913–931. https://doi.org/10.1002/nbm.2833

161. Aran D, Camarda R, Odegaard J, Paik H, Oskotsky B, Krings G, Goga A, Sirota M, Butte AJ (2017) Comprehensive analysis of normal adjacent to tumor transcriptomes. Nat Commun 8(1):1077. https://doi.org/10.1038/s41467-017-01027-z

162. Lemee JM, Clavreul A, Menei P (2015) Intratumoral heterogeneity in glioblastoma: don't forget the peritumoral brain zone. Neuro-Oncology 17(10):1322–1332. https://doi.org/10.1093/neuonc/nov119

163. Gilkes DM, Semenza GL, Wirtz D (2014) Hypoxia and the extracellular matrix: drivers of tumour metastasis. Nat Rev Cancer 14(6):430–439. https://doi.org/10.1038/nrc3726

164. Steeg PS (2016) Targeting metastasis. Nat Rev Cancer 16(4):201–218. https://doi.org/10.1038/nrc.2016.25

165. Christen T, Pannetier NA, Ni WW, Qiu D, Moseley ME, Schuff N, Zaharchuk G (2014) MR vascular fingerprinting: a new approach to

compute cerebral blood volume, mean vessel radius, and oxygenation maps in the human brain. NeuroImage 89:262–270. https://doi.org/10.1016/j.neuroimage.2013.11.052

166. Digernes I, Bjornerud A, Vatnehol SAS, Lovland G, Courivaud F, Vik-Mo E, Meling TR, Emblem KE (2017) A theoretical framework for determining cerebral vascular function and heterogeneity from dynamic susceptibility contrast MRI. J Cereb Blood Flow Metab 37(6):2237–2248. https://doi.org/10.1177/0271678x17694187

167. Hernandez-Torres E, Kassner N, Forkert ND, Wei L, Wiggermann V, Daemen M, Machan L, Traboulsee A, Li D, Rauscher A (2017) Anisotropic cerebral vascular architecture causes orientation dependency in cerebral blood flow and volume measured with dynamic susceptibility contrast magnetic resonance imaging. J Cereb Blood Flow Metab 37(3):1108–1119. https://doi.org/10.1177/0271678x16653134

168. Lemasson B, Pannetier N, Coquery N, Boisserand LS, Collomb N, Schuff N, Moseley M, Zaharchuk G, Barbier EL, Christen T (2016) MR vascular fingerprinting in stroke and brain tumors models. Sci Rep 6:37071. https://doi.org/10.1038/srep37071

169. Semmineh NB, Xu J, Skinner JT, Xie J, Li H, Ayers G, Quarles CC (2015) Assessing tumor cytoarchitecture using multiecho DSC-MRI derived measures of the transverse relaxivity at tracer equilibrium (TRATE). Magn Reson Med 74(3):772–784. https://doi.org/10.1002/mrm.25435

Chapter 14

MRI Morphometry in Brain Tumors: Challenges and Opportunities in Expert, Radiomic, and Deep-Learning-Based Analyses

Marco C. Pinho, Kaustav Bera, Niha Beig, and Pallavi Tiwari

Abstract

Morphometry refers to the quantitative study of form, which has gained popularity in neurosciences for non-invasive in vivo evaluation of the normal and aging brain through the use of neuroimaging data, and hence designated as brain morphometry. In the rapidly evolving field of neuro-oncology, morphological evaluation provided by neuroimaging studies has been a cornerstone for the initial diagnosis, classification, management, and post-treatment follow-up of brain tumors. However, it has historically relied on predominantly subjective and qualitative observations made by imaging experts based on clinical experience. The wealth of knowledge obtained through visual inspection of tumor imaging has made remarkable contributions to the field and enhanced our understanding of tumor biology and natural history; however, further developments have been hampered by the lack of robust methods for more automated and quantitative evaluation. These methods are becoming more readily available and have been fueled by breakthrough developments in imaging post-processing and artificial intelligence. In this chapter, we review past contributions and evolution of the field of brain tumor morphological evaluation as it evolves into more automated computerized methods including radiomics and deep learning.

Key words Brain tumor, MRI, Glioma, Morphometry, Radiomics, Radiogenomics, Deep learning

Abbreviations

2D	2-dimensional
3D	3-dimensional
AA	Anaplastic astrocytoma
ANN	Artificial neural network
BBB	blood-brain barrier
BraTs	Brain tumor segmentation
CE	Contrast enhancement
CNS	Central nervous system
CoLIAGe	Co-occurrence of local anisotropic gradient orientations
CSF	cerebrospinal fluid
CT	Computed tomography
DTI	Diffusion tensor imaging

Giorgio Seano (ed.), *Brain Tumors*, Neuromethods, vol. 158,
https://doi.org/10.1007/978-1-0716-0856-2_14, © Springer Science+Business Media, LLC, part of Springer Nature 2021

DWI	Diffusion-weighted imaging
GBCAs	Gadolinium-based contrast agents
GBM	Glioblastoma multiforme, or glioblastoma
GLCM	Gray-level co-occurrence matrix
GLRLM	Gray-level run length matrix
GRE	Gradient echo
ICA	Iodinated contrast agent
IDH	Isocitrate dehydrogenase
IV	Intravenous
KPS	Karnofski Performance Scale
LBP	Local binary patterns
LGG	Low-grade glioma
MRI	Magnetic resonance imaging
MRS	Magnetic resonance spectroscopy
nCET	Non-contrast-enhancing tumor
NEX	Number of excitations
NSCLC	Non-small cell lung cancer
OS	Overall survival
PET	Positron emission tomography
PFS	Progression-free survival
PTE	Peritumoral edema
PWI	Perfusion-weighted imaging
ROI	Region of interest
SE	Spin echo
SFTA	Segmentation-based fractal texture analysis
SWI	Susceptibility-weighted imaging
T1W	T1-weighted
T1W+C	T1-weighted post contrast
T2W	T2-weighted
T2W-FLAIR	T2-weighted fluid-attenuated inversion recovery
TCGA	The Cancer Genome Atlas
TCIA	The Cancer Imaging Archive
TN	Tumor necrosis
TSE	Turbo spin echo
VASARI	Visually AcceSAble Rembrandt Images
WHO	World Health Organization
XRT	Chemoradiation

1   Introduction

The brain parenchyma has the potential to harbor a large and heterogeneous group of benign and malignant neoplasms across different age groups. Primary brain neoplasms arise from diverse cell types and are most often of neuroectodermal origin, with gliomas representing the most clinically significant group of tumors in adults due to their overall frequency and morbi-mortality [1]. Secondary tumors can involve the brain parenchyma by hematogenous dissemination (most commonly), cerebrospinal fluid (CSF) spread, and direct extension. Despite being relatively rare in comparison to their counterparts in other body regions in the adult population [1], brain tumors account for a disproportional amount of overall morbidity, disability, and death. One of the main contributing factors for the distinctively ominous nature of these neoplasms is the close proximity/involvement of brain regions which are essential for survival or maintenance of basic neurological skills for a functional living. The often-infiltrative nature of these tumors makes them inseparable from the surrounding brain parenchyma and significantly limits options for curative surgical and nonsurgical ablative therapies [2–4]. However, there are other complicating factors for timely and curative treatment, including the often delayed and nonspecific nature of symptoms at presentation [5–9], and presence of an effective blood-brain barrier (BBB) which limits the delivery of most chemotherapeutic agents to the tumor bed at effective levels [10, 11]. Moreover, recent scientific developments in cell biology and genetics have disclosed additional challenging biological characteristics for some specific tumor subtypes, such as gliomas, and particularly for the one tumor type which is both the most common and deadly, i.e., glioblastoma (GBM). Intratumoral heterogeneity, clonal selection/evolution, and presence of pluripotential stem cell reservoirs represent additional layers of complexity for medical treatment of these neoplasms [12–15]. Nonsurprisingly, despite significant technological advancements and improvements in understanding of glial cell biology over the recent decades, only minor improvements in clinically significant outcome variables have been achieved [5–7]. For GBM, overall prognosis remains dismal and post-treatment quality of life in the few long-term survivors is poor [5, 15].

Alongside micro-neurosurgical techniques and technical improvements in radiation therapy, cross-sectional neuroimaging has undoubtedly been one of the most impactful developments in the field of neuro-oncology since its inception [16–19]. The use of computed tomography (CT) revolutionized modern neurosurgery and for the first time allowed neurosurgeons to non-invasively identify and localize a brain tumor reliably [20, 21]. However, since the advent of magnetic resonance imaging (MRI) [22], it has been largely supplanted. Due to its powerful contrast capabilities, versatility, inherent multiplanar capabilities, and ongoing technical advances, MRI has become an indispensable tool in modern neuro-oncology for diagnostic confirmation, localization, histology prediction, surgical planning, and treatment monitoring [23–26]. Among the breadth of information provided by MRI, the evaluation of tumor morphology/characteristics, location, and relationship to adjacent brain structures represent the cornerstones of imaging diagnosis and allow clinicians to make complex, informed decisions about best clinical practices. The study of tumor imaging morphology in particular has evolved tremendously in a few decades. From initial anecdotal and subjective inferences about tumor biology, to small case series of radiology-pathology correlation and exploration of imaging features predictive of clinical outcomes, the field of imaging morphometry has made meaningful progress, but is nonetheless still in its early years. More recently, an explosive number of new technological tools and methodologies coupled with groundbreaking developments in tumor genetic and epigenetic pathways have substantially accelerated the pace of new discoveries and will pave the way for more data-driven, objective, and certainly impactful results. In this chapter, we discuss the evolution of different methods for the analysis of brain tumor morphology and their results with a main focus on glial neoplasms and MRI. We start with a review of early qualitative, operator-dependent approaches discussing their contributions and limitations, and then switch to subsequent sections exploring new quantitative, computerized, and automated approaches. These latter sections particularly focus on the fields of Radiomics/Radiogenomics, with a detailed description of methods, processing steps, feature selection, classifier construction, and clinical applications.

2   Current Role of Clinical MRI in the Management of Brain Tumors

Due to the often nonspecific clinical presentation of most brain tumors, cross-sectional imaging is usually required for confirmation. CT is sometimes initially ordered in patients presenting to emergency rooms or with low clinical suspicion, and can confirm the presence of a brain tumor, however lacks significantly in overall accuracy when compared to MRI. The main reason underlying the overall supremacy of MRI for evaluation of brain tumors (as well as most intracranial disorders) is its exquisite ability to generate a number of contrast types between normal and abnormal tissues, which aids in both detection and characterization of intracranial masses. Therefore, CT has been mostly supplanted by MRI and currently has a limited role in the evaluation of brain tumors (used sometimes for detection of calcifications and associated osseous abnormalities). Morphology is the most important finding to raise initial suspicion for a brain neoplasm in any given patient. Tumors are identified on MRI as expansive areas of signal abnormality which may be focal or diffuse, solitary or multiple. Focal and solitary lesions usually allow a more straightforward diagnosis, while multiple, diffuse, or ill-defined lesions sometimes can be more challenging, due to overlapping presentations with inflammatory, infectious, vascular, or other disease processes [27–30]. Gliomas, which are tumors originated from glial cells (and which will be the main focus of this review), constitute the overall majority of primary intra-axial brain tumors [1], and can be confidently identified in most patients, due to some unique morphological attributes. The most common subtypes of glioma (which include oligoden-drogliomas, diffuse astrocytomas, anaplastic astrocytomas, and glioblastomas) have an intrinsic ability to infiltrate the brain paren-chyma expanding it, distorting it, and growing along with both white and gray matter structures. Therefore, the signature mor-phological trait is that of masses with irregular shapes and ill-defined margins, which tend to overall mimic the morphology of the brain region of origin. Shape is the most reliable feature to differentiate a glial neoplasm from the other most important and frequent group of intra-axial brain tumors: metastatic lesions. MRI is a powerful tool not only to identify a glioma, but also to suggest its subtype and important biological traits (such as proliferative potential and aggressiveness, i.e., grade). Contrast-enhancement (CE) on T1-weighted (T1W) sequences due to leakage of gadolinium-based contrast agents (GBCAs) is one of the most sig-nificant and reliable predictive findings of aggressiveness in diffuse astrocytomas (the most common subtypes of gliomas), and results from tumor-induced disruption of the blood-brain barrier.

There are several other imaging features, which have been extensively explored in terms of morphological characterization for histology prediction, which include MRI signal, shape and margin characteristics, edema, heterogeneity, necrosis/cysts, vascularity, calcifications, and hemorrhage, among others [31–37]. Although it is a general consensus that MRI findings alone are not sufficient to definitely diagnose a particular brain tumor and that histopatho-logical evaluation is ultimately required to establish tumor type and grade, brain tissue sampling is an invasive, risk-associated proce-dure [38, 39], and the management decision to sample a tumor or not is essentially made by a combination of clinical and MRI data. Most clinicians agree, for example, that a "watch and wait" approach without tissue sampling can be an appropriate option for a presumable low-grade glioma (LGG) diagnosed solely by MRI morphological evaluation, particularly in an oligo-symptomatic patient or in a surgically challenging brain region. In addition to its invasive nature, histopathology has other limitations in brain tumor characterization, such as incomplete sampling in stereotactic or surgical biopsies, which may result in mischaracterization of tumor type or grade [40, 41]. Partial sampling errors further highlight the importance of imaging, since MRI provides information about the entire tumor mass, instead of small fragments.

Additionally, MRI constitutes the single most important ancil-lary diagnostic modality in all other steps of management of a brain tumor, including conservative surveillance to assess for interval changes [42–44], structural and functional pre-treatment planning for surgical, radiation or other focal (i.e., laser) treatment modalities [45], and lifelong post-treatment surveillance, since diffuse glial neoplasms are usually not curable, and at best controllable [5–7]. It is not uncommon, therefore, for patients with glial neoplasms to undergo dozens of MRI studies in their lifetime, sometimes on an annual basis, as in lower-grade gliomas on MRI surveillance [42–44], or as often as every 2–3 months during the early post-chemoradiation (XRT) stage of malignant gliomas.

Despite its crucial role, it is generally believed that the vast amount of information provided by MRI is not fully used clinically, due to the subjective nature of imaging interpretation provided by radiologists and other clinicians involved in neuro-oncology care. Over the last decade, a significant body of research has been done as an attempt to replace the cognitive process of morphological evaluation of brain tumor MRI features by more measureable, morphometric approaches, which have the potential to be more standardized, accurate, and reproducible. The following sections will provide an overview about methods and results, evolving from predominantly qualitative, expert-dependent, and subjective approaches to more recent automated, quantitative, and data-driven methods. The focus will be on imaging data obtained from standard clinical MRI sequences, and Table 14.1 summarizes the main strengths and weaknesses of each method. "Advanced" functional and physiological approaches available clinically with MRI such as perfusion, permeability, diffusion, and spectroscopy, for example, will be the subject of other chapters.

3   Expert-Based Approaches for Tumor Morphometry—Overview of Methods, Advances in Knowledge, and Limitations

3.1   General Overview of Methods

There is an abundance of imaging literature exploring methods to characterize brain tumors morphologically on routine structural MRI based on expert interpretations. The overarching goal is to gain insights into important biological properties and clinically relevant measures. The individual methodologies used in these predominantly qualitative studies vary substantially, however, there are some common denominators. The vast majority of studies are small- to medium-sized observational studies with tens [31, 32, 33, 36, 46, 47, 48, 49–54] to several hundred [33, 51, 55, 56–59] consecutive or non-consecutive patients from a single [46, 47, 55, 57, 58, 60] or multiple [56, 51] institutions during a several-year period. Patients are usually selected retrospectively through search of clinical or pathological databases, and limitations of study design [61, 62] and reporting [63, 64] need to be taken into account for interpretation of results. Inclusion/exclusion criteria are usually based on age (studies typically separate adult from pediatric cohorts), histology (most studies are selective for the most common diffuse glioma types such as astrocytoma or oligodendroglioma), or degree of malignancy (some studies focus on all glioma grades while others focus on low- [33, 65, 66] or

Table 14.1
Comparison of MRI Brain Tumor Evaluation Methods

Expert-based evaluation		Radiomics		Deep learning	
Pros	Cons	Pros	Cons	Pros	Cons
Observation-driven Low computational requirements Simple to perform Abundant historical literature	Labor-intensive Mostly qualitative Low statistical power Intra and inter-reader variability Poor reproducibility	Involves extracting hand-crafted features using machine learning which often provide some degree of biological interpretability	Large sources of variance in imaging parameters introduced across sites and scanners Often dependent on segmentation of tumor habitat Often based on small retrospective, single-site, and relatively biased data	Domain-agnostic Does not require segmentation of tumor habitat	Deep learning features are known as "black-box" due to limited explanatory capacity of the deep features Limited by the relative sparsity of training samples, not always suited for applications with limited availability of well-curated samples.

high- [48, 51, 52, 57, 58, 60, 67, 68–71] grade gliomas). MRI acquisition methods also vary substantially in terms of technical specifications, including magnetic field strength (from early low field scanners to clinically available 3 T, research 7 T magnets, and beyond), variable gradient coil specifications with significant improvements over time, and available MRI pulse sequences. Most available studies, however, have used clinical scanners in the range of 0.5–3 T and routinely available clinical sequences, which are usually 2-dimensional (2D), multi-slice, whole-brain coverage imaging utilizing spin-echo (SE), or turbo SE (TSE) acquisitions. In addition to greater availability across institutions, the main advantage of the use of routine clinical images instead of custom modified research protocols is the potential for a more straight-forward translation of research findings to clinical practice. MRI contrast types almost invariably include T1-weighted (T1W) before and after the intravenous (IV) administration of a GBCA, conventional T2-weighted (T2W) TSE, T2W fluid-attenuated inversion recovery (T2W-FLAIR), and T2*, as either standard GRE or susceptibility-weighted imaging (SWI) more recently. Some of the available studies attempted to study multiple imaging characteristics as explanatory variables [31, 32, 34–36, 47, 51, 54], while some focused on a single imaging trait, such as CE [33, 46, 56, 57, 72–77], edema [50, 78–82], necrosis [49, 52, 83–85], cysts [55, 68, 70, 71, 86, 87], tumor location, shape, or margins [2, 88–91]. From a statistical standpoint, these imaging features have been classified in different ways by investigators, with variables sometimes being dichotomous (presence vs absence of CE), nominal (patterns of CE, such as faint/patchy, nodular, or ring-like), ordinal (degree of CE, from mild to intense), or con-tinuous (volume of CE measured by manual or semi-automatic outlining). The main outcome variables of clinical interest on the literature include pathological grade [31, 32, 36, 57, 73], resect-ability [65, 92] progression-free survival (PFS), and OS [33, 55, 57, 58, 68, 71]. For studies correlating MRI features with clinical outcomes, an important and sometimes overlooked step is correc-tion for non-imaging measures known to be correlated with the outcomes of interest during multivariate analysis, such as age, per-formance status, extent of surgical resection, non-surgical treat-ment modalities such as radiation and/or chemotherapy, and molecular markers [93], which are becoming increasingly avail-able. A common denominator for all of these early imaging stud-ies is the method of data analysis, with reliance on one or more expert readers for image interpretation. Visual subjective interpre-tation of images fueled radiology research over the past century because it took advantage of highly trained human brains to quickly visualize an enormous amount of data, separate normal from abnormal findings, and classify those findings according to some pre-determined categories. It has been a powerful tool to quickly identify meaningful observations, confirm associations and trends, which have largely advanced the imaging literature. However, expert interpretation of medical images has a number of inherent, and difficult to address limitations. Despite being simple to implement, it is usually very labor-intensive, and often-times non-practical when evaluating large datasets (on the order of several thousands of patients), which are becoming increasingly available in the age of medical Big Data [94, 95]. Interpretations are largely qualitative and rely on categorical variables, which have limited statistical power. Notwithstanding methodological efforts in training and standardization, expert reading is inherently sub-jective and affected by psychological factors, including personal experience and a number of interpretation biases [96]. Therefore, inter- and even intra-reader variability are significant issues, limit-ing the reproducibility of individual studies and possibly contrib-uting to conflicting results on studies designed to answer a similar question [55, 68, 70, 71]. Even when continuous variables are used (linear measurements, manual outlining of tumor volumes), issues in variability may still persist [97–100]. Despite these limi-tations, expert-based studies have made extremely impactful con-tributions to brain tumor research, and will continue to have a significant role in the foreseeable future, as computerized auto-mated methods are still in development. The following subsec-tions will review the main morphological features that have been explored by expert-based studies exploring glioma morphological features to date, with a focus in meaningful contributions to clini-cal knowledge.

3.2 Contrast Enhancement (CE) in Brain Tumors

The presence of abnormal post-CE in infiltrative gliomas has been conclusively been determined in numerous studies as an overall reliable marker of tumor aggressiveness, poor prognosis, and short overall survival [31, 32, 36, 51, 57, 101]. Butler et al., in one of the earliest imaging-pathology correlation studies, analyzed 84 patients with histologically confirmed astrocytomas and proposed an index of CE (ICE) based on visual evaluation of pre- and post-contrast CT images by readers blinded to clinical and pathological data. They were able to demonstrate significant correlations of CE with tissue markers of aggressiveness, such as cellularity, vascularity, pleomorphism, necrosis, and a composite malignancy index [102]. These findings have been confirmed by multiple other studies [2, 3, 34, 103–105]. More recently, larger studies employing more standardized imaging evaluation tools have again confirmed this relationship [51]. Instead of focusing on the presence vs absence of CE, some studies have recognized the importance of enhancement morphology, demonstrating that some specific patterns of CE (such as marked and heterogeneous, ring-like, and centrally necrotic) are more predictive of higher grade [52], creating com-posite scores with higher weights to more worrisome patterns.

However, clinical experience has demonstrated that the significant of post-CE is much more complex, and depends on several factors including subtype of tumor histology. For example, it is well known that many types of gliomas in the pediatric age group often demonstrate prominent CE, unrelated to malignancy [106]. The best example in this category is the group of pilocytic astrocytomas, which almost invariably demonstrate nodular or patchy areas of intense CE despite being World Health Organization (WHO) grade I tumors, with an indolent biological behavior and high rate of cure by surgical resection. Similar features are seen in other gliomas found in children and young adults, such as in other focal astrocytomas, neuronal, and mixed neuronal-glial neoplasms [107]. Fortunately, these tumors can usually be distinguished from their malignant counterparts by demographic, as well as more insidious clinical presentation, ancillary morphological features, and sometimes typical sites of origin [106, 107].

3.2.1 CE in Low-Grade Gliomas

Even for classic diffuse grade II gliomas, the association between CE and malignant features is not entirely sensitive or specific. In a large retrospective study in a French glioma group, Pallud and colleagues evaluated clinical images of 927 consecutive WHO grade II gliomas (two blinded expert readers) and demonstrated the presence of CE in 143 (15.9%). They also demonstrated the value of morphological analysis of CE, with all patients demonstrating either patchy/faint (65%) or nodular-like CE (35%). No patients demonstrated centrally necrotic, ring-like CE, which has been consistently reported as much more specific for malignancy [53]. Of note, despite the absence of worse overall survival in the LGGs with any pattern of CE, subgroups of nodular-like or progressive post-contrast enhancement did demonstrate reduced survival, suggesting more aggressive subtypes. This again highlights the importance of morphological evaluation of MRI and clinical findings in addition to histology results, especially in patients with limited tissue sampling (stereotactic or surgical biopsy rather than resections) given the potential for sampling bias [40]. Chaichana et al. performed a large, single-center retrospective analysis of all supratentorial diffuse grade II glioma patients undergoing resection at their institution (with the specific exclusion of biopsy cases) [33]. From 189 identified patients, 64 (34%) demonstrated some degree of CE, which was classified with a slightly different scheme (heterogeneous, nodular, or ring-enhancement). Despite the absence of differences in clinical and treatment-related variables between patients with and without CE, the presence of CE was independently associated with reduced survival and increased recurrence on both univariate and multivariate analysis. Patterns of CE were nodular in 31 (48%) of patients, heterogeneous in 26 (41%), and ring in 7 (11%), but subgroups of enhancement patterns were not evaluated separately. An impactful observation of this study was the fact that patients with CE tumors that had a gross total resection (versus subtotal resection) had both improved OS and PFS, highlighting a potential role of MRI for treatment decision making in grade II patients despite histology. Their literature review revealed only five studies with at least 100 patients for comparison of results [72, 108–111], but all were deemed to have significant limitations and/or design flaws, such as selection criteria (use of non-WHO grade II gliomas), use of limited biopsy specimens, lack of uniform histological criteria, and omission of variables known to affect OS in multivariate analysis (such as extent of resection, tumor size, and radiation therapy). The combined results of these studies do indicate that the presence of CE does not necessarily predict malignancy for a glioma, and that morphological patterns of CE have prognostic value.

3.2.2 Non-CE Tumors

Conversely, the absence of contrast enhancement (which is typically seen in grade II lesions) does not guarantee a benign biological behavior, as has been repeatedly demonstrated [59, 112–115]. An early impactful study published by Chamberlain et al. in 1988 evaluating CT imaging characteristics of 229 consecutive gliomas demonstrated a lack of post-contrast enhancement in a large proportion (40%) of anaplastic astrocytomas (51/126 patients) and even in 4 out of 93 diagnosed glioblastomas. It could be argued that the inferior sensitivity of CT to contrast enhancement could explain the lack of detection of BBB disruption in these malignant gliomas; however, Ginsberg et al [46] evaluated preoperative MR imaging of 40 consecutive supratentorial gliomas with absent CE and noted that 16 (40%) of the non-CE lesions had anaplastic features, confirming those results. Okamoto and colleagues published an interesting small case series in which they were able to capture the early appearance and evolution of malignant gliomas fortuitously while reviewing MRI images performed a few months before admission for a malignant glioma diagnosis. These initial MRI images showed a number of non-enhancing lesions which were interpreted as non-neoplastic processes, and which became more typical ring-enhancing masses in a few subsequent months, demonstrating that malignant, rapidly progressing gliomas have a dynamic appearance and that imaging at an early stage can be deceptive. In an attempt to estimate the true practical significance of these observations, Scott et al. [59] performed a population analysis in the Alberta Cancer Registry from multiple institutions and reviewed all medical records for patients diagnosed with supratentorial gliomas from 1994 to 1999. Of a total of 314 patients, 54 (18%) lacked post-contrast enhancement, and one-third of these tumors were malignant by histology, with the most significant predictive risk factor for malignancy being advanced age. There is, therefore, robust evidence that patients with newly discovered non-enhancing masses cannot be assumed to have a LGG, and that histopathological diagnosis or at least close image surveillance is needed to determine the biological behavior of such lesions, particularly in adults and elderly patients, who are statistically at much greater risk of harboring an aggressive lesion. Of note, there is a well-defined role for advanced MRI techniques such as perfusion, diffusion, and spectroscopy in this setting, which are beyond the scope of this review [116], but which may offer additional insights about the malignant potential of a non-enhancing lesion or provide more reliable biopsy targets [117] to avoid sampling bias.

3.3 Tumor-Related Edema

Edema is a common and distinctive pathological finding commonly seen in association with multiple types of brain tumors, including gliomas. It is loosely defined as a combination of abnormal water accumulation in the tumor and peritumoral brain parenchyma and swelling, which can be both reliability identified on MRI as prolongation of T1 and T2 values associated with volumetric enlargement of the brain parenchyma. Other distinctive morphological features of peritumoral edema (PTE) include its fairly homogeneous, near CSF signal, geographical morphology assuming the shape of the brain parenchyma without rounded morphology or convex margins, and tendency to spread along white matter tracts with finger-like projections sparing cortical and deep gray matter structures [54]. These are important features to distinguish PTE from other non-enhancing tumor components, such as tumor necrosis, cysts, and non-contrast-enhancing tumor (nCET). Despite being common and easily identifiable, there are still controversies regarding the mechanism of formation and significance of PTE in gliomas. Theories to explain the etiology and molecular mechanisms of tumor-related edema are complex and beyond the scope of this text, and the reader is referred to specific articles on the subject for further reading [118–120]. In short, current belief is that rather than the result of tumor secretions, PTE in the brain is mostly vasogenic, and results from molecularly mediated increased BBB permeability [120]. However, the term "vasogenic edema" can be misleading in clinical practice and suggest benignity, while it has been well demonstrated that peritumoral edema in gliomas contains variable numbers of neoplastic infiltrative cells [121–123]. These satellite cells are thought to have a key role as determinants of disease recurrence and ultimately treatment failures and mortality. This is in contrast to the purely vasogenic edema associated with non-infiltrative brain tumors, such as meningiomas and metastases [79]. As such, PTE is an important treatment target in gliomas, and usually included in radiotherapy treatment fields to reduce the rate of marginal recurrences [124]. Therefore, despite the well demonstrated vascular mechanisms associated with peritumoral edema in glial neoplasms, the term "tumor-infiltrated edema" may be more appropriate and accurate than "vasogenic edema."

Interestingly, the degree of glioma PTE varies substantially in different patients, and it is common knowledge in clinical practice that the degree of PTE correlates well with tumor grade and aggressiveness. To that effect, it has been demonstrated that, as a general rule, the degree of PTE in malignant gliomas parallels the degree of disruption of the BBB and tumor contrast enhancement [47]. This is further supported by the correlation of PTE with molecular markers of microvascular extravasation, such as VEGF [125]. However, even in patients with tumors of same histology and grade, the degree of edema can vary markedly, which has led to research interest to explore this tumor morphological feature as a prognostic factor. The degree of PTE has been shown to be independent from tumor size [81], but appears to be positively correlated with deeper tumor locations in the cerebral white matter. Several retrospective studies of malignant gliomas have identified that the amount of edema correlates negatively with overall survival [50, 53, 78, 82]. However, other studies have failed to demonstrate such correlation [126], or a prognostic significance of edema only in a subset of tumors, such as GBMs with a methylated MGMT promoter [127]. A recent systematic review by Liu [80] analyzing the prognostic significance of PTE had inconclusive results, suggestive that controversy remains in regards to its true prognostic value. One of the main limitations to compare the different existing studies are the variable methods of edema assessment, which is reliant on blinded independent expert readers and different grading methods. For example, studies by Pierallini [36] and Hammoud et al [53] utilized simple three-tiered scores based on visual inspection of the volume of edema compared to the overall tumor volume. Pierallini and colleagues utilized two-way ANOVA for repeated measures to demonstrate that MR features, and not readers had a main effect on results. They also validated their visual method with a semi-automated computerized volumetric method based on signal intensity thresholding [36]. In a multi-feature qualitative study exploring MRI imaging correlates of survival in high-grade gliomas, Pope et al identified PTE as one of the few statistically significant imaging prognostic indicators of survival [54]. They utilized a slightly different score, measuring the extent of edema beyond the tumor margins semi-quantitatively (greater or less than 1 cm), a method which has been adapted by other authors in more recent studies [50, 78]. Most importantly, this study by Pope and colleagues was one of the first that explicitly attempted to make a formal discrimination between PTE and nCET. Despite the fact that separation of these two distinct tumor subregions on routine imaging can be challenging by visual inspection, it is crucial, since they reflect markedly distinct pathological processes. In fact, failing to distinguish PTE and nCET may be one of the major factors behind the current conflicting

PTE literature results, given the fact that nCET is typically seen in less aggressive subtypes of high-grade gliomas, such as anaplastic astrocytoma (AA) and secondary GBM [128]. This likely explains the favorable prognostic significance of nCET demonstrated originally by Pope et al [54]. In summary, PTE is an important and informative component of brain tumors and overall associated with worsened outcomes, both due to direct detrimental intracranial effects and also as a marker of more aggressive tumor characteristics; however, current expert-based visual methods to separate PTE from other tumor components have significant limitations in need of more accurate approaches.

3.4   Tumor Necrosis

Tumor necrosis (TN) is a hallmark of aggressive brain neoplasms, conventionally thought to result mechanistically from an imbalance between rapid tumor growth and availability of oxygen and nutrients. It can be reliably identified micro and macroscopically on tumor sections as well as on cross-sectional imaging, and differs fundamentally from apoptosis, which is a form of signal-mediated, programmed cell death [129]. Alongside microvascular proliferation, necrosis is a defining histological feature for GBM, the most common and aggressive subtype of glioma, and therefore, crucial for traditional tumor grading according to WHO criteria. It is typically defined as "pseudopalisading," due to the morphology of the peri-necrotic tumor cells, a subset of particularly hypoxic, and invasive tumor cells, migrating away from the necrotic centers and likely accounting for microscopic tumor spread beyond gross tumor margins [130, 131]. The imaging literature has explored the presence of TN as an important morphological feature since its early days [102], demonstrating accurate correlations between histology-defined necrosis with central areas of hypodensity or fluid signal surrounded by infiltrative tumor cells in stereotactic specimens [3]. Subsequently proposed scoring systems for grading and outcome prediction derived from expert-defined morphological features have all incorporated TN as one of the main defining features [32, 34–36]. Additionally, the presence of TN identified on pretreatment MRI has been demonstrated to independently impact survival even in tumors of similar histological grade. For example, Hammoud et al. in 1996 scored 48 supratentorial GBMs for their degree of TN and demonstrated it to be strongest prognostic variable, with median survivals of 42, 24, 15, and 12 months on multivariate analysis [53]. On a larger study involving 416 consecutive patients, Lacroix and colleagues [58] identified five independent predictors of survival in GBM patients: age, Karnofsky Performance Scale (KPS) score, extent of resection, TN, and CE. Necrosis was selected as one of three factors (alongside age and KPS, which are the most robust clinical prognostic factors) on a proposed outcome scale to identify patients with longer survival. These findings have been replicated in other studies utilizing different grading criteria [50, 78, 126]. However, Gutman et al reported a study which was unable to find an association of TN with poorer outcome in GBM patients, conflicting with prior results. However, they speculated that the particular percentage cut-offs used in that study may have lacked sensitivity to detect an association in their relatively small patient population. More recently, other approaches have been used to capture richer morphometric features of necrosis utilizing more complex descriptors. For example, Liu et al [49] utilized fractal dimension and "lacunarity" analysis of segmented MRI images to demonstrate the predictive potential of necrosis on survival.

3.5 Tumor Cysts

One of the main confounders for morphological identification of TN on MRI images is the presence of tumoral cysts, since both may present with some overlapping features on MRI, notably as a well-circumscribed area of non-contrast-enhancing fluid signal. Cysts can be formed as the result of confluent necrosis in aggressive neoplasms (so-called cystic necrosis); however, there are other presumptive mechanisms for cyst formation, such as the confluence of microcystic spaces occupied by leaking plasma fluid or extracellular fluid secreted by tumor cells [86, 132]. Of more significant practical importance, tumor cysts unrelated to necrosis are particularly common in biologically indolent glial tumors, such as pilocytic astrocytoma, ganglioglioma, low-grade astrocytoma, and oligodendroglioma. Cystic features in these subtypes of gliomas are associated with significantly better outcomes [66, 133]. Therefore it is imperative to distinguish cystic necrosis in agressive brain tumors from true tumor cysts. Morphological features in favor of cystic necrosis include irregular morphology of the cystic spaces, anfractuous margins and thick, solid enhancing walls, while true cysts are typically spherical or oval-shaped, well-demarcated, and demonstrate smooth, non-enhancing, or minimally enhancing walls. However, this distinction is not straightforward, and may account for some of the variability of results in the literature regarding the significance of glioma cysts. For example, Maldaun et al. [71] evaluated 22 GBM patients for large tumor cysts (> 50% of tumor volume), matched for other prognostically significant outcome variables, and found improved recurrence-free and overall survival in patients with large tumor cysts. Utsuki et al [70] also compared cystic and non-cystic GBMs utilizing similar criteria in 37 patients with histological evaluation of tumor margins and demonstrated that cystic GBMs were less infiltrative, had a better defined tumor-parenchymal interface and less than 2-cm-thick peripheral edema compared to non-cystic lesions. More importantly, they observed prolonged median survival time (19.8 vs 12.8 months) and 2-year survival rates (50 vs 17%) in cystic versus non-cystic GBM. These observations, as well as the finding of significantly younger patients in the cystic group, led the authors to believe that cystic GBMs could represent the malignant transformation of lower-grade gliomas (secondary GBMs), thus explaining the improved outcomes. However, a larger and more recent study performed by Kaur et al. [68] indicated conflicting results, finding no survival advantage for patients with cystic GBMs compared to non-cystic tumors matched by age and other patient characteristics. Moreover, Sarmiento et al. [55] more recently evaluated another large patient cohort of 351 consecutive GBM patients and demonstrated a lack of prognostic implication for cystic versus non-cystic tumors as well as a lack of association of cystic features with molecular markers of secondary glioblastoma. These more recent and better-powered studies seem to convincingly dispute findings from the earlier literature.

3.6 MRI Scores and Morphological Feature Sets

A review of the currently available (predominantly qualitative and expert-driven) morphological studies of brain tumors demonstrates that the significant variability in methods used for feature scoring remains one of the main barriers for interpretation and knowledge advancement. Grading systems are arbitrary, usually derived from anecdotal experience and not data-driven, and highly subjective to reader bias. Few are evaluated for repeatability and almost all result from single-center experience, which limits generalizability. As an attempt to overcome some of these limitations and provide more reproducible readings, a group of investigators [51] proposed the use of a standardized set of visible subjective MRI imaging features associated with brain tumors, better known as VASARI (Visually AcceSAble Rembrandt Images) [134]. These image features were selected by consensus opinion of a panel of neuroradiologist experts from multiple institutions based on the existing literature, and acknowledge the importance of four main tumor subregions commonly explored by MRI: contrast-enhanced tumor, non-enhanced tumor, necrosis, and edema. The proposed "standardized" vocabulary in this system was created with the intent to describe tumor features in a "reproducible, clinically meaningful, and biologically relevant" manner. To achieve multi-institutional validation, the VASARI group of investigators performed a comprehensive evaluation of these expert-made subjective assessments and compared the results to gene expression data and patient outcomes available for the publically available "The Cancer Genome Atlas (TCGA)" dataset. One of the important and most enlightening components of this exploratory study was the formal evaluation of reliability measures for the multiple readers. The interrater agreement coefficients were classified as moderate to high, but the actual numbers (ranging from approximately 0.5–0.6) demonstrate the existing challenges to reach very high reproducibility utilizing nominal or ordinal scales, even when a formalized, standardized feature set is utilized. The standardization is an important step to reduce the large amount of assessment variability currently seen, but the results highlight the importance of more user-independent, automated, and quantitative methods for feature identification and classification, some of which will be discussed in the subsequent sections.

4   Radiomics: Quantitative and Non-invasive Phenotyping of Tumors on Imaging

The growing availability of radiological data in oncology (>1 million publicly available images on The Cancer Imaging Archive [TCIA]), coupled with increased computational resources, has led to the development of a number of radiomics and advanced machine-learning techniques for disease characterization. Radiomics is the high-throughput extraction of quantitative features from radiological medical images particularly from routinely acquired radiographic scans (i.e., CT, positron emission tomography [PET], and MRI) for disease characterization. Lambin et al. [135] first coined the term in 2012, and radiomics has since spanned across various oncology disciplines, including breast, prostate, lung, and brain tumors, for the clinical purposes of personalizing treatment decisions in diagnosis, prognosis, as well as treatment response. Radiomics is known to non-invasively characterize the subtle attributes of the lesion, which are often difficult to be appreciated by the naked eye, even by an experienced radiologist. These radiomics-derived image biomarkers capture the inherent heterogeneity in the tumor and its immediate surroundings (also known as "tumor habitat") in order to characterize the disease phenotype. Crucially, radiomics could have significant clinical implications in improving patient outcome by enabling modification and personalization of treatment based on patient's quantitative tumor phenotype. For instance, a detailed depiction of post-treatment lesions via radiomics could allow for more reliable evaluation and stratification of patients as having complete response (eradication of disease), partial response (significant reduction in tumor extent), stable disease (minor reduction in tumor, no new tumor), or progressive disease (increase in tumor bulk, new foci of disease). This detailed post-treatment analysis could offer significant advantages over the current clinical evaluation via imaging, which is typically limited to the response evaluation criteria in solid tumors (RECIST [136], PERCIST [137]). These response criteria are based on measuring changes in tumor dimensions or volume, limiting their ability in spatially locating localized tissue response to treatment for clinical purposes [138, 139]. Radiomics overcomes this limitation by capturing different quantitative measurements of the tumor, such as (1) shape-related attributes, (2) semantic features, (3) statistical co-occurrence [140] (Haralick), or convolutional filter responses [141, 142] that capture intra-lesion heterogeneity, and (4) peri-lesional heterogeneity by capturing features from the tumor surroundings, just outside of the tumor, also known as "habitat" features. These radiomic features have shown a great deal of success in comprehensive, quantitative, and pixel-wise characterization of disease appearance for the purposes of predicting response as well as disease aggressiveness, using routine PET, CT, or MRI scans.

5   Building a Quantitative Radiomics-Based Model for Brain Tumors

Radiomic applications in neuro-oncology especially in the context of brain tumors have largely included (a) predicting progression-free or overall survival using pre-treatment routine MRI scans, (b) predicting or monitoring response to treatment following surgery, (c) differentiating pseudo-progression/radiation necrosis from tumor recurrence for patients undergoing chemoradiation following surgery, (d) predicting treatment response to novel treatments including targeted therapy or anti-angiogenic therapies, and (e) investigating the molecular and genetic characteristics of brain tumors using radiomic approaches, also known as radiogenomic analysis.

Figure 14.1 describes a typical radiomic workflow, which constitutes the following steps: (a) image acquisition and data curation, (b) pre-processing, (c) identifying and annotating the region of interest, (d) radiomic feature extraction, (e) feature selection to identify the most discriminative features, (f) building a machine learning classifier using training cohort, and preferably testing it on an independent validation set, h) statistical analysis of the performance of the classifier on the training and validation cohorts toward solving the clinical problem at hand (i.e., benign versus malignant, responder versus non-responder, pseudo-progression versus tumor recurrence).

5.1   Image Acquisition

The preferred modality of choice in neuro-oncology over the last few decades has been multiparametric MRI including contrast-enhanced T1W, T2W, and T2W-FLAIR scans. Neuroradiologists rely on gadolinium-enhanced MRIs for evaluation of brain tumors due to it enabling a high-contrast images of soft tissues. Standard of care MRI scans (<2 mm slice thickness) not only allow radiologists to visualize the tumor but also helps them regarding surgical staging and gradation. Similarly, T2W and T2W-FLAIR imaging sequences signals capture interstitial leakage and low cellular density, reflecting edema. Other advanced techniques often used for diagnosis and treatment evaluation include perfusion-weighted imaging (PWI), diffusion-weighted imaging (DWI), diffusion tensor imaging (DTI), and magnetic resonance spectroscopy (MRS). However, their utility is limited due to considerable parameter

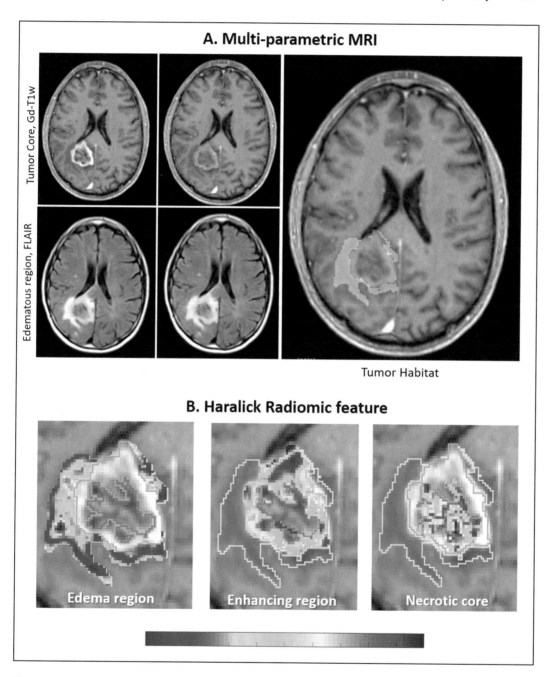

A. Multi-parametric MRI

Tumor Core, Gd-T1w

Edematous region, FLAIR

Tumor Habitat

B. Haralick Radiomic feature

Edema region

Enhancing region

Necrotic core

Fig. 14.1 Typical radiomics workflow for brain tumor MRI analysis

variations in imaging acquisitions across platforms and scanners, cost, and unavailability at some clinical sites.

5.2 Pre-Processing Pipeline for Radiomic Analysis

A key consideration with the use of radiomic analysis is to account for wide-scale differences across the acquired radiographic images differing across scanners and institutions. Scanner variabilities [143] include different slice thicknesses [144], reconstruction kernels [145], and processing steps [146]. Savio et al. [147] found statistically significant differences when using radiomic features to classify multiple sclerosis plaques when changing slice thicknesses from 1 to 3 mm. Similarly, Buch et al. [148] showed that texture was found to be significantly dependent on scanner platforms and number of excitations (NEX).

Intensity non-standardness refers to the issue of MR image "intensity drift" across different imaging acquisitions, and results in MR image intensities lacking tissue-specific numeric meaning within the same MRI protocol, for the same body region, or for images of the same patient obtained on the same scanner. Therefore, considerable pre-processing needs to be performed before the radiographic images are suitable for radiomic analysis. The initial pre-processing pipeline includes registration of the MRI protocols to a single reference anatomical template. Different modes of registration are implemented including rigid (linear, affine) as well as non-rigid or deformable transformations (B-spline, quadratic). Registration also regularizes the spatial resolution of the images where each slice is re-sampled to have a defined uniform pixel spacing and then interpolated to a fixed slice thickness. Following this, usually all the sequences are resampled to an isotropic 3D volume. In order to prevent the skull enhancement on a T1W+C scan to interfere with texture analysis, the skull is stripped from all the image slices during pre-processing. Following skull-stripping [149, 150], corrections are applied on the skull stripped, resampled, and registered image in order to rectify the common artifacts including acquisition-based intensity artifacts, bias field inhomogeneity, and intensity non-standardness. Bias field artifacts [151] include standard variations of signal intensity across the MRI images. Several methods have been published to tackle the issue of intensity non-standardness including several variants of matching image histograms with a standard [152, 153] or with certain landmarks within the image [154, 155]. Meanwhile, methods [151, 156] to correct bias field artifacts include filtering methods like high-pass or homomorphic filtering algorithms, parametric surface fitting, and nonparametric nonuniform intensity normalization (N3) [157] algorithms.

5.3 Volume of Interest Identification and Segmentation

The identification and annotation of the region of interest (ROI) is a crucial step in radiomic analysis and in the case of brain tumors, there is often the need to identify pathologically distinct regions due to tumor heterogeneity, within the MRI scans. For instance, in GBMs, the neuroradiologist needs to identify and mark the regions of edema, enhancement, non-enhancing lesion, necrosis, and inflammation, depending on the type of lesion (i.e., pre-treatment or post-treatment lesion). This often requires the neuroradiologist to visualize different types of MR image contrast, such as T1W, T2W, and T2W-FLAIR scans, in conjunction. The pre-processed images can be viewed in open-source and freely available software, which allows experienced neuroradiologists and experts to place a designated boundary around the ROI across various slices. Figure 14.2a shows the identified ROI of tumor habitat (necrosis, enhancing tumor, and edema) in GBM using multiparametric MRI scans.

Following identification of the ROI(s), the identified tumor is segmented into its various sub-compartments, which is a crucial step [158, 159] in the radiomics pipeline. The segmentation step is crucial because the radiomic features that are finally used for building the model are extracted from the segmented volume itself. While the features could very well be extracted from the entire image, minimizing the volume of interest from the entire image to the segmented boundaries eases the processing time as well as maximizes the signal-to-noise ratio. Studies often use multiple radiologists to increase reliability by having a consensus opinion on the segmentation; however, there is often a segmentation discordance across studies [160–163] using measures to report inter-observer segmentation variability including the Dice-Sorensen coefficient and Jaccard Index.

While human reader segmentation obtained via consensus is currently the gold standard [160] for radiomic analysis and ensures high accuracy, several automated [161, 162] segmentation approaches using deep learning interfaces as well as semi-automatic [160] seeding-based algorithms have found popularity in segmenting the tumor and associated regions of interest. While some analyses might include only the tumor and the regions surrounding the tumor, sophisticated radiomic analysis has often focused on defining tumor sub-compartments including enhancement, non-enhancement, necrosis as well as specific sub-compartments in the immediate periphery of the tumor. Over the last few years, Deep Learning models have especially shown tremendous potential in segmenting brain tumors on MRIs. A convolutional neural network-based model [162] obtained impressive performances in the Brain Tumor Segmentation (BraTS) Challenge in 2013 and 2015. Since then, several other deep learning algorithms including U-Net, Conv-Net, Transfer Learning, and Deep Hourglass approaches have been used for the segmentation of brain tumors on the BraTS dataset [164, 165].

5.4 Quantitative Features

Once the tumor sub-compartments are segmented, radiomic analysis could be performed across each of these sub-compartments individually. The most commonly employed radiomic features that are utilized in brain tumor analysis include:

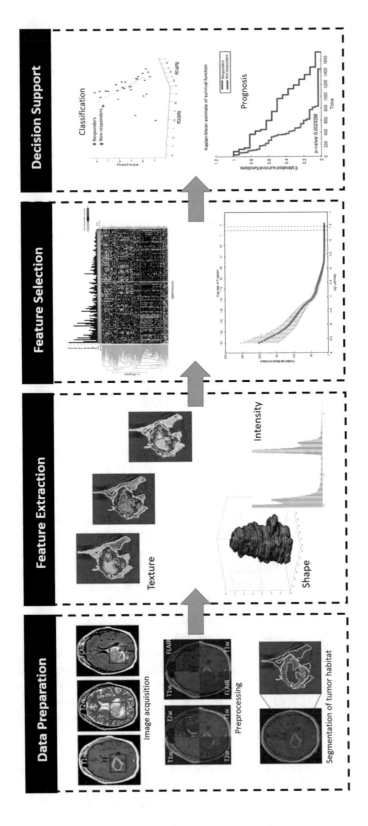

Fig. 14.2 (**a**) Identification of tumor habitat using multi-parametric MRI, where red represents the necrotic core, green is the enhancing region of the tumor, and yellow represents the edematous region of GBM. (**b**) Tumor sub-compartment based Haralick feature extraction. This implies that the edema region of the repre-sentative GBM has a more homogeneous arrangement of pixels when compared to the enhancing and necrotic core of the tumor

1. Semantic features—These include qualitative features of the tumor as described by the neuroradiologist analyzing the MRI images. A large-scale study was employed to the The Cancer Genome Atlas (TCGA) data of brain tumors (low-grade gliomas and glioblastomas) to define semantic features that characterize the tumors and discover their associations with genomic features. This led to the founding of a comprehensive radiologist defined feature set known as VASARI [166], that normalized the grading of MRI imaging features, with the ensuing goal to minimize inter-observer discordance. The 30 distinctive VASARI features [167] were based on the four commonly analyzed tumor sub-compartments (non-enhancing tumor, enhancing tumor, necrosis, and edema), and included features such as location of the lesion, morphology, margin, lesion vicinity, and remote alterations.

2. First -order statistics—These are a set of features that use basic statistics to define the intensity distribution of each pixel in the volume of interest. These features include mean, median, standard deviation, kurtosis, variance, and skewness. Kong et al [168] found a significant correlation between first-order statistical features extracted from segmented GBMs with a different molecular class of GBM, thus predicting the molecular phenotype of the tumor. First-order statistics have shown moderate to good stability, calculated using intra-class correlation coefficient (ICC), using 45 patients from the publicly available TCIA dataset, in a study published by Lee et al [163].

3. Shape-related cues—Shape and surface deformations induced due to uneven tumor growth can be attributed to irregular and aggressive tumor infiltration. Therefore, a quantitative approach to assessing the shape features can also be useful. Global features that characterize the segmentation and capture the tumor volume, i.e., the number of voxels, the major and minor axis of the shape of the segmented compartments, and elongation of the ROI (ratio between major and minor axes of the ROI) are some of the common shape features that are currently evaluated. Additionally, local features that correspond to the curvature values computed for the surface of each 3D segmented compartment, voxel-by-voxel, are also evaluated using first-order statistics including mean, median, mode, and skewness. Ismail et al [169] recently demonstrated that 3D shape features acquired from post-treatment lesions on MRI can distinguish radiographically similar pathologies of pseudo-progression and tumor progression.

4. Texture features—Usually higher-order statistical features, the gamut of texture features is often the most studied feature set in the brain tumor space [158, 170, 171]. For a given ROI, the differing texture features serve to distinguish the differences

between voxel intensities in the volume of interest. They usually represent the heterogeneity in the patterns both related to intensity as well as spatial orientation, which usually defines malignancy in its heterogeneous enhancement within the tumor boundaries both in the enhancing and non-enhancing region. The different texture features that are frequently used in radiomic analysis include:

(a) *Local binary patterns (LBP)* [172, 173]—A simple form of texture analysis, first defined in 1994 [172] involves creating a feature vector to describe the image by calculating and comparing the intensities of surrounding pixels, on a patch-by-patch basis. In a 3×3 array of pixels, the center-pixel would then be compared to the eight surrounding ones, and a binary 8-digit decimal value would be computed based on whether each of the eight pixels is greater in intensity (1) to the surrounding pixel or lesser (0). This would then be repeated for the entire image, with each pixel given a decimal number to indicate its local binary pattern. A histogram is then computed with the binary patterns across the volume of interest, and the normalized histogram serves as the LBP of the ROI. Sachdeva et al. [174] used LBP as one of the features in an artificial neural network (ANN) to classify brain tumors including GBMs, medulloblastomas, meningiomas, and metastatic disease.

(b) *Gray-level co-occurrence matrix (GLCM)* [140, 175]—GLCM is a relatively popular feature extraction method in medical image analysis as it allows for capturing variations in gray-level image characteristics via second-order intensity statistics (e.g., angular second moment, contrast, and differential entropy). For a given local neighborhood, these second-order texture features measure statistical distances between textures with varying properties [176]. This is represented in the form of a matrix which depends on the spatial orientation between two pixels in the image. Each row and each column represent a single gray value, building a matrix which represents the spatial relationships between the pixels in the image. The method was first described by Haralick and Shanmugam in a seminal paper [140] and the features are now commonly referred as Haralick features. Haralick features have often been the most representative features in radiomic analysis. In a study [175] involving 120 patients to classify between normal MRIs, patients with multiple sclerosis (MS) and brain tumors, an ANN based on Haralick features successfully identified all the tumor images without exception, while having a 92.86% accuracy in classifying MS images. Figure 14.2b shows the Haralick feature extraction from the different sub-compartments of the GBM tumor habitat.

(c) *Gray-level run length matrix (GLRLM)*—In comparison with GLCM features, GLRLM [177] features look to analyze the pixel runs instead of pairs of pixels. A pixel run includes the number of pixels of a specific gray value that are in a right direction, in the right sequence. While the rows of the matrix still represent gray levels, the columns represent run lengths. Yang et al. [178] extracted 44 GLRLM features from 82 patients with GBM and found the GLRLM features on T2W-FLAIR to be most predictive of classifying neural subtype of GBM as well as predicting 12-month overall survival (OS) with an area under the receiver operating curve (AUC) of 0.72.

(d) *Laws features*—Kenneth Laws [179] described a set of texture energy descriptors to characterize the heterogeneity within the image. The Laws measures define the texture of the image using various parameters including spot, edge, ripple, and level surfaces. Prasanna et al [180] found that Laws energy descriptors were the most distinguishing features from the peritumoral zone in GBMs and used it to predict overall survival in a set of 65 patients with GBM.

5. Wavelet features—These are features that utilize different wavelengths, amplitudes, and frequencies to recognize attributes across a wide range of scales within the image.

(a) *Gabor filters*—Dennis Gabor first defined a novel way how an image was represented in the visual cortex [181]. This wavelet feature, which is now called Gabor features, describes the unique textural pattern within an image by capturing gradients across varying frequencies and wavelengths. The Gabor filters utilized are tuned according to various scales, rotations, and window sizes. Iv et al. [182] used Gabor features to classify medulloblastomas based off MRI images from 109 children into different molecular subgroups with an AUC consistently over 0.70.

(b) *Co-occurrence of local anisotropic gradient orientations (CoLlAGe)*—A recently introduced radiomics wavelet descriptor [183], CoLIAGe, builds on the existing Gabor filter and captures the apparent disorder in gradient orientations down to a pixel basis within an image. CoLIAGe has been shown to be a highly discriminative feature and has been shown to distinguish between tumor recurrence and radiation necrosis in GBMs following chemo-radiation, based off MRI scans alone [170].

6. Fractal-based features—Fractal-based imaging features have also been recently used as part of the radiomic toolset in brain tumors. These use dimensional analysis to quantify image complexity. Hausdorff's fractal dimension of the tumor area is a way

to measure these complexities as well as capture regions with high variations in intensity.

5.5   Feature Selection

Very often, the number of radiomic features extracted from the various sub-compartments (edema, enhancing, non-enhancing, and necrosis regions) ranges in the thousands. It is thus essential to trim the number of features, keeping in mind the patient sample size is much lower. In order to alleviate this so-called "curse of dimensionality" [184], which describes the situation when the number of features is many times higher than the number of training images, only the most discriminative features are selected for radiomic analysis. Usually, an order of magnitude defined by the sample size is used to prune the features by various feature selection algorithms [185, 186] before the model can be constructed. For example, using an order of 0.1 means that a sample size of 100 patients would mean a maximum of 10 features could be used to build a classifier. This can be done in either in a univariate or multivariate way:

Univariate—The univariate feature selection methods include statistical tests like Pearson and Spearman's correlation tests, chi-square tests, and Wilcoxon rank sum test which calculate a score which defines the strength of how each feature is related to the outcome. All these methods determine the effectiveness of feature and report statistical significance, i.e., whether a specific feature could differentiate between the variable studied with a minimal degree of error. The p value reported by these tests is then used to select the features. Lower the p-value, lower the error and better the quality of the feature. Univariate feature selection methods are simpler to utilize but are limited by failing to measure inter-feature associations and thus are prone to selecting features which might be similar or redundant.

Multivariate—Multivariate feature selection algorithms serve to solve the limitations of univariate models by not only considering the degree of correlation between the feature and the outcome of interest but also the mutual correlation between the features as well thus avoiding redundancy. Ding et al. [187] proposed a multivariate feature selection method to maximize association while minimizing redundancy, referred as mRMR framework which has found use in radiomic feature analysis across different disease types. Cho et al. [188] used a mRMR feature selection algorithm to select the optimum radiomic features in order to grade gliomas into high and low grades in 285 cases based off pre-treatment MRI. Other prominent multivariate feature selection methods include joint mutual information (JMI) [189], variable importance on projection measure for principal component analysis (PCA-VIP) [190], conditional mutual information (CMI) [191], and quadratic programming [192], among others. A few machine learning classifiers such as LASSO [193] and elastic net regression [194] have an embedded feature selection algorithm as a part of the classification methodology.

5.6 Classifier Construction

The final step in creating a machine learning computational model to predict outcome or evaluate response involves building a machine learning classifier with the most stable and discriminative features post feature selection. A classifier leverages machine learning techniques to categorize and classify various datasets according to defined labels. In case of medical image analysis, the label is usually a diagnosis or response to therapy, e.g., a classifier will divide patients with GBM (dataset) treated with targeted therapy into those who respond and those who do not respond to therapy (label). For example, a study involving predicting short- versus long-term survival in GBM patients treated with chemo-radiation using pre-treatment MRI scans would be considered as a binary classification problem. Broadly categorizing, classifiers can be divided into supervised and unsupervised methods.

Supervised—The supervised classification model utilizes a predefined set of labels, representing the outcomes of interest. Broadly, supervised classifiers can be divided into linear and non-linear classifiers. Linear models make the decision to categorize the class which it belongs to by a linear combination of its attributes or "features." These include simple mathematical classifications like linear discriminant analysis (LDA), Bayes classifiers or more discriminative models including logistic regression, quadratic discriminant analysis (QDA), and support vector machines (SVM). Besides linear models, other methods include boosting, decision trees like Random forests and kernel estimation functions like k-nearest neighbor (k-NN).

Unsupervised—An unsupervised approach, also known as clustering, is employed when the target labels or the precise outcomes of interest are not known *apriori*. Unsupervised clustering methods include principal component analysis (PCA), Bayesian-, hierarchical-, and partitioning based approaches. Unsupervised clustering approaches might lead to the discovery of previously unknown categories during the clustering optimization. In brain tumor radiomics, unsupervised learning approaches have been used in automatizing brain tumor segmentation. A fuzzy c-means clustering algorithm was used by Clark et al. [161] to automatically segment GBMs on MRI using T1W+C, T2W, and T2W-FLAIR MRI.

6   Recent Advent of Deep Learning Methods

Unlike traditional radiomics methodology that relies on accurate segmentation and selecting hand-crafted features, deep learning methods [195–199] are domain-agnostic, and do not require accurate segmentation. A deep learning model [199, 200] differs from traditional radiomics as it generates features via multiple layers or abstractions to solve a classification problem. Deep Learning networks can be seen as an artificial representation of the complicated human neural architecture in their multiple branching patterns, networks, and chains. Like machine learning methods, deep learning models can be both supervised and unsupervised. Convolutional neural networks (CNNs) are the most common deep learning models used in medical image analysis. A CNN, much like a neuron in the visual cortex, has an input layer and an output layer while the computations are performed within multiple hidden layers, which include convolutional layers. A convolutional layer, from where the CNN derives the name, is the primary layer from where features are extracted from the image, using small patches of image data, representing each patch by a matrix of pixel values. This matrix is essentially what is known as a feature detector or filter. Altering the filter matrix allows one to compute several tasks on the image including detecting edges, color, sharpening, blurring, and focusing. The CNN, through the process of training, determines the usefulness of these filters in recognizing subtle visual signatures hidden in the image. A CNN via transfer learning, which is a machine learning method to leverage pre-trained methods for a new task, in addition to traditional hand-crafted features was used by Lao et al. [201] in predicting survival in GBM. The model was trained on 75 patients and validated independently on a hold-out set of 37 patients. The top features were identified as deep-learned features from the CNN, and yielded a C-index of 0.71 in predicting overall survival. Unlike the hand-crafted machine learning features which often provide some degree of biological interpretability, deep learning features are more of a "black-box." These networks are limited in their explanatory capacity of the deep features with neither a set of diagnostic rules nor an insight into the results [198]. Moreover, deep learning models are limited by the relative sparsity of training samples in medical image analysis, and hence may not always be suited for the specific application.

7   Analysis of the Constructed Model

Performance of a supervised classification method to define a binary endpoint is analyzed by a receiver operating characteristic (ROC) curve. For instance, a binary endpoint classification could be identifying which GBMs tend to recur versus which ones do not recur following treatment based on radiomics analysis. A ROC curve is constructed by plotting the classified true positives against the false positives while simultaneously varying the decision threshold [202]. The performance is then reported by the area under the ROC curve (AUC) with a higher AUC representing improved performance. Statistical measures like sensitivity and specificity obtained from the ROC curve as well as accuracy, reliability, true and false positive rates are also used to quantify performance of the classifier. In comparison, it is more onerous to evaluate the performance of an unsupervised classification approach, due to the lack of pre-defined outcomes of interest to compare the performance against. External validity measures [203, 204] are often used to measure the clustering performance by human comparison or comparing with a set standard. Some of these measures include purity, Rand index, F measures, and normalized mutual information [202, 205]. Meanwhile, internal quality measures include having a low inter-cluster similarity while maintaining a high intra-cluster similarity [206]. Internal measures do not need a gold standard or a human-defined value to compare against unlike external standards.

Survival analysis includes evaluating the performance model on a time to event basis rather than a binary endpoint. This involves measures including the univariate Kaplan-Meier analysis, which employs a single model to predict whether there is a statistically significant separation between the two predicted classes or categories. The significance is defined by an accepted $p\text{-value} < 0.05$ which means a Type I error of less than 5%. A more powerful way of survival analysis includes the Cox proportional hazards analysis, which is a multivariate model, which can combine multiple models including clinical parameters to produce a hazard ratio (HR). The Cox HR is defined by the ratio of hazard rates corresponding to the two different outcomes of interest. For example, if the radiomic model had an HR of 2 in predicting overall survival (OS) in GBMs, it would mean that increased radiomic values in a patient have twice high a chance of poor OS as compared to those with low radiomic values. Other endpoints beside OS include progression-free survival (PFS), and disease free-survival (DFS).

8   Applications of Radiomics in Brain Tumors

Figure 14.3 illustrates some of the most common radiomic applications for brain tumor evaluation.

8.1   Diagnosis and Categorization of Brain Tumors

The histopathological gradation of gliomas into low (WHO Class II) and high (III and IV) as well as morphological subtypes (ependymomas, oligodendrogliomas, astrocytomas, brainstem, and mixed) are clinically relevant as they govern management options,

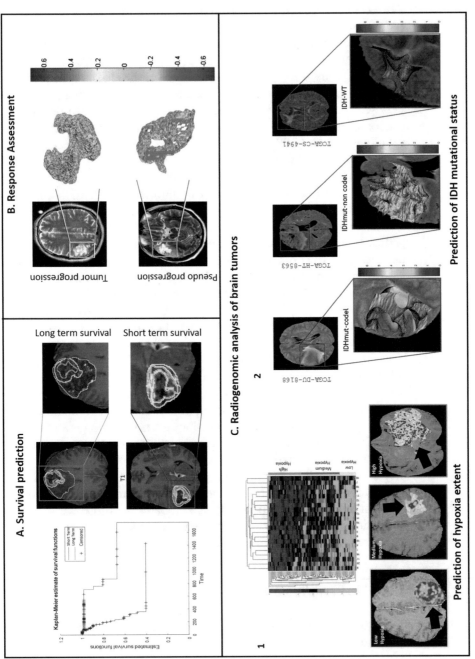

Fig. 14.3 Applications of radiomics in brain tumors (**a**) Radiomic features extracted from the tumor habitat of GBM were prognostic of survival on MRI ($p < 0.0015$). (**b**) Shape features can discriminate true progression from pseudoprogression. (**c**) (1) Radiomic correlates of the hypoxia signaling pathway were found to be prognostic of survival ($p < 0.01$) (2) Radiogenomic analysis revealed radiomic features predictive of IDH mutation status. *Figure **c**(1) modified with permission from Beig N et al., Radiogenomic analysis of hypoxia pathway is predictive of overall survival in Glioblastoma, Sci Rep. 2018;8(1):7 under the terms of the Creative Commons license (http://creativecommons.org/licenses/by/4.0/)

such as the doses and relative frequencies of chemotherapy and radiation following resection. In addition to the histopathological grading, gliomas have distinct molecular phenotypes (isocitrate dehydrogenase [IDH], 1p19qcodel) which can help in personalizing treatment decisions. Several radiomic studies have tackled this problem by identifying distinct radiomic phenotypes of gliomas on imaging. For instance, Zacharaki et al. [171] deployed an SVM classifier to categorize the different classes of brain tumors based on radiomic analysis of conventional and perfusion MRI. The SVM model yielded a sensitivity of 87% and specificity of 79% in distinguishing brain metastasis from gliomas, and 85% sensitivity and 96% specificity in distinguishing grade II from grade III and IV gliomas. Sasikala et al. [207] similarly used a set of randomized heuristic search techniques using wavelet features to obtain an accuracy of 98% in differentiating between normal, benign, or malignant brain tumors (n = 100). Meanwhile, an unsupervised k-means clustering approach differentiated between metastasis from GBMs with a sensitivity of 92% and a specificity of 71% (n = 50) using features extracted from a combination of conventional MRI as well as perfusion-weighted MRI [208]. Gutierrez et al. [209] expanded radiomic classification to pediatric posterior fossa tumors as well in a retrospective analysis in 40 children (17 medulloblastomas, 7 ependymomas, and 16 pilocytic astrocytomas). Using conventional T1W + C and T2W images along with ADC maps, an SVM classifier had an accuracy of over 91.4% in correctly assigning the type of posterior fossa tumors. Meanwhile, a total of 109 pediatric patients was used by Iv et al. to classify medulloblastomas into the four molecular subgroups (WNT, Sonic Hedgehog, Group 3, Group 4), using radiomic features including Gabor and tumor edge-sharpness features. They found that the edge sharpness feature was most discriminative in differentiating between Sonic hedgehog and Group 4 tumors [182].

8.2 Survival Prediction and Risk of Recurrence in Brain Tumors

Radiomics has found its most potent application in the survival analysis of brain tumors. Several supervised and unsupervised machine learning models have utilized a plethora of features including shape, texture, wavelet, intensity, as well as "deep learned" features for risk prediction. Kickingereder et al. [210] retrospectively analyzed 119 patients with $n = 79$ for training and $n = 40$ for independent validation and used a supervised principal component analysis to stratify patients according to OS and PFS. The top selected features belonged to Haralick texture family. On the validation set, the radiomics model had a HR—2.43 ($p = 0.002$) in predicting PFS. Meanwhile, McGarry et al. [211] used a radiomic profile (RP)-based approach by combining T1W+C and T2W-FLAIR pre-surgical MRIs of 81 patients. Out of the 81 generated RPs, 5 of them were found to be highly correlated with OS using Cox regression analysis ($p < 0.0005$). These RPs were then correlated with the histological samples from the biopsy. Macyszyn et al. [212] used a dataset of 105 GBM patients and intensity-based features teamed up with an SVM classifier to distinguish between (a) patients surviving more than 6 months or less (short term); (b) patients surviving less/more than 12 months (long term). The two classifiers were then combined to create a single index which they called the survival prediction index (SPI). Higher SPI predicted longer survival and vice versa. The developed predictor was tested on a prospective cohort of 29 new patients and yielded an HR of 10.64 on predicting short-term vs long-term survival. They also used the radiomic model to classify the GBM into the four molecular subtypes with a 76% accuracy.

The concept of tumor habitat on multi-parametric MRI has been used by Lee et al. and Zhou et al. to define clinical outcomes in GBM patients. Tumor habitats were chosen by dividing regions within the tumor according to their varying voxel intensities by 2D and 3D histogram analysis, wherein they formed two classes spatially into high-and low using an intensity threshold. Lee et al. [213] used 14 spatial diversity features to have an AUC of 0.76 in predicting survival status among 74 GBM patients ($\leq$12 months vs >12 months). Zhou et al. [214] meanwhile used 32 patients with GBM from TCIA and applied the tumor habitat spatial analysis, using an SVM classifier to have an 81.25% accuracy in determining survival status (<400 and >400 days). Meanwhile, Prasanna et al. [180] looked at the peritumoral zone, rather than looking conventionally inside the tumor and used ten radiomic features on 65 patients, including the novel CoLIAGe and Haralick features, in a Random Forest classifier to differentiate between short-term and long-term survival (p = 1.47 x 10–5). A larger study by Bae et al. [215] involving 217 patients (n = 163 for training and n = 54 for testing) used a combination of shape and Haralick textural features in a Random Forest classifier to predict OS (HR—2.58; p = 0.03) and PFS (HR—3.24; p = 0.01) on the testing set.

8.3 Treatment-Related Changes and Response Assessment in Brain Tumors

Current standard of care in GBMs includes surgical resection followed by concurrent chemoradiation. One of the most common problems confronting radiation oncologists in the GBM treatment paradigm is to reliably distinguish treatment-related changes, i.e., radiation necrosis or pseudo-progression from true tumor recurrence. While it is extremely difficult for even expert radiologists to distinguish between true and pseudo-progression using conventional MRIs, radiomics has shown promising applications in this field and demonstrated impressive accuracy in distinguishing between the two conditions. Ismail et al. [169] used a novel 3D shape-based radiomic classifier on 105 GBM patients from two different institutions (n = 59 for training and n = 46 for testing) to differentiate true progression from pseudo-progression. With the top 5 shape features embedded in an SVM classifier, an accuracy of

90.2% was reported in correctly classifying pseudo-progression from tumor recurrence. Meanwhile, Tiwari et al. [170] had previously shown that an SVM classifier using an mRMR feature selection algorithm on a dataset of $n = 43$ for training and $n = 15$ for validation of post-chemoradiation MRI scans, could successfully distinguish cerebral radionecrosis from tumor recurrence (AUC—0.79; $p < 0.05$). Interestingly, this study also reported that while the radiomic model had an accuracy of 80%, they employed two expert neuro-radiologists who had an accuracy of 47% and 53%, in distinguishing radiation necrosis and tumor recurrence. With a smaller subset of GBM patients ($n = 31$), Hu et al. [216] had an AUC of 0.94 in identifying pseudo-progression using voxel-based intensity-based features from multiparametric MRI, using an SVM classifier. Lohmann et al. [217] meanwhile used Haralick features from 18F-FET PET images of 47 patients with brain metastases post-radiotherapy, along with standard PET parameters including tumor-to-brain ratios (TBRs) of uptake had an accuracy of 85% in correctly differentiating between radiation injury and recurrent brain metastasis. Radiomic analysis has also shown promise in stratifying response to novel anti-angiogenic agents like bevacizumab, which has shown increasing potential in recurrent GBMs. A study [218] in 172 patients (divided into training and validation in a 2:1 ratio) with multi-parametric pre-treatment MRIs and a supervised PCA analysis using shape, texture, and intensity features successfully stratified outcome to bevacizumab on the testing set based on OS (HR = 2.60; $p = 0.001$).

8.4 Radiogenomic Analysis of Brain Tumors

Radiogenomics is an evolving discipline to elucidate a deeper biological meaning of imaging phenotype, i.e., radiomics by finding correlations with the genetic signature of the disease ("genomics"), including its gene expression, mutations, or downstream signaling pathways. The association between the imaging phenotype and the genetic patterns of the disease can be computed using simple correlations with amplified or downregulated genes or with a specific genetic subset or a combined genetic signature. Radiogenomics in the field of brain tumors can be leveraged to predict treatment response and underpin the malignant biological processes that drive most of the GBMs to poor prognosis. Radiogenomic applications in this space have led to the restructuring of the WHO classification of diffuse gliomas with the 2016 update [219], focusing on mutations and chromosomal alterations (e.g., IDH mutant and wild type, 1p/19q status). An imaging surrogate to determine the molecular subtyping of gliomas based on IDH and 1p/19q genotypes [220] can potentially avoid invasive interventions and sampling errors on genetic analysis. Zhang et al. [221] found that texture and shape features have the highest predictive value and predicted an accuracy of 89% (AUC = 0.9231) in the validation cohort. Similarly, epidermal growth factor receptor variant III (EGFRvIII) mutation has been considered to be driver mutation and therapeutic target in GBM. Bakas et al. [222] found that the heterogeneity of GBM can be captured using perfusion temporal dynamics from the peritumoral edema on MRI. They also reported that the EGFRvIII mutation can also be predicted using radiomics. Beig et al. [223] used a radiomic signature on treatment naïve multiparametric MRIs (n = 115) to find correlations with a Hypoxia Enrichment Score (HES) from 21 hypoxia-associated genes implicated in GBM aggressiveness (see Fig. 14.3c). This HES-based radiomic signature was successful in distinguishing GBM patients with short-term vs long-term survival (p = 0.003). Meanwhile, Levner et al. [224] used an L1-regularized neural network using texture features to predict MGMT promoter methylation status in 59 GBM patients with an average 87.7% accuracy. MGMT has also been shown to be associated with chemotherapy benefit but its promoter methylation status is difficult to determine, which again brings the value of radiomics analysis as a surrogate. Kickingereder et al. [225] used pre-operative multiparametric MRI scans of 152 patients and used a number of classifiers (RF, Lasso and Elastic net Generalized model, Gradient boosting, etc.) to link semantic and volume/perfusion-based imaging parameters with corresponding molecular and genetic signatures. In univariate analysis, associations were found with EGFR amplification and loss of CDKN2A. The machine learning classifiers were also successful in predicting EGFR status and RTK II status. Hu et al. [226] used textural features from multiparametric MRI (conventional, diffusion tensor imaging (DTI), DSC-pMRI) in a PCA to correlate with the genetic signature for six driver genes from image-guided biopsies of enhancing and non-enhancing regions of the tumor (n = 13). Interestingly, the radiogenomic model found the highest accuracy for PDGFRA (77.1%) from within the enhancing region, while CDKN2A and RB1 showed an accuracy of 87.5% but from the peritumoral region.

9 Future Scope and Potential of Radiomics as a Novel Biomarker for Brain Tumors

Recent developments in brain tumor research clearly demonstrate that the field will continue to see a transition from predominantly qualitative, expert-based methods of morphological assessment to more automated and quantitative methods of tumor evaluation. In this setting, the emerging field of radiomics has already shown tremendous promise across multiple applications in oncology including in diagnosis, prognosis, as well evaluating response to treatment in brain tumors. Radiomics, as it stands, needs to find its place as a decision support tool that could be employed routinely in day-to-day clinical applications by neuroradiologists, neurosurgeons, oncologists, and radiation oncologists, toward benefitting clinical and therapeutic decision making. However, its adoption as a part of routine clinical workflow will be dependent on overcoming certain limitations. A significant limitation is that most radiomic findings are often based on small retrospective, single-site, and relatively biased data. Similarly, multi-institutional studies need to account for large sources of variance in imaging parameters introduced across sites and scanners. This can be accomplished by large-scale, multi-institutional prospective validation studies that involve extensive validation of the stability of radiomic features across sites and scanners. Similarly, clinical adoption will also hinge on improved understanding of the morphological and biological basis of the radiomic features that ultimately drive disease phenotyping. With increasing interest in the discipline, the time is opportune for radiomic methods to be prospectively validated in large multi-site clinical trials which can pave the way for radiomics to be accepted as standard of care in clinical decision making. This also calls for a collaborative enterprise between oncologists, radiologists, physicians, and computational scientists, working in synergy from tool development to validation.

Acknowledgements

Research reported in this publication was supported by The Ohio Third Frontier Technology Validation Fund and The Wallace H. Coulter Foundation Program in the Department of Biomedical Engineering at Case Western Reserve University.

Department of Defense Peer Reviewed Cancer Research Program (PRCRP) Career Development Award.

Dana Foundation David Mahoney Neuroimaging Program.

The content is solely the responsibility of the authors and does not necessarily represent the official views of the National Institutes of Health, the U.S. Department of Veterans Affairs, the Department of Defense, or the United States Government.

References

1. Ostrom QT, Gittleman H, Truitt G, Boscia A, Kruchko C, Barnholtz-Sloan JS (2018) CBTRUS statistical report: primary brain and other central nervous system tumors diagnosed in the United States in 2011–2015. Neuro-Oncology 20(suppl_4):iv1–iv86. https://doi.org/10.1093/neuonc/noy131

2. Burger PC, Heinz ER, Shibata T, Kleihues P (1988) Topographic anatomy and CT correlations in the untreated glioblastoma multiforme. J Neurosurg 68(5):698–704. https://doi.org/10.3171/jns.1988.68.5.0698

3. Kelly PJ, Daumas-Duport C, Scheithauer BW, Kall BA, Kispert DB (1987) Stereotactic histologic correlations of computed tomography- and magnetic resonance imaging-defined abnormalities in patients with glial neoplasms. Mayo Clin Proc 62(6):450–459

4. Burger PC (1983) Pathologic anatomy and CT correlations in the glioblastoma multiforme. Appl Neurophysiol 46(1–4):180–187

5. Lapointe S, Perry A, Butowski NA (2018) Primary brain tumours in adults. Lancet (London, England) 392(10145):432–

446. https://doi.org/10.1016/s0140-6736(18)30990-5

6. Sasaki H, Yoshida K (2017) Treatment recommendations for adult patients with diffuse gliomas of grades II and III according to the New WHO Classification in 2016. Neurol Med Chir 57(12):658–666. https://doi.org/10.2176/nmc.ra.2017-0071

7. Perkins A, Liu G (2016) Primary brain tumors in adults: diagnosis and treatment. Am Fam Physician 93(3):211–217

8. Weller M, Wick W, Aldape K, Brada M, Berger M, Pfister SM, Nishikawa R, Rosenthal M, Wen PY, Stupp R, Reifenberger G (2015) Glioma. Nat Rev Dis Primers 1:15017. https://doi.org/10.1038/nrdp.2015.17

9. Ricard D, Idbaih A, Ducray F, Lahutte M, Hoang-Xuan K, Delattre JY (2012) Primary brain tumours in adults. Lancet (London, England) 379(9830):1984–1996. https://doi.org/10.1016/s0140-6736(11)61346-9

10. Dong X (2018) Current strategies for brain drug delivery. Theranostics 8(6):1481–1493. https://doi.org/10.7150/thno.21254

11. Omidi Y, Barar J (2012) Impacts of blood-brain barrier in drug delivery and targeting of brain tumors. Bioimpacts 2(1):5–22. https://doi.org/10.5681/bi.2012.002

12. Clarke J, Penas C, Pastori C, Komotar RJ, Bregy A, Shah AH, Wahlestedt C, Ayad NG (2013) Epigenetic pathways and glioblastoma treatment. Epigenetics 8(8):785–795. https://doi.org/10.4161/epi.25440

13. Friedmann-Morvinski D (2014) Glioblastoma heterogeneity and cancer cell plasticity. Crit Rev Oncog 19(5):327–336

14. Pointer KB, Clark PA, Zorniak M, Alrfaei BM, Kuo JS (2014) Glioblastoma cancer stem cells: biomarker and therapeutic advances. Neurochem Int 71:1–7. https://doi.org/10.1016/j.neuint.2014.03.005

15. Veliz I, Loo Y, Castillo O, Karachaliou N, Nigro O, Rosell R (2015) Advances and challenges in the molecular biology and treatment of glioblastoma-is there any hope for the future? Ann Transl Med 3(1):7. https://doi.org/10.3978/j.issn.2305-5839.2014.10.06

16. Henson JW, Gaviani P, Gonzalez RG (2005) MRI in treatment of adult gliomas. Lancet Oncol 6(3):167–175. https://doi.org/10.1016/s1470-2045(05)01767-5

17. Henson JW, Gonzalez RG (2004) Neuroimaging in glioma therapy. Expert Rev Neurother 4(4):665–671. https://doi.org/10.1586/14737175.4.4.665

18. Lemort M, Canizares-Perez AC, Van der Stappen A, Kampouridis S (2007) Progress in magnetic resonance imaging of brain tumours. Curr Opin Oncol 19(6):616–622. https://doi.org/10.1097/CCO.0b013e3282f076b2

19. Rees J (2003) Advances in magnetic resonance imaging of brain tumours. Curr Opin Neurol 16(6):643–650. https://doi.org/10.1097/01.wco.0000102626.38669.b9

20. Brismar J, Stromblad LG, Salford LG (1978) Impact of CT in the neurosurgical management of intracranial tumors. Neuroradiology 16:506–509

21. Baker HL Jr, Houser OW, Campbell JK (1980) National Cancer Institute study: evaluation of computed tomography in the diagnosis of intracranial neoplasms. I. Overall results. Radiology 136(1):91–96. https://doi.org/10.1148/radiology.136.1.7384529

22. Damadian R (1971) Tumor detection by nuclear magnetic resonance. Science (New York, NY) 171(3976):1151–1153

23. Iv M, Yoon BC, Heit JJ, Fischbein N, Wintermark M (2018) Current clinical state of advanced magnetic resonance imaging for brain tumor diagnosis and follow up. Semin Roentgenol 53(1):45–61. https://doi.org/10.1053/j.ro.2017.11.005

24. Villanueva-Meyer JE, Mabray MC, Cha S (2017) Current clinical brain tumor imaging. Neurosurgery 81(3):397–415. https://doi.org/10.1093/neuros/nyx103

25. Mabray MC, Barajas RF Jr, Cha S (2015) Modern brain tumor imaging. Brain Tumor Res Treat 3(1):8–23. https://doi.org/10.14791/btrt.2015.3.1.8

26. Castillo M (2014) History and evolution of brain tumor imaging: insights through radiology. Radiology 273(2 Suppl):S111–S125. https://doi.org/10.1148/radiol.14140130

27. Anderson MD, Colen RR, Tremont-Lukats IW (2014) Imaging mimics of primary malignant tumors of the central nervous system (CNS). Curr Oncol Rep 16(8):399. https://doi.org/10.1007/s11912-014-0399-8

28. Huisman TA (2009) Tumor-like lesions of the brain. Cancer Imaging 9(Spec A):S10–S13. https://doi.org/10.1102/1470-7330.2009.9003

29. Okamoto K, Furusawa T, Ishikawa K, Quadery FA, Sasai K, Tokiguchi S (2004) Mimics of brain tumor on neuroimaging: part II. Radiat Med 22(3):135–142

30. Okamoto K, Furusawa T, Ishikawa K, Quadery FA, Sasai K, Tokiguchi S (2004) Mimics of brain tumor on neuroimaging: part I. Radiat Med 22(2):63–76

31. Asari S, Makabe T, Katayama S, Itoh T, Tsuchida S, Ohmoto T (1993) Astrocytic gliomas: MRI and pathological grade. Acta Med Okayama 47(6):383–389. https://doi.org/10.18926/amo/31566

32. Asari S, Makabe T, Katayama S, Itoh T, Tsuchida S, Ohmoto T (1994) Assessment

of the pathological grade of astrocytic gliomas using an MRI score. Neuroradiology 36(4):308–310

33. Chaichana KL, McGirt MJ, Niranjan A, Olivi A, Burger PC, Quinones-Hinojosa A (2009) Prognostic significance of contrast-enhancing low-grade gliomas in adults and a review of the literature. Neurol Res 31(9):931–939. https://doi.org/10.1179/174313209x395454

34. Dean BL, Drayer BP, Bird CR, Flom RA, Hodak JA, Coons SW, Carey RG (1990) Gliomas: classification with MR imaging. Radiology 174(2):411–415. https://doi.org/10.1148/radiology.174.2.2153310

35. Mihara F, Numaguchi Y, Rothman M, Sato S, Fiandaca MS (1995) MR imaging of adult supratentorial astrocytomas: an attempt of semi-automatic grading. Radiat Med 13(1):5–9

36. Pierallini A, Bonamini M, Bozzao A, Pantano P, Stefano DD, Ferone E, Raguso M, Bosman C, Bozzao L (1997) Supratentorial diffuse astrocytic tumours: proposal of an MRI classification. Eur Radiol 7(3):395–399. https://doi.org/10.1007/s003300050173

37. Upadhyay N, Waldman AD (2011) Conventional MRI evaluation of gliomas. Br J Radiol 84(Spec 2):S107–S111. https://doi.org/10.1259/bjr/65711810

38. Malone H, Yang J, Hershman DL, Wright JD, Bruce JN, Neugut AI (2015) Complications following stereotactic needle biopsy of intracranial tumors. World Neurosurg 84(4):1084–1089. https://doi.org/10.1016/j.wneu.2015.05.025

39. Yong RL, Lonser RR (2013) Safety of closed brain biopsy: population-based studies weigh in. World Neurosurg 79(1):53–54. https://doi.org/10.1016/j.wneu.2012.05.016

40. Jackson RJ, Fuller GN, Abi-Said D, Lang FF, Gokaslan ZL, Shi WM, Wildrick DM, Sawaya R (2001) Limitations of stereotactic biopsy in the initial management of gliomas. Neuro-Oncology 3(3):193–200. https://doi.org/10.1093/neuonc/3.3.193

41. Kelly PJ (1992) Stereotactic resection and its limitations in glial neoplasms. Stereotact Funct Neurosurg 59(1–4):84–91. https://doi.org/10.1159/000098922

42. Jakola AS, Skjulsvik AJ, Myrmel KS, Sjavik K, Unsgard G, Torp SH, Aaberg K, Berg T, Dai HY, Johnsen K, Kloster R, Solheim O (2017) Surgical resection versus watchful waiting in low-grade gliomas. Ann Oncol 28(8):1942–1948. https://doi.org/10.1093/annonc/mdx230

43. Jakola AS, Unsgard G, Myrmel KS, Kloster R, Torp SH, Losvik OK, Lindal S, Solheim O (2013) Surgical strategy in grade II astrocy-

toma: a population-based analysis of survival and morbidity with a strategy of early resection as compared to watchful waiting. Acta Neurochir 155(12):2227–2235. https://doi.org/10.1007/s00701-013-1869-8

44. Whittle IR (2010) What is the place of conservative management for adult supratentorial low-grade glioma? Adv Tech Stand Neurosurg 35:65–79

45. Wang LL, Leach JL, Breneman JC, McPherson CM, Gaskill-Shipley MF (2014) Critical role of imaging in the neurosurgical and radiotherapeutic management of brain tumors. Radiographics 34(3):702–721. https://doi.org/10.1148/rg.343130156

46. Ginsberg LE, Fuller GN, Hashmi M, Leeds NE, Schomer DF (1998) The significance of lack of MR contrast enhancement of supratentorial brain tumors in adults: histopathological evaluation of a series. Surg Neurol 49(4):436–440

47. Pronin IN, Holodny AI, Petraikin AV (1997) MRI of high-grade glial tumors: correlation between the degree of contrast enhancement and the volume of surrounding edema. Neuroradiology 39(5):348–350

48. Hawighorst H, Schreiber W, Knopp MV, Essig M, Engenhart-Cabilic R, Brix G, van Kaick G (1996) Macroscopic tumor volume of malignant glioma determined by contrast-enhanced magnetic resonance imaging with and without magnetization transfer contrast. Magn Reson Imaging 14(10):1119–1126

49. Liu S, Wang Y, Xu K, Wang Z, Fan X, Zhang C, Li S, Qiu X, Jiang T (2017) Relationship between necrotic patterns in glioblastoma and patient survival: fractal dimension and lacunarity analyses using magnetic resonance imaging. Sci Rep 7(1):8302. https://doi.org/10.1038/s41598-017-08862-6

50. Wu CX, Lin GS, Lin ZX, Zhang JD, Liu SY, Zhou CF (2015) Peritumoral edema shown by MRI predicts poor clinical outcome in glioblastoma. World J Surg Oncol 13:97. https://doi.org/10.1186/s12957-015-0496-7

51. Gutman DA, Cooper LA, Hwang SN, Holder CA, Gao J, Aurora TD, Dunn WD Jr, Scarpace L, Mikkelsen T, Jain R, Wintermark M, Jilwan M, Raghavan P, Huang E, Clifford RJ, Mongkolwat P, Kleper V, Freymann J, Kirby J, Zinn PO, Moreno CS, Jaffe C, Colen R, Rubin DL, Saltz J, Flanders A, Brat DJ (2013) MR imaging predictors of molecular profile and survival: multi-institutional study of the TCGA glioblastoma data set. Radiology 267(2):560–569. https://doi.org/10.1148/radiol.13120118

52. Pierallini A, Bonamini M, Pantano P, Palmeggiani F, Raguso M, Osti MF, Anaveri G, Bozzao L (1998) Radiological assess-

ment of necrosis in glioblastoma: variability and prognostic value. Neuroradiology 40(3):150–153

53. Hammoud MA, Sawaya R, Shi W, Thall PF, Leeds NE (1996) Prognostic significance of preoperative MRI scans in glioblastoma multiforme. J Neuro-Oncol 27(1):65–73

54. Pope WB, Sayre J, Perlina A, Villablanca JP, Mischel PS, Cloughesy TF (2005) MR imaging correlates of survival in patients with high-grade gliomas. AJNR Am J Neuroradiol 26(10):2466–2474

55. Sarmiento JM, Nuno M, Ortega A, Mukherjee D, Fan X, Black KL, Patil CG (2014) Cystic glioblastoma: an evaluation of IDH1 status and prognosis. Neurosurgery 74(1):71–75.: discussion 75-76. https://doi.org/10.1227/neu.0000000000000200

56. Pallud J, Capelle L, Taillandier L, Fontaine D, Mandonnet E, Guillevin R, Bauchet L, Peruzzi P, Laigle-Donadey F, Kujas M, Guyotat J, Baron MH, Mokhtari K, Duffau H (2009) Prognostic significance of imaging contrast enhancement for WHO grade II gliomas. Neuro-Oncology 11(2):176–182. https://doi.org/10.1215/15228517-2008-066

57. Wang Y, Wang K, Wang J, Li S, Ma J, Dai J, Jiang T (2016) Identifying the association between contrast enhancement pattern, surgical resection, and prognosis in anaplastic glioma patients. Neuroradiology 58(4):367–374. https://doi.org/10.1007/s00234-016-1640-y

58. Lacroix M, Abi-Said D, Fourney DR, Gokaslan ZL, Shi W, DeMonte F, Lang FF, McCutcheon IE, Hassenbusch SJ, Holland E, Hess K, Michael C, Miller D, Sawaya R (2001) A multivariate analysis of 416 patients with glioblastoma multiforme: prognosis, extent of resection, and survival. J Neurosurg 95(2):190–198. https://doi.org/10.3171/jns.2001.95.2.0190

59. Scott JN, Brasher PM, Sevick RJ, Rewcastle NB, Forsyth PA (2002) How often are nonenhancing supratentorial gliomas malignant? A population study. Neurology 59(6):947–949

60. Chaichana KL, Jusue-Torres I, Lemos AM, Gokaslan A, Cabrera-Aldana EE, Ashary A, Olivi A, Quinones-Hinojosa A (2014) The butterfly effect on glioblastoma: is volumetric extent of resection more effective than biopsy for these tumors? J Neuro-Oncol 120(3):625–634. https://doi.org/10.1007/s11060-014-1597-9

61. Chan K, Bhandari M (2011) Three-minute critical appraisal of a case series article. Indian J Orthop 45(2):103–104. https://doi.org/10.4103/0019-5413.77126

62. Westhoff CL (1995) Epidemiologic studies: pitfalls in interpretation. Dialogues Contracept 4(5):5–6. 8

63. Kempen JH (2011) Appropriate use and reporting of uncontrolled case series in the medical literature. Am J Ophthalmol 151(1):7–10.e11. https://doi.org/10.1016/j.ajo.2010.08.047

64. von Elm E, Altman DG, Egger M, Pocock SJ, Gotzsche PC, Vandenbroucke JP (2008) The strengthening the reporting of observational studies in epidemiology (STROBE) statement: guidelines for reporting observational studies. J Clin Epidemiol 61(4):344–349. https://doi.org/10.1016/j.jclinepi.2007.11.008

65. Talos IF, Zou KH, Ohno-Machado L, Bhagwat JG, Kikinis R, Black PM, Jolesz FA (2006) Supratentorial low-grade glioma resectability: statistical predictive analysis based on anatomic MR features and tumor characteristics. Radiology 239(2):506–513. https://doi.org/10.1148/radiol.2392050661

66. Keles GE, Lamborn KR, Berger MS (2001) Low-grade hemispheric gliomas in adults: a critical review of extent of resection as a factor influencing outcome. J Neurosurg 95(5):735–745. https://doi.org/10.3171/jns.2001.95.5.0735

67. Tortosa A, Vinolas N, Villa S, Verger E, Gil JM, Brell M, Caral L, Pujol T, Acebes JJ, Ribalta T, Ferrer I, Graus F (2003) Prognostic implication of clinical, radiologic, and pathologic features in patients with anaplastic gliomas. Cancer 97(4):1063–1071. https://doi.org/10.1002/cncr.11120

68. Kaur G, Bloch O, Jian BJ, Kaur R, Sughrue ME, Aghi MK, McDermott MW, Berger MS, Chang SM, Parsa AT (2011) A critical evaluation of cystic features in primary glioblastoma as a prognostic factor for survival. J Neurosurg 115(4):754–759. https://doi.org/10.3171/2011.5.Jns11128

69. Bohman LE, Swanson KR, Moore JL, Rockne R, Mandigo C, Hankinson T, Assanah M, Canoll P, Bruce JN (2010) Magnetic resonance imaging characteristics of glioblastoma multiforme: implications for understanding glioma ontogeny. Neurosurgery 67(5):1319–1327.; discussion 1327-1318. https://doi.org/10.1227/NEU.0b013e3181f556ab

70. Utsuki S, Oka H, Suzuki S, Shimizu S, Tanizaki Y, Kondo K, Tanaka S, Kawano N, Fujii K (2006) Pathological and clinical features of cystic and noncystic glioblastomas. Brain Tumor Pathol 23(1):29–34. https://doi.org/10.1007/s10014-006-0195-8

71. Maldaun MV, Suki D, Lang FF, Prabhu S, Shi W, Fuller GN, Wildrick DM, Sawaya R (2004) Cystic glioblastoma multiforme: survival outcomes in 22 cases. J Neurosurg

100(1):61–67. https://doi.org/10.3171/jns.2004.100.1.0061

72. Lote K, Egeland T, Hager B, Skullerud K, Hirschberg H (1998) Prognostic significance of CT contrast enhancement within histological subgroups of intracranial glioma. J Neuro-Oncol 40(2):161–170

73. Sharma S, Jain SK, Sinha VD (2017) Use of preoperative ependymal enhancement on magnetic resonance imaging brain as a marker of grade of glioma. J Neurosci Rural Pract 8(4):545–550. https://doi.org/10.4103/jnrp.jnrp_78_17

74. Reyes-Botero G, Dehais C, Idbaih A, Martin-Duverneuil N, Lahutte M, Carpentier C, Letouze E, Chinot O, Loiseau H, Honnorat J, Ramirez C, Moyal E, Figarella-Branger D, Ducray F (2014) Contrast enhancement in 1p/19q-codeleted anaplastic oligodendrogliomas is associated with 9p loss, genomic instability, and angiogenic gene expression. Neuro-Oncology 16(5):662–670. https://doi.org/10.1093/neuonc/not235

75. White ML, Zhang Y, Kirby P, Ryken TC (2005) Can tumor contrast enhancement be used as a criterion for differentiating tumor grades of oligodendrogliomas? AJNR Am J Neuroradiol 26(4):784–790

76. Earnest F, Kelly PJ, Scheithauer BW, Kall BA, Cascino TL, Ehman RL, Forbes GS, Axley PL (1988) Cerebral astrocytomas: histopathologic correlation of MR and CT contrast enhancement with stereotactic biopsy. Radiology 166(3):823–827. https://doi.org/10.1148/radiology.166.3.2829270

77. Graif M, Bydder GM, Steiner RE, Niendorf P, Thomas DG, Young IR (1985) Contrast-enhanced MR imaging of malignant brain tumors. AJNR Am J Neuroradiol 6(6):855–862

78. Wu CX, Lin GS, Lin ZX, Zhang JD, Chen L, Liu SY, Tang WL, Qiu XX, Zhou CF (2015) Peritumoral edema on magnetic resonance imaging predicts a poor clinical outcome in malignant glioma. Oncol Lett 10(5):2769–2776. https://doi.org/10.3892/ol.2015.3639

79. Min ZG, Niu C, Rana N, Ji HM, Zhang M (2013) Differentiation of pure vasogenic edema and tumor-infiltrated edema in patients with peritumoral edema by analyzing the relationship of axial and radial diffusivities on 3.0T MRI. Clin Neurol Neurosurg 115(8):1366–1370. https://doi.org/10.1016/j.clineuro.2012.12.031

80. Liu SY, Mei WZ, Lin ZX (2013) Preoperative peritumoral edema and survival rate in glioblastoma multiforme. Onkologie 36(11):679–684. https://doi.org/10.1159/000355651

81. Seidel C, Dorner N, Osswald M, Wick A, Platten M, Bendszus M, Wick W (2011) Does age matter?—a MRI study on peritumoral edema in newly diagnosed primary glioblastoma. BMC Cancer 11:127. https://doi.org/10.1186/1471-2407-11-127

82. Schoenegger K, Oberndorfer S, Wuschitz B, Struhal W, Hainfellner J, Prayer D, Heinzl H, Lahrmann H, Marosi C, Grisold W (2009) Peritumoral edema on MRI at initial diagnosis: an independent prognostic factor for glioblastoma? Eur J Neurol 16(7):874–878. https://doi.org/10.1111/j.1468-1331.2009.02613.x

83. Raza SM, Fuller GN, Rhee CH, Huang S, Hess K, Zhang W, Sawaya R (2004) Identification of necrosis-associated genes in glioblastoma by cDNA microarray analysis. Clin Cancer Res 10(1 Pt 1):212–221

84. Raza SM, Lang FF, Aggarwal BB, Fuller GN, Wildrick DM, Sawaya R (2002) Necrosis and glioblastoma: a friend or a foe? A review and a hypothesis. Neurosurgery 51(1):2–12. discussion 12-13

85. Barker FG 2nd, Davis RL, Chang SM, Prados MD (1996) Necrosis as a prognostic factor in glioblastoma multiforme. Cancer 77(6):1161–1166

86. Lohle PN, Verhagen IT, Teelken AW, Blaauw EH, Go KG (1992) The pathogenesis of cerebral gliomatous cysts. Neurosurgery 30(2):180–185

87. Afra D, Norman D, Levin VA (1980) Cysts in malignant gliomas. Identification by computerized tomography. J Neurosurg 53(6):821–825. https://doi.org/10.3171/jns.1980.53.6.0821

88. Zhou M, Chaudhury B, Hall LO, Goldgof DB, Gillies RJ, Gatenby RA (2017) Identifying spatial imaging biomarkers of glioblastoma multiforme for survival group prediction. J Magn Reson Imaging 46(1):115–123. https://doi.org/10.1002/jmri.25497

89. Johnson DR, Diehn FE, Giannini C, Jenkins RB, Jenkins SM, Parney IF, Kaufmann TJ (2017) Genetically defined oligodendroglioma is characterized by indistinct tumor borders at MRI. AJNR Am J Neuroradiol 38(4):678–684. https://doi.org/10.3174/ajnr.A5070

90. Chaddad A, Desrosiers C, Hassan L, Tanougast C (2016) A quantitative study of shape descriptors from glioblastoma multiforme phenotypes for predicting survival outcome. Br J Radiol 89(1068):20160575. https://doi.org/10.1259/bjr.20160575

91. Lee JW, Wen PY, Hurwitz S, Black P, Kesari S, Drappatz J, Golby AJ, Wells WM 3rd, Warfield SK, Kikinis R, Bromfield EB (2010) Morphological characteristics of

brain tumors causing seizures. Arch Neurol 67(3):336–342. https://doi.org/10.1001/archneurol.2010.2

92. Scherer M, Jungk C, Younsi A, Kickingereder P, Muller S, Unterberg A (2016) Factors triggering an additional resection and determining residual tumor volume on intraoperative MRI: analysis from a prospective single-center registry of supratentorial gliomas. Neurosurg Focus 40(3):E4. https://doi.org/10.3171/2015.11.Focus15542

93. Siegal T (2016) Clinical relevance of prognostic and predictive molecular markers in gliomas. Adv Tech Stand Neurosurg 43:91–108. https://doi.org/10.1007/978-3-319-21359-0_4

94. Morris MA, Saboury B, Burkett B, Gao J, Siegel EL (2018) Reinventing radiology: big data and the future of medical imaging. J Thorac Imaging 33(1):4–16. https://doi.org/10.1097/rti.0000000000000311

95. Lee CH, Yoon HJ (2017) Medical big data: promise and challenges. Kidney Res Clin Pract 36(1):3–11. https://doi.org/10.23876/j.krcp.2017.36.1.3

96. Kaptchuk TJ (2003) Effect of interpretive bias on research evidence. BMJ 326(7404):1453–1455. https://doi.org/10.1136/bmj.326.7404.1453

97. Henker C, Kriesen T, Glass A, Schneider B, Piek J (2017) Volumetric quantification of glioblastoma: experiences with different measurement techniques and impact on survival. J Neuro-Oncol 135(2):391–402. https://doi.org/10.1007/s11060-017-2587-5

98. Shah GD, Kesari S, Xu R, Batchelor TT, O'Neill AM, Hochberg FH, Levy B, Bradshaw J, Wen PY (2006) Comparison of linear and volumetric criteria in assessing tumor response in adult high-grade gliomas. Neuro-Oncology 8(1):38–46. https://doi.org/10.1215/s1522851705000529

99. Smedley NF, Ellingson BM, Cloughesy TF, Hsu W (2018) Longitudinal patterns in clinical and imaging measurements predict residual survival in glioblastoma patients. Sci Rep 8(1):14429. https://doi.org/10.1038/s41598-018-32397-z

100. Kanaly CW, Ding D, Mehta AI, Waller AF, Crocker I, Desjardins A, Reardon DA, Friedman AH, Bigner DD, Sampson JH (2011) A novel method for volumetric MRI response assessment of enhancing brain tumors. PLoS One 6(1):e16031. https://doi.org/10.1371/journal.pone.0016031

101. Ellingson BM, Wen PY, Cloughesy TF (2018) Evidence and context of use for contrast enhancement as a surrogate of disease burden and treatment response in malignant glioma.

Neuro-Oncology 20(4):457–471. https://doi.org/10.1093/neuonc/nox193

102. Butler AR, Horii SC, Kricheff II, Shannon MB, Budzilovich GN (1978) Computed tomography in astrocytomas. A statistical analysis of the parameters of malignancy and the positive contrast-enhanced CT scan. Radiology 129(2):433–439. https://doi.org/10.1148/129.2.433

103. Burger PC, Dubois PJ, Schold SC Jr, Smith KR Jr, Odom GL, Crafts DC, Giangaspero F (1983) Computerized tomographic and pathologic studies of the untreated, quiescent, and recurrent glioblastoma multiforme. J Neurosurg 58(2):159–169. https://doi.org/10.3171/jns.1983.58.2.0159

104. Lilja A, Bergstrom K, Spannare B, Olsson Y (1981) Reliability of computed tomography in assessing histopathological features of malignant supratentorial gliomas. J Comput Assist Tomogr 5(5):625–636

105. Lewander R, Bergstrom M, Boethius J, Collins VP, Edner G, Greitz T, Willems J (1978) Stereotactic computer tomography for biopsy of gliomas. Acta Radiol Diagn 19(6):867–888

106. Panigrahy A, Bluml S (2009) Neuroimaging of pediatric brain tumors: from basic to advanced magnetic resonance imaging (MRI). J Child Neurol 24(11):1343–1365. https://doi.org/10.1177/0883073809342129

107. Shin JH, Lee HK, Khang SK, Kim DW, Jeong AK, Ahn KJ, Choi CG, Suh DC (2002) Neuronal tumors of the central nervous system: radiologic findings and pathologic correlation. Radiographics 22(5):1177–1189. https://doi.org/10.1148/radiographics.22.5.g02se051177

108. Leighton C, Fisher B, Bauman G, Depiero S, Stitt L, MacDonald D, Cairncross G (1997) Supratentorial low-grade glioma in adults: an analysis of prognostic factors and timing of radiation. J Clin Oncol Off J Am Soc Clin Oncol 15(4):1294–1301. https://doi.org/10.1200/jco.1997.15.4.1294

109. Kreth FW, Faist M, Rossner R, Volk B, Ostertag CB (1997) Supratentorial World Health Organization grade 2 astrocytomas and oligoastrocytomas. A new pattern of prognostic factors. Cancer 79(2):370–379

110. Philippon JH, Clemenceau SH, Fauchon FH, Foncin JF (1993) Supratentorial low-grade astrocytomas in adults. Neurosurgery 32(4):554–559

111. Bauman G, Lote K, Larson D, Stalpers L, Leighton C, Fisher B, Wara W, MacDonald D, Stitt L, Cairncross JG (1999) Pretreatment factors predict overall survival for patients with low-grade glioma: a recursive partition-

ing analysis. Int J Radiat Oncol Biol Phys 45(4):923–929

112. Cohen-Gadol AA, DiLuna ML, Bannykh SI, Piepmeier JM, Spencer DD (2004) Non-enhancing de novo glioblastoma: report of two cases. Neurosurg Rev 27(4):281–285. https://doi.org/10.1007/s10143-004-0346-5

113. Okamoto K, Ito J, Takahashi N, Ishikawa K, Furusawa T, Tokiguchi S, Sakai K (2002) MRI of high-grade astrocytic tumors: early appearance and evolution. Neuroradiology 44(5):395–402. https://doi.org/10.1007/s00234-001-0725-3

114. Moore-Stovall J, Venkatesh R (1993) Serial nonenhancing magnetic resonance imaging scans of high grade glioblastoma multiforme. J Natl Med Assoc 85(2):122–128

115. Chamberlain MC, Murovic JA, Levin VA (1988) Absence of contrast enhancement on CT brain scans of patients with supratentorial malignant gliomas. Neurology 38(9):1371–1374

116. Shukla G, Alexander GS, Bakas S, Nikam R, Talekar K, Palmer JD, Shi W (2017) Advanced magnetic resonance imaging in glioblastoma: a review. Chin Clin Oncol 6(4):40. https://doi.org/10.21037/cco.2017.06.28

117. Maia AC Jr, Malheiros SM, da Rocha AJ, Stavale JN, Guimaraes IF, Borges LR, Santos AJ, da Silva CJ, de Melo JG, Lanzoni OP, Gabbai AA, Ferraz FA (2004) Stereotactic biopsy guidance in adults with supratentorial nonenhancing gliomas: role of perfusion-weighted magnetic resonance imaging. J Neurosurg 101(6):970–976. https://doi.org/10.3171/jns.2004.101.6.0970

118. Carlson MR, Pope WB, Horvath S, Braunstein JG, Nghiemphu P, Tso CL, Mellinghoff I, Lai A, Liau LM, Mischel PS, Dong J, Nelson SF, Cloughesy TF (2007) Relationship between survival and edema in malignant gliomas: role of vascular endothelial growth factor and neuronal pentraxin 2. Clin Cancer Res 13(9):2592–2598. https://doi.org/10.1158/1078-0432.Ccr-06-2772

119. Badie B, Schartner JM, Hagar AR, Prabakaran S, Peebles TR, Bartley B, Lapsiwala S, Resnick DK, Vorpahl J (2003) Microglia cyclooxygenase-2 activity in experimental gliomas: possible role in cerebral edema formation. Clin Cancer Res 9(2):872–877

120. Lin ZX (2013) Glioma-related edema: new insight into molecular mechanisms and their clinical implications. Chin J Cancer 32(1):49–52. https://doi.org/10.5732/cjc.012.10242

121. Ruiz-Ontanon P, Orgaz JL, Aldaz B, Elosegui-Artola A, Martino J, Berciano MT, Montero JA, Grande L, Nogueira L, Diaz-Moralli S, Esparis-Ogando A, Vazquez-Barquero A, Lafarga M, Pandiella A, Cascante M, Segura V, Martinez-Climent JA, Sanz-Moreno V, Fernandez-Luna JL (2013) Cellular plasticity confers migratory and invasive advantages to a population of glioblastoma-initiating cells that infiltrate peritumoral tissue. Stem Cells 31(6):1075–1085. https://doi.org/10.1002/stem.1349

122. Yamahara T, Numa Y, Oishi T, Kawaguchi T, Seno T, Asai A, Kawamoto K (2010) Morphological and flow cytometric analysis of cell infiltration in glioblastoma: a comparison of autopsy brain and neuroimaging. Brain Tumor Pathol 27(2):81–87. https://doi.org/10.1007/s10014-010-0275-7

123. Chen J, Li Y, Yu TS, McKay RM, Burns DK, Kernie SG, Parada LF (2012) A restricted cell population propagates glioblastoma growth after chemotherapy. Nature 488(7412):522–526. https://doi.org/10.1038/nature11287

124. Choi SH, Kim JW, Chang JS, Cho JH, Kim SH, Chang JH, Suh CO (2017) Impact of including peritumoral edema in radiotherapy target volume on patterns of failure in glioblastoma following temozolomide-based chemoradiotherapy. Sci Rep 7:42148. https://doi.org/10.1038/srep42148

125. Strugar JG, Criscuolo GR, Rothbart D, Harrington WN (1995) Vascular endothelial growth/permeability factor expression in human glioma specimens: correlation with vasogenic brain edema and tumor-associated cysts. J Neurosurg 83(4):682–689. https://doi.org/10.3171/jns.1995.83.4.0682

126. Pierallini A, Bonamini M, Osti MF, Pantano P, Palmeggiani F, Santoro A, Maurizi Enrici R, Bozzao L (1996) Supratentorial glioblastoma: neuroradiological findings and survival after surgery and radiotherapy. Neuroradiology 38(Suppl 1):S26–S30

127. Carrillo JA, Lai A, Nghiemphu PL, Kim HJ, Phillips HS, Kharbanda S, Moftakhar P, Lalaezari S, Yong W, Ellingson BM, Cloughesy TF, Pope WB (2012) Relationship between tumor enhancement, edema, IDH1 mutational status, MGMT promoter methylation, and survival in glioblastoma. AJNR Am J Neuroradiol 33(7):1349–1355. https://doi.org/10.3174/ajnr.A2950

128. Ohgaki H, Kleihues P (2013) The definition of primary and secondary glioblastoma. Clin Cancer Res 19(4):764–772. https://doi.org/10.1158/1078-0432.Ccr-12-3002

129. Oliver L, Olivier C, Marhuenda FB, Campone M, Vallette FM (2009) Hypoxia and the malignant glioma microenvironment: regulation and implications for therapy. Curr Mol Pharmacol 2(3):263–284

130. Brat DJ, Castellano-Sanchez AA, Hunter SB, Pecot M, Cohen C, Hammond EH, Devi SN, Kaur B, Van Meir EG (2004) Pseudopalisades in glioblastoma are hypoxic, express extracellular matrix proteases, and are formed by an actively migrating cell population. Cancer Res 64(3):920–927

131. Rong Y, Durden DL, Van Meir EG, Brat DJ (2006) 'Pseudopalisading' necrosis in glioblastoma: a familiar morphologic feature that links vascular pathology, hypoxia, and angiogenesis. J Neuropathol Exp Neurol 65(6):529–539

132. Adn M, Saikali S, Guegan Y, Hamlat A (2006) Pathophysiology of glioma cyst formation. Med Hypotheses 66(4):801–804. https:// doi.org/10.1016/j.mehy.2005.09.048

133. Laws ER Jr, Taylor WF, Clifton MB, Okazaki H (1984) Neurosurgical management of low-grade astrocytoma of the cerebral hemispheres. J Neurosurg 61(4):665–673. https://doi.org/10.3171/jns.1984.61.4.0665

134. Wiki for the VASARI feature set. Updated May 25, 2012. https://wiki.nci.nih.gov/display/CIP/VASARI

135. Lambin P, Rios-Velazquez E, Leijenaar R, Carvalho S, van Stiphout RG, Granton P, Zegers CM, Gillies R, Boellard R, Dekker A (2012) Radiomics: extracting more information from medical images using advanced feature analysis. Eur J Cancer 48(4):441–446

136. Eisenhauer EA, Therasse P, Bogaerts J, Schwartz LH, Sargent D, Ford R, Dancey J, Arbuck S, Gwyther S, Mooney M, Rubinstein L, Shankar L, Dodd L, Kaplan R, Lacombe D, Verweij J (2009) New response evaluation criteria in solid tumours: revised RECIST guideline (version 1.1). Eur J Cancer 45(2):228–247. https://doi.org/10.1016/j.ejca.2008.10.026

137. Wahl RL, Jacene H, Kasamon Y, Lodge MA (2009) From RECIST to PERCIST: evolving considerations for PET response criteria in solid tumors. J Nucl Med 50(Suppl 1):122s–150s. https://doi.org/10.2967/jnumed.108.057307

138. Gwyther SJ (2006) Current standards for response evaluation by imaging techniques. Eur J Nucl Med Mol Imaging 33(Suppl 1):11–15. https://doi.org/10.1007/s00259-006-0130-6

139. Kurland BF, Gerstner ER, Mountz JM, Schwartz LH, Ryan CW, Graham MM, Buatti JM, Fennessy FM, Eikman EA, Kumar V, Forster KM, Wahl RL, Lieberman FS (2012) Promise and pitfalls of quantitative imaging in oncology clinical trials. Magn Reson Imaging 30(9):1301–1312. https://doi.org/10.1016/j.mri.2012.06.009

140. Haralick RM, Shanmugam K (1973) Textural features for image classification. IEEE Trans Syst Man Cybern 6:610–621

141. Ou X, Pan W, Zhang X, Xiao P (2016) Skin image retrieval using Gabor wavelet texture feature. Int J Cosmet Sci 38(6):607–614. https://doi.org/10.1111/ics.12332

142. Dilger SK, Uthoff J, Judisch A, Hammond E, Mott SL, Smith BJ, Newell JD Jr, Hoffman EA, Sieren JC (2015) Improved pulmonary nodule classification utilizing quantitative lung parenchyma features. J Med Imaging (Bellingham) 2(4):041004. https://doi.org/10.1117/1.Jmi.2.4.041004

143. Mackin D, Fave X, Zhang L, Fried D, Yang J, Taylor B, Rodriguez-Rivera E, Dodge C, Jones AK, Court L (2015) Measuring computed tomography scanner variability of radiomics features. Investig Radiol 50(11):757–765. https://doi.org/10.1097/RLI.0000000000000180

144. Lu L, Ehmke RC, Schwartz LH, Zhao B (2016) Assessing agreement between radiomic features computed for multiple CT imaging settings. PLoS One 11(12):e0166550. https://doi.org/10.1371/journal.pone.0166550

145. Kim H, Park CM, Lee M, Park SJ, Song YS, Lee JH, Hwang EJ, Goo JM (2016) Impact of reconstruction algorithms on CT radiomic features of pulmonary tumors: analysis of intra- and inter-reader variability and inter-reconstruction algorithm variability. PLoS One 11(10):e0164924. https://doi.org/10.1371/journal.pone.0164924

146. Labby ZE, Straus C, Caligiuri P, MacMahon H, Li P, Funaki A, Kindler HL, Armato SG (2013) Variability of tumor area measurements for response assessment in malignant pleural mesothelioma. Med Phys 40(8):081916

147. Savio SJ, Harrison LC, Luukkaala T, Heinonen T, Dastidar P, Soimakallio S, Eskola HJ (2010) Effect of slice thickness on brain magnetic resonance image texture analysis. Biomed Eng Online 9:60. https://doi.org/10.1186/1475-925X-9-60

148. Buch K, Kuno H, Qureshi MM, Li B, Sakai O (2018) Quantitative variations in texture analysis features dependent on MRI scanning parameters: a phantom model. J Appl Clin Med Phys 19(6):253–264. https://doi.org/10.1002/acm2.12482

149. Segonne F, Dale AM, Busa E, Glessner M, Salat D, Hahn HK, Fischl B (2004) A hybrid approach to the skull stripping problem in MRI. NeuroImage 22(3):1060–1075. https://doi.org/10.1016/j.neuroimage.2004.03.032

150. Speier W, Iglesias JE, El-Kara L, Tu Z, Arnold C (2011) Robust skull stripping of clinical glioblastoma multiforme data. Med

Image Comput Comput Assist Interv 14(Pt 3):659–666

151. Juntu J, Sijbers J, Van Dyck D, Gielen J (2005) Bias field correction for MRI images. In: Kurzyński M, Puchała E, Woźniak M, żołnierek A (eds) Advances in soft computing. Springer, Berlin Heidelberg, pp 543–551

152. Wang L, Lai H-M, Barker GJ, Miller DH, Tofts PS (1998) Correction for variations in MRI scanner sensitivity in brain studies with histogram matching. Magn Reson Med 39(2):322–327. https://doi.org/10.1002/mrm.1910390222

153. Nyúl LG, Udupa JK (1999) On standardizing the MR image intensity scale. Magn Reson Med 42(6):1072–1081. https://doi.org/10.1002/(SICI)1522-2594(199912)42:6<1072::AID-MRM11>3.0.CO;2-M

154. Madabhushi A, Udupa JK (2006) New methods of MR image intensity standardization via generalized scale. Med Phys 33(9):3426–3434. https://doi.org/10.1118/1.2335487

155. Madabhushi A, Udupa JK, Moonis G (2006) Comparing MR image intensity standardization against tissue characterizability of magnetization transfer ratio imaging. J Magn Reson Imaging 24(3):667–675. https://doi.org/10.1002/jmri.20658

156. Gispert JD, Reig S, Pascau J, Vaquero JJ, García-Barreno P, Desco M (2004) Method for bias field correction of brain T1-weighted magnetic resonance images minimizing segmentation error. Hum Brain Mapp 22(2):133–144. https://doi.org/10.1002/hbm.20013

157. Tustison NJ, Avants BB, Cook PA, Zheng Y, Egan A, Yushkevich PA, Gee JC (2010) N4ITK: improved N3 bias correction. IEEE Trans Med Imaging 29(6):1310–1320. https://doi.org/10.1109/TMI.2010.2046908

158. Gillies RJ, Kinahan PE, Hricak H (2015) Radiomics: images are more than pictures, they are data. Radiology 278(2):563–577. https://doi.org/10.1148/radiol.2015151169

159. Zhao B, Tan Y, Tsai W-Y, Qi J, Xie C, Lu L, Schwartz LH (2016) Reproducibility of radiomics for deciphering tumor phenotype with imaging. Sci Rep 6(1). https://doi.org/10.1038/srep23428

160. Parmar C, Rios Velazquez E, Leijenaar R, Jermoumi M, Carvalho S, Mak RH, Mitra S, Shankar BU, Kikinis R, Haibe-Kains B, Lambin P, Aerts HJWL (2014) Robust radiomics feature quantification using semi-automatic volumetric segmentation. PLoS One 9(7). https://doi.org/10.1371/journal.pone.0102107

161. Clark MC, Hall LO, Goldgof DB, Velthuizen R, Murtagh FR, Silbiger MS (1998) Automatic tumor segmentation using knowledge-based techniques. IEEE Trans Med Imaging 17(2):187–201. https://doi.org/10.1109/42.700731

162. Pereira S, Pinto A, Alves V, Silva CA (2016) Brain tumor segmentation using convolutional neural networks in MRI images. IEEE Trans Med Imaging 35(5):1240–1251. https://doi.org/10.1109/TMI.2016.2538465

163. Lee M, Woo B, Kuo MD, Jamshidi N, Kim JH (2017) Quality of radiomic features in glioblastoma multiforme: impact of semiautomated tumor segmentation software. Korean J Radiol 18(3):498–509. https://doi.org/10.3348/kjr.2017.18.3.498

164. Benson E, Pound MP, French AP, Jackson AS, Pridmore TP. Deep Hourglass for Brain Tumor Segmentation. In: Crimi A, Bakas S, Kuijf H, Keyvan F, Reyes M, van Walsum T, editors. Brainlesion: Glioma, Multiple Sclerosis, Stroke and Traumatic Brain Injuries. Cham: Springer International Publishing; 2019. p. 419–428. doi:10.1007/978-3-030-11726-9_37

165. Bakas S, Reyes M, Jakab A, Bauer S, et al. Identifying the Best Machine Learning Algorithms for Brain Tumor Segmentation, Progression Assessment, andOverall Survival Prediction in the BRATS Challenge. arXiv:181102629 [cs, stat]. 2019 Apr 23;

166. VASARI Research Project—The cancer imaging archive (TCIA) public access—Cancer Imaging Archive Wiki

167. Mazurowski MA, Desjardins A, Malof JM (2013) Imaging descriptors improve the predictive power of survival models for glioblastoma patients. Neuro-Oncology 15(10):1389–1394. https://doi.org/10.1093/neuonc/nos335

168. Kong D-S, Kim J, Ryu G, You H-J, Sung JK, Han YH, Shin H-M, Lee I-H, Kim S-T, Park C-K, Choi SH, Choi JW, Seol HJ, Lee J-I, Nam D-H (2018) Quantitative radiomic profiling of glioblastoma represents transcriptomic expression. Oncotarget 9(5):6336–6345. https://doi.org/10.18632/oncotarget.23975

169. Ismail M, Hill V, Statsevych V, Huang R, Prasanna P, Correa R, Singh G, Bera K, Beig N, Thawani R, Madabhushi A, Aahluwalia M, Tiwari P (2018) Shape features of the lesion habitat to differentiate brain tumor progression from pseudoprogression on routine multiparametric MRI: a multisite study. Am J Neuroradiol. https://doi.org/10.3174/ajnr.A5858

170. Tiwari P, Prasanna P, Wolansky L, Pinho M, Cohen M, Nayate AP, Gupta A, Singh G, Hatanpaa KJ, Sloan A, Rogers L, Madabhushi

A (2016) Computer-extracted texture features to distinguish cerebral radionecrosis from recurrent brain tumors on multiparametric MRI: a feasibility study. Am J Neuroradiol 37(12):2231–2236. https://doi.org/10.3174/ajnr.A4931

171. Zacharaki EI, Wang S, Chawla S, Yoo DS, Wolf R, Melhem ER, Davatzikos C (2009) Classification of brain tumor type and grade using MRI texture and shape in a machine learning scheme. Magn Reson Med 62(6):1609–1618. https://doi.org/10.1002/mrm.22147

172. Ojala T, Pietikainen M, Harwood D. Performance evaluation of texture measures with classification based on Kullback discrimination of distributions. In: Proceedings of 12th International Conference on Pattern Recognition, October 1994 1994. pp 582–585 vol.581. doi:https://doi.org/10.1109/ICPR.1994.576366

173. Ojala T, Pietikainen M, Maenpaa T (2002) Multiresolution gray-scale and rotation invariant texture classification with local binary patterns. IEEE Trans Pattern Anal Mach Intell 24(7):971–987

174. Sachdeva J, Kumar V, Gupta I, Khandelwal N, Ahuja CK (2013) Segmentation, feature extraction, and multiclass brain tumor classification. J Digit Imaging 26(6):1141–1150. https://doi.org/10.1007/s10278-013-9600-0

175. Jafarpour S, Sedghi Z, Amirani MC (2012) A robust brain MRI classification with GLCM features. Int J Comput Appl 37(12), 1–5.

176. Gnep K, Fargeas A, Gutiérrez-Carvajal RE, Commandeur F, Mathieu R, Ospina JD, Rolland Y, Rohou T, Vincendeau S, Hatt M, Acosta O, Rd C (2017) Haralick textural features on T2-weighted MRI are associated with biochemical recurrence following radiotherapy for peripheral zone prostate cancer. J Magn Reson Imaging 45(1):103–117. https://doi.org/10.1002/jmri.25335

177. Galloway MM (1975) Texture analysis using gray level run lengths. Comput Graph Image Process 4(2):172–179. https://doi.org/10.1016/S0146-664X(75)80008-6

178. Yang D, Rao G, Martinez J, Veeraraghavan A, Rao A (2015) Evaluation of tumor-derived MRI-texture features for discrimination of molecular subtypes and prediction of 12-month survival status in glioblastoma. Med Phys 42(11):6725–6735. https://doi.org/10.1118/1.4934373

179. Laws KI Rapid texture identification. In: 1980, 1980. International Society for Optics and Photonics, pp 376–382

180. Prasanna P, Patel J, Partovi S, Madabhushi A, Tiwari P (2017) Radiomic features from the peritumoral brain parenchyma on treatment-naïve multi-parametric MR imaging predict long versus short-term survival in glioblastoma multiforme: preliminary findings. Eur Radiol 27(10):4188–4197

181. Marĉelja S (1980) Mathematical description of the responses of simple cortical cells. JOSA 70(11):1297–1300

182. Iv M, Zhou M, Shpanskaya K, Perreault S, Wang Z, Tranvinh E, Lanzman B, Vajapeyam S, Vitanza NA, Fisher PG, Cho YJ, Laughlin S, Ramaswamy V, Taylor MD, Cheshier SH, Grant GA, Poussaint TY, Gevaert O, Yeom KW (2018) MR imaging–based radiomic signatures of distinct molecular subgroups of medulloblastoma. Am J Neuroradiol. https://doi.org/10.3174/ajnr.A5899

183. Prasanna P, Tiwari P, Madabhushi A (2016) Co-occurrence of local anisotropic gradient orientations (CoLlAGe): a new radiomics descriptor. Sci Rep 6. (1 SRC – BaiduScholar). https://doi.org/10.1038/srep37241

184. Jain AK, Duin RPW, Mao J (2000) Statistical pattern recognition: a review. IEEE Trans Pattern Anal Mach Intell 22(1):4–37

185. Guyon I, Elisseeff A (2003) An introduction to variable and feature selection. J Mach Learn Res 3(Mar):1157–1182

186. Saeys Y, Inza I, Larranaga P (2007) A review of feature selection techniques in bioinformatics. Bioinformatics 23(19):2507–2517. https://doi.org/10.1093/bioinformatics/btm344

187. Ding C, Peng H (2005) Minimum redundancy feature selection from microarray gene expression data. J Bioinform Comput Biol 3(2):185–205. https://doi.org/10.1142/s0219720005001004 PMID: 15852500

188. Cho H-H, Lee S-H, Kim J, Park H (2018) Classification of the glioma grading using radiomics analysis. PeerJ 6:e5982

189. Yang HH, Moody J (1999) Feature Selection Based on Joint Mutual Information. In Proceedings of International ICSC Symposium on Advances in Intelligent DataAnalysis. pp. 22–25.

190. Ginsburg SB, Viswanath SE, Bloch BN, Rofsky NM, Genega EM, Lenkinski RE, Madabhushi A (2015) Novel PCA-VIP scheme for ranking MRI protocols and identifying computer-extracted MRI measurements associated with central gland and peripheral zone prostate tumors. J Magn Reson Imaging 41(5):1383–1393

191. Fleuret F (2004) Fast binary feature selection with conditional mutual information. J Mach Learn Res 5:1531–1555

192. Frank M, Wolfe P (1956) An algorithm for quadratic programming. Naval Res Logistics Quarterly 3(1–2):95–110

193. Tibshirani R (1996) Regression shrinkage and selection via the lasso. J R Stat Soc Series B Methodol 58:267–288

194. Zou H, Hastie T (2005) Regularization and variable selection via the elastic net. J R Stat Soc Series B Stat Method 67(2):301–320

195. LeCun Y, Bengio Y, Hinton G (2015) Deep learning. Nature 521(7553):436–444

196. Schmidhuber J (2015) Deep learning in neural networks: an overview. Neural Netw 61:85–117

197. Litjens G, Kooi T, Bejnordi BE, Setio AAA, Ciompi F, Ghafoorian M, Van Der Laak JA, Van Ginneken B, Sánchez CI (2017) A survey on deep learning in medical image analysis. Med Image Anal 42:60–88

198. Shen D, Wu G, Suk H-I (2017) Deep learning in medical image analysis. Annu Rev Biomed Eng 19:221–248

199. Deng L (2014) Deep learning: methods and applications. Foundations Trends Signal Process 7(3–4):197–387. https://doi.org/10.1561/2000000039

200. Deng L (2000) Deep learning: methods and applications. Found Trends Signal Process (3–4). 7197387 SRC – BaiduScholar. https://doi.org/10.1561/2000000039

201. Lao J, Chen Y, Li Z-C, Li Q, Zhang J, Liu J, Zhai G (2017) A deep learning-based radiomics model for prediction of survival in glioblastoma multiforme. Sci Rep 7(1):10353

202. Bera K, Velcheti V, Madabhushi A (2018) Novel quantitative imaging for predicting response to therapy: techniques and clinical applications. Am Soc Clin Oncol Educ Book 38:1008–1018

203. Rand WM (1971) Objective criteria for the evaluation of clustering methods. J Am Stat Assoc 66(336):846–850

204. Halkidi M, Batistakis Y, Vazirgiannis M (2002) Cluster validity methods: part I. ACM SIGMOD Rec 31(2):40–45

205. Rand WM, J. (1971) Objective criteria for the evaluation of clustering methods. Stat Assoc 66. (336 SRC - BaiduScholar):846–850

206. Abundez I, Arizmendi A, Quiroz EMJ, Rendón E (2011) Internal versus external cluster validation indexes. Int Commun 5. (1 SRC - BaiduScholar):27–34

207. Sasikala M, Kumaravel N (2008) A wavelet-based optimal texture feature set for classification of brain tumours. J Med Eng Technol 32(3):198–205

208. Mouthuy N, Cosnard G, Abarca-Quinones J, Michoux N (2012) Multiparametric magnetic resonance imaging to differentiate high-grade gliomas and brain metastases. J Neuroradiol 39(5):301–307

209. Gutierrez DR, Awwad A, Meijer L, Manita M, Jaspan T, Dineen RA, Grundy RG, Auer DP (2014) Metrics and textural features of MRI diffusion to improve classification of pediatric posterior Fossa tumors. Am J Neuroradiol 35(5):1009–1015. https://doi.org/10.3174/ajnr.A3784

210. Kickingereder P, Burth S, Wick A, Götz M, Eidel O, Schlemmer H-P, Maier-Hein KH, Wick W, Bendszus M, Radbruch A, Bonekamp D (2016) Radiomic profiling of glioblastoma: identifying an imaging predictor of patient survival with improved performance over established clinical and radiologic risk models. Radiology 280(3):880–889

211. McGarry SD, Hurrell SL, Kaczmarowski AL, Cochran EJ, Connelly J, Rand SD, Schmainda KM, LaViolette PS (2016) Magnetic resonance imaging-based Radiomic profiles predict patient prognosis in newly diagnosed Glioblastoma before therapy. Tomography (Ann Arbor, Mich) 2(3):223–228

212. Macyszyn L, Akbari H, Pisapia JM, Da X, Attiah M, Pigrish V, Bi Y, Pal S, Davuluri RV, Roccograndi L, Dahmane N, Martinez-Lage M, Biros G, Wolf RL, Bilello M, O'Rourke DM, Davatzikos C (2015) Imaging patterns predict patient survival and molecular subtype in glioblastoma via machine learning techniques. Neuro-Oncology 18(3):417–425. https://doi.org/10.1093/neuonc/nov127

213. Lee J, Narang S, Martinez J, Rao G, Rao A (2015) Spatial habitat features derived from multiparametric magnetic resonance imaging data are associated with molecular subtype and 12-month survival status in glioblastoma multiforme. PLoS One 10(9):e0136557

214. Zhou M, Hall L, Goldgof D, Russo R, Balagurunathan Y, Gillies R, Gatenby R (2014) Radiologically defined ecological dynamics and clinical outcomes in glioblastoma multiforme: preliminary results. Transl Oncol 7(1):5–13

215. Bae S, Choi YS, Ahn SS, Chang JH, Kang S-G, Kim EH, Kim SH, Lee S-K (2018) Radiomic MRI Phenotyping of glioblastoma: improving survival prediction. Radiology 289(3):797–806

216. Hu X, Wong KK, Young GS, Guo L, Wong ST (2011) Support vector machine multiparametric MRI identification of pseudoprogression from tumor recurrence in patients with resected glioblastoma. J Magn Reson Imaging 33(2):296–305

217. Lohmann P, Stoffels G, Ceccon G, Rapp M, Sabel M, Filss CP, Kamp MA, Stegmayr C, Neumaier B, Shah NJ, Langen K-J, Galldiks N (2017) Radiation injury vs. recurrent brain metastasis: combining textural feature radiomics analysis and standard parameters may increase ^{18}F-FET PET accuracy without dynamic scans. Eur Radiol 27(7):2916–2927

218. Kickingereder P, Götz M, Muschelli J, Wick A, Neuberger U, Shinohara RT, Sill M, Nowosielski M, Schlemmer H-P, Radbruch A, Wick W, Bendszus M, Maier-Hein KH, Bonekamp D (2016) Large-scale radiomic profiling of recurrent glioblastoma identifies an imaging predictor for stratifying anti-angiogenic treatment response. Clin Cancer Res 22(23):5765–5771

219. Louis DN, Perry A, Reifenberger G, von Deimling A, Figarella-Branger D, Cavenee WK, Ohgaki H, Wiestler OD, Kleihues P, Ellison DW (2016) The 2016 World Health Organization classification of tumors of the central nervous system: a summary. Acta Neuropathol 131(6):803–820

220. Lu C-F, Hsu F-T, Hsieh KL-C, Kao Y-CJ, Cheng S-J, Hsu JB-K, Tsai P-H, Chen R-J, Huang C-C, Yen Y, Chen C-Y (2018) Machine learning-based radiomics for molecular subtyping of gliomas. Clin Cancer Res 24(18):4429–4436

221. Zhang B, Chang K, Ramkissoon S, Tanguturi S, Bi WL, Reardon DA, Ligon KL, Alexander BM, Wen PY, Huang RY (2017) Multimodal MRI features predict isocitrate dehydrogenase genotype in high-grade gliomas. Neuro-Oncology 19(1):109–117

222. Bakas S, Akbari H, Pisapia J, Martinez-Lage M, Rozycki M, Rathore S, Dahmane N, O'Rourke DM, Davatzikos C (2017) In Vivo detection of EGFRvIII in glioblastoma via perfusion magnetic resonance imaging signature consistent with deep peritumoral infiltration: the φ-index. Clin Cancer Res 23(16):4724–4734

223. Beig N, Patel J, Prasanna P, Hill V, Gupta A, Correa R, Bera K, Singh S, Partovi S, Varadan V (2018) Radiogenomic analysis of hypoxia pathway is predictive of overall survival in glioblastoma. Sci Rep 8(1):7

224. Levner I, Drabycz S, Roldan G, De Robles P, Cairncross JG, Mitchell R (2009) Predicting MGMT methylation status of glioblastomas from MRI texture. In: Yang G-Z, Hawkes D, Rueckert D, Noble A, Taylor C (eds) Medical image computing and computer-assisted intervention—MICCAI 2009, vol 5762. Springer, Berlin, Heidelberg, pp 522–530

225. Kickingereder P, Bonekamp D, Nowosielski M, Kratz A, Sill M, Burth S, Wick A, Eidel O, Schlemmer H-P, Radbruch A, Debus J, Herold-Mende C, Unterberg A, Jones D, Pfister S, Wick W, von Deimling A, Bendszus M, Capper D (2016) Radiogenomics of glioblastoma: machine learning–based classification of molecular characteristics by using multiparametric and multiregional MR imaging features. Radiology 281(3):907–918. https://doi.org/10.1148/radiol.2016161382

226. Hu LS, Ning S, Eschbacher JM, Baxter LC, Gaw N, Ranjbar S, Plasencia J, Dueck AC, Peng S, Smith KA, Nakaji P, Karis JP, Quarles CC, Wu T, Loftus JC, Jenkins RB, Sicotte H, Kollmeyer TM, O'Neill BP, Elmquist W, Hoxworth JM, Frakes D, Sarkaria J, Swanson KR, Tran NL, Li J, Mitchell JR (2017) Radiogenomics to characterize regional genetic heterogeneity in glioblastoma. Neuro-Oncology 19(1):128–137

INDEX

Giorgio Seano (ed.), *Brain Tumors*, Neuromethods, vol. 158,
https://doi.org/10.1007/978-1-0716-0856-2, © Springer Science+Business Media, LLC, part of Springer Nature 2021

Printed in the United States
by Baker & Taylor Publisher Services